Spatially Oriented Behavior

Edited by
Alan Hein and
Marc Jeannerod

With 112 Figures

Springer-Verlag
New York Berlin Heidelberg Tokyo

Alan Hein
Department of Psychology
Massachusetts Institute of Technology
Cambridge, Massachusetts 02139, U.S.A.

Marc Jeannerod
INSERM Unité 94
F-69500 Bron, France

Library of Congress Cataloging in Publication Data
Main entry under title:
Spatially oriented behavior.
    Bibliography: p.
    Includes index.
    1. Spatial behavior. I. Hein, Alan. II. Jeannerod,
Marc. [DNLM: 1. Space perception. 2. Spatial behavior.
BF 469 H468s]
BF469.S69 1982   153.7'52   82-19497

Typeset by Ms Associates, Champaign, Illinois.

9 8 7 6 5 4 3 2 1

ISBN-13:978-1-4612-5490-4   e-ISBN-13:978-1-4612-5488-1
DOI: 10.1007/978-1-4612-5488-1

This volume is dedicated to
Hans Lukas Teuber
1916–1977

Preface

This volume is the outcome of a Symposium held in Lyon, France. The meeting was organized under the auspices of the Institut National de la Santé et de la Recherche Médicale (INSERM, Paris). We are grateful to the Université Claude-Bernard which allowed us to use the house of the Brothers Lumiere for the site of the meeting. We would also like to acknowledge the generosity of the Fondation Merieux (Lyon) which provided us with a reception at the house where Claude Bernard was born.

In addition to the authors of this volume we wish to thank the following individuals for their contributions to the success of the Symposium: Christine Baleydier, Simon Faugier-Grimaud, Francoise Girardet, Jacqueline Jeannerod, Henry Kennedy, Michele Magnin, Claude Prablanc, Katherine Page, Lawrence Stark and Francois Vital-Durand.

Support from the Office of Naval Research (Contract # N00014-80-K-0243), the National Eye Institute (Grant # 1 P30-EY02621), the Institut National de la Santé et de la Recherche Médicale (Paris) and Sherin Stahl, a participant in the Undergraduate Research Opportunities Program of the Massachusetts Institute of Technology made this volume possible.

Alan Hein
Marc Jeannerod

Contents

Chapter 1 How Do We Direct Our Actions in Space? 1
Marc Jeannerod

Different Modes of Processing Visual Spatial
Information.. 2
Role of the Extraretinal Signal in Perceptual
Stability of Visual Space............................... 4
The Paralyzed Eye Situation as a Cue for
Understanding Visuomotor Localization.............. 6
The Role of Registered vs. Actual Gaze Position in
Guiding Behavior...................................... 8

**Chapter 2 Maintenance of Equilibrium During
Movement** 15
*M. Dufossé, J. M. Macpherson, J. Massion, and
A. Polit*

How Is Body Posture Established?..................... 15
How Do Movement and Posture Interact? 16
Coordination Between Posture and Movement in the
Quadruped... 18
Central Organization of Postural Changes 27
Conclusions.. 31

**Chapter 3 Neural Mechanisms of Visual Orientation in
Rodents: Targets Versus Places** 35
Melvin A. Goodale

Visually Guided Movements Toward Targets 36
Visually Guided Movements Toward Places......... 50
Conclusions.. 57

Chapter 4 Role of the Monkey Superior Colliculus in the Spatial Localization of Saccade Targets 63
David L. Sparks and Lawrence E. Mays

Role of the Superior Colliculus in Saccadic Eye Movements 64
Models of the Saccadic Eye Movement System 67
Neural Representations of Saccade Targets 72

Chapter 5 Interface of Visual Input and Oculomotor Command for Directing the Gaze on Target 87
John Schlag and Madeleine Schlag-Rey

Internal Medullary Lamina 88
Experimental Examination of the Hypothesis of Adequate Response 91
Visual Responses to Target Location in Space 94
Discussion 98

Chapter 6 Coordination of Eye-Head Movements in Alert Monkeys: Behavior of Eye-Related Neurons in the Brain Stem 105
Francis Lestienne, Doug Whittington, and Emilio Bizzi

Experimental Method 107
Behavior of PRF Preoculomotor Neurons During Eye-Head Coordination 108
Coordination of Head and Eyes in the Saccadic Changes of Gaze 114
Conclusions 116

Chapter 7 Contribution of Eye Movement to the Representation of Space 119
Alan Hein and Rhea Diamond

How Is Visual Direction Given? 120
What Is the Source of Knowledge of Eye Position? .. 128
Conclusions 132

Chapter 8 Control of the Optokinetic Reflex by the Nucleus of the Optic Tract in the Cat 135
Klaus P. Hoffman

Functional Anatomy of NOT 136
Relationship Between NOT Output and OKN 141
Conclusions 151

Chapter 9 Development of Optokinetic Nystagmus and Effects of Abnormal Visual Experience During Infancy 155
Janice R. Naegele and Richard Held

Human OKN Development 156
Other Studies 169
Future Research 171

Chapter 10 Spatially Determined Visual Activity in Early Infancy 175
Marshall M. Haith

Appreciation of Spatial Relations 176
Visual Anticipation of Spatiotemporal Events 184
Spatial Sensitivity at Birth 188
Conclusions 192

Chapter 11 Spatial Sense of the Human Infant 197
Peter C. Dodwell

Experiments in Reaching 199
Experiments in Intermodal Spatial Coordination 201
Orienting to Touch 206
Conclusions 209

Chapter 12 Space, the Organism and Objects, Their Cognitive Elaboration in the Infant 215
André Bullinger

The Development of Organized Movements for Tracking a Moving Visual Object 216

Chapter 13 Motion Parallax Sensitivity and Space Perception 223
Ken Nakayama

Space Perception 223
Psychophysical Observations in Motion Parallax 229
Implications for Space Perception 238

Chapter 14 Perceptual Consequences of Experimental Extraocular Muscle Paralysis 243
Leonard Matin, John K. Stevens, and Evan Picoult

The Kernel Observation under Partial Paralysis: Perceived Change in Elevation 245
Quantitative Measurements of Perceived Eye-Level Horizontal 247

Illusory Changes in Visual Localization of the
Median Plane 250
Influence of Level of Paralysis 250
Three-Parameter Model of the Influence of
Paralysis on Visual Localization and Oculomotor
Control 251
Illumination Versus Darkness 255
Auditory-Visual Matches 256
EEPI and Cancellation Are Not Suppressed in
Illumination; Visual Localization and Intersensory
Localization Are Guided by Cancellation in
Darkness 258
A Paradox and Its Resolution 259
Source of EEPI 261

Chapter 15 Mechanisms of Space Constancy **263**
Bruce Bridgeman

Beyond Corollary Discharge 265
Rules of Motion Perception During Saccades
Versus During Fixation 266
Exploring Space Constancy with Figure-Ground
Reversible Images 270
Mechanisms of Stabilization 274

**Chapter 16 Visual Information Processing for Saccadic
Eye Movements** **281**
John M. Findlay

Processes Involved in Saccade Generation 282
Retinal Eccentricity and Saccade Direction 286
Saccade Latency 289
Saccade Amplitude 292
Conclusions 299

**Chapter 17 Optic Ataxia: A Specific Disorder in
Visuomotor Coordination** **305**
M. Thérèse Perenin and Alain Vighetto

Case Reports 306
Neuropsychological Symptoms 310
Proximal and Distal Deficits in Optic Ataxia 310
Comparison with Findings from Experiments in
Monkeys 315
Effects of Altering Spatial Cues on Reaching
Behavior in Optic Ataxia 317
Discussion 320

Chapter 18 Multimodal Structure of the Extrapersonal Space 327
Otto-Joachim Grüsser

Compartments of Extrapersonal Space 327
Coordinates of Extrapersonal Space 331
Disturbances in the Perception of Objects and
Space Following Parietal and Occipital Corex
Lesions 337
Multimodal Control of Eye Movements 344
Conclusions 350

Author Index 353

Subject Index 363

Contributors

Emilio Bizzi, Department of Psychology, Massachusetts Institute of Technology, Cambridge, Massachusetts 02139, U.S.A.

André Bullinger, F.P.S.E., University of Geneva, CH-1211 Geneva 4, Switzerland

Bruce Bridgeman, Departments of Psychology and Psychobiology, University of California, Santa Cruz, California 95064, U.S.A.

Rhea Diamond, Department of Psychology, Massachusetts Institute of Technology, Cambridge, Massachusetts 02139, U.S.A.

Peter C. Dodwell, Department of Psychology, Queens University, Kingston, Ontario, Canada

M. Dufossé, Department de Neurophysiologie generale, C.N.R.S.-I.N.P., B.P. 71, 13266 Marseille Cedex 9, France

John M. Findlay, Department of Psychology, University of Durham, Durham Dh1 3LE, England

Melvin A. Goodale, Department of Psychology, University of Western Ontario, London, Ontario Canada N6A 5C2

Otto-Joachim Grüsser, Physiologisches Institut, Freie Universität, 100 Berlin 33, West Germany

Marshall M. Haith, Department of Psychology, Queens University, Kingston, Ontario, Canada

Alan Hein, Department of Psychology, Massachusetts Institute of Technology, Cambridge, Massachusetts 02139, U.S.A.

Richard Held, Department of Psychology, Massachusetts Institute of Technology, Cambridge, Massachusetts 02139, U.S.A.

Klaus P. Hoffmann, Abteilung für Vergleichende Neurobiologie, Universität Ulm, D-7900 Ulm, West Germany

Marc Jeannerod, INSERM Unité 94, F-69500 Bron, France

Francis Lestienne, Department of Psychology, Massachusetts Institute of Technology, Cambridge, Massachusetts 02139, U.S.A.

J. M. Macpherson, Department de Neurophysiologie generale, C.N.R.S.-I.N.P., B.P. 71, 13266 Marseille Cedex 9, France

J. Massion, Department de Neurophysiologie generale, C.N.R.S.-I.N.P., B.P.71, 13266 Marseille Cedex 9, France

Leonard Matin, Department of Psychology, Columbia University, New York, New York, 10027, U.S.A.

Lawrence E. Mays, Department of Psychology and Neuroscience Program, University of Alabama—Birmingham, Birmingham, Alabama 35294, U.S.A.

Janice R. Naegele, Department of Psychology, Massachusetts Institute of Technology, Cambridge, Massachusetts 02139, U.S.A.

Ken Nakayama, Smith-Kettlewell Institute of Visual Sciences, San Francisco, California 94115, U.S.A.

M. Thérèse Perenin, Laboratoire de Neurophysiologie Expérimentale, INSERM Unité 94, F-69500 Bron, France

Evan Picoult, Department of Psychology, Columbia University, New York, New York 10027, U.S.A.

A. Polit, Department de Neurophysiologie generale, C.N.R.S.-I.N.P., B.P. 71, 13266 Marseille Cedex 9, France

John Schlag, Department of Anatomy, University of California—Los Angeles, Brain Research Institute, Los Angeles, California 90024, U.S.A.

Madeleine Schlag-Rey, Department of Anatomy, University of California—Los Angeles, Brain Research Institute, Los Angeles, California 90024, U.S.A.

David L. Sparks, Department of Psychology and Neuroscience Program, University of Alabama—Birmingham, Birmingham, Alabama 32594, U.S.A.

John K. Stevens, Department of Physiology and Playfair Neuroscience Unit, University of Toronto, Toronto, Ontario, Canada N5T 2S8

Alain Vighetto, Laboratoire de Neurophysiologie Expérimentale, INSERM Unité 94, F-69500, Bron, France

Doug Whittington, Department of Psychology, Massachusetts Institute of Technology, Cambridge, Massachusetts 02139, U.S.A.

Chapter 1

How Do We Direct Our Actions in Space?

Marc Jeannerod

Spatially oriented behavior requires that the goals toward which actions are directed be localized with respect to a body reference. This requirement would obviously not be a problem for an ideal organism in which sensory receptors would be kept in constant relation to the body reference. In that case, the retinal locus to which a visual object projected would have unambiguous spatial value. In most animals, however, and certainly in man, the relationship of retinal coordinates to body axis varies with movements of the eye and head. Therefore, behavior directed at an object located in extrapersonal space cannot rely solely on the position of its image on the retina. As stressed by Miles and Evarts (1979), reconstruction of the position of an object in space relative to the body requires summation of three signals:

1. The position of the image on the retina
2. The position of the eye in the orbit relative to the head
3. The position of the head relative to the body.

It could be argued that the reconstruction of object position in space would be greatly simplified by first nulling some of the signals involved. For instance, bringing the image automatically to the fovea, via the fixation reflex, would result in nulling signal 1. Similarly, keeping the eye axis automatically superimposed on the head axis, via the vestibulo-ocular reflex, would bring signal 2 to zero. Finally, a fairly good reconstruction of object position relative to the body would be obtained by monitoring only signal 3.

These requirements do not obtain, however. Reaching behavior is not organized serially. It has recently been demonstrated that the neural commands for eye movements and to neck and arm muscles, as tested by the latency of electromyographic activation, are released synchronously, even though the appearance of eye, head, and arm movements is that of a sequence (Biguer, Jeannerod, & Prablanc, 1982).

This result could be interpreted as evidence for the existence of a common spatial reconstructor in which the transformation from retinal to spatial coordinates could occur. The reconstruction would make available to the various motor command structures information about target location on a body-centered map of visual space. This design would imply a constant updating in the reconstructor by signals describing the present position of the eye, head, and arm with respect to the common body reference.

In fact, it is now well known that the signal about target location received by oculomotor structures is coded in spatial, not in retinal, coordinates (see Robinson, 1975). Evidence on this point comes from experiments by several authors. Hallet and Lightstone (1976), for example, showed that a saccade can be accurately directed at a target light briefly presented just before execution of a prior saccade which was directed at another target. The subject first executes the saccade to the location of the first target, and from there directs his eyes to the location of the second. The difference between the location of the first target and the location of the second, which is covered by the second saccade, is not equivalent to the difference between the position of the eyes at the time the second target was shown and the location of that target. This result indicates that the second saccade was not programed to cover the difference between the initial position of the eye and the second target (the amplitude of the retinal signal), but to cover the difference between the two targets (the amplitude of the spatial difference). Similar observations were made by Prablanc, Massé, and Echallier (1978). Sparks and Mays report (Chapter 4, this volume) even more dramatic results in the monkey. Their results show that if a perturbing eye deviation is produced by electrical stimulation of the superior colliculus immediately prior to, or during, a target-oriented saccade, the eyes will nevertheless reach their initial goal, again indicating that the program for the movement was to reach a point in the orbit and not on the retina (see also Mays & Sparks, 1980).

Different Modes of Processing Visual Spatial Information

The hypothesis of a spatially coded reconstructor offers a plausible explanation for the processing of visual input at the visuomotor level but not for other, more perceptual, levels of processing. Evidence for such a dissociation between different levels of processing has been found in several experimental and clinical observations. An extreme case can be found in patients with limited lesions within the geniculo-striate pathways. These lesions produce scotoma, the loss of perceptual experience from the part of the visual field controlled by the destroyed

area. The scotoma, however, is only a relatively blind zone, from which
response to visual stimulation can still be obtained. The important
point is that whether the subject can respond depends on the requirement of the task. If the task demands a verbal response based on subjective experience of the stimulus, using the usual perceptual or cognitive
mode of processing, none can be given. If, alternately, the subject is
forced to give a purely visuomotor response (e.g., by directing the
gaze or by pointing with the hand) in the direction of the "unseen"
stimulus, he can produce accurate movements (Perenin & Jeannerod,
1978, 1979; Poeppel, Held, & Frost, 1973; Weiskrantz, Warrington,
Sanders, & Marshall, 1974). This result clearly indicates that describing
the localization of an object in space verbally is quite a different task
from that of reaching to the same object and involves quite different
neural structures.

Evidence for different task-dependent modes of processing for localization of visual targets can also be demonstrated in normal subjects.
The experiments cited here address the intriguing problem of perceptual stability of the visual environment during movements of the eyes.
As pointed out by Matin (1972),

> [The] combination of eye, head and bodily movements continually brings
> about variations in the retinal loci of images corresponding to stationary
> objects, and questions of how we perceive these objects in stable positions
> and do not confuse them with moving objects are thus central to any
> understanding of space perception and sensorimotor coordination.

However, comparison of the results of experiments testing "space perception" with those of experiments testing "sensorimotor coordination" during voluntary eye movements clearly shows that the two
processes can be dissociated. If a subject is required to localize with
respect to each other two point sources of light briefly presented in succession in temporal association with an ocular saccade (i.e., by reporting
whether the second light appeared to the right or left of the first), he
may eventually make large spatial errors.[1] These errors are observed not
only during the saccade itself but, in fact, during a span of time extending from about 200 msec prior to the saccade, up to several hundred
milliseconds after its completion. Finally, analysis of the spatial distribution of the errors indicates that the perceived space is contracted in
the direction of the eye movement (Matin, 1972; Matin & Pearce,
1965). In contrast, if the subject in the same experimental situation is

---

[1] A more casual observation related to this point is the jerky appearance of the
visual field when voluntary eye movements are performed under stroboscopic illumination (Gregory, 1958). This observation confirms that the stabilization of the
visual field is incomplete under conditions of restricted illumination, which were
also used in experiments by Matin's group.

required to strike at the test light with a hammer he will do so with remarkable accuracy in spite of having the impression of making large errors (Skavenski & Hansen, 1978). The implication of these experiments is that there is no one simple correspondence between the perceived location of a source of light and the spatial information used by the visuomotor system to localize it with respect to the body (see Steinman, 1975).

Role of the Extraretinal Signal in Perceptual Stability of Visual Space

At this point it is necessary to focus on the postulated mechanisms for perceptual stability on the one hand, and for visuomotor localization on the other. Comparison between the two may help to reveal differences and similarities. Perceptual stability during eye movements is currently explained by the contribution, at the level of the visual system, of an extraretinal signal related to the generation and/or execution of the eye movement. Several complementary functions have been ascribed to the extraretinal signal. Although the function of maintaining perceptual invariance, by canceling the effects of eye movement on visual perception, is considered essential by psychologists (see Matin, 1972), others have emphasized the role of the extraretinal signal in controlling behavioral responses to visual motion. Thus, the basic problem is that of differentiating a visual movement occurring independently within the outside world, from a self-generated movement. A theoretical assumption to account for this distinction is that the "efference copy" related to a self-generated movement nulls the interpretation of visual motion inflow to the sensory neurons each time this inflow is a direct result of behavior (Holtz & Mittelstaedt, 1950). Neuronal responses fulfilling the criterion for an efference copy type of mechanism have been recorded from visual neurons in elegant experiments conducted in crustaceans (Wiermsa & Yamaguchi, 1967), in insects (Palka, 1969), and, more recently, in mammals. In the monkey superior colliculus, Robinson and Wurtz (1976) found neurons that responded when a visual stimulus rapidly crossed their receptive field, but did not when an eye movement swept the receptive field across the same but stationary stimulus. Neurons in other parts of the central visual system, including the visual cortex, have also been shown to receive eye movement related, extraretinal signals (see review in Jeannerod, Kennedy, & Magnin, 1979).

In the behavioral context, perceptual stabilization may also be essential for stabilization of the whole animal, particularly in those animal species whose behavior is strongly driven by visual motion signals. Experiments have been reported that evidence the effects of efference copy on an animal's behavior. Sperry, in 1943, observed that fish with inverted vision caused by surgical 180° eye rotation tend to turn continuously in circles, quite in the same way as normal fish when stimulated by a visual surround moving at constant velocity. In a later paper Sperry (1950) interpreted the circling behavior due to inverted vision as the result of a disharmony between the retinal input generated by movement of the animal and the compensatory mechanism for maintaining the stability of the visual field. Surgical eye rotation of 180°, by making the compensatory mechanism in diametric disharmony with the retinal input, "would therefore cause accentuation rather than cancellation of the illusory outside movement." The mechanism postulated by Sperry is a centrally arising discharge that reaches the visual centers as a corollary of "any excitation pattern that normally results in a movement" and is "specific for each movement with regard to its direction and speed." It should be noted that "efference copy" and "corollary" discharges are quite germane concepts that are included in, but do not completely overlap with, the more general concept of an extraretinal signal (also see Teuber, 1960).

The previously mentioned experiments designed for testing the role of extraretinal signals in perceptual stability had, in fact, shown the relative inappropriateness of this mechanism. From the Matin and Pearce experiment (1965) the extraretinal signal would appear to be too weak (subjects make localization errors) and not properly timed to the eye movement. If, as argued by Matin (1972), visual stabilization nevertheless occurs in normal conditions, it is because other mechanisms add to the extraretinal signal. Purely retinal mechanisms could combine for masking undesirable visual effects of saccadic eye movements (see Campbell & Wurtz, 1978; see also Chapter 15, this volume).

There are few data available about the information contained in the extraretinal signal. As emphasized by Sperry (1950), a requirement for adequate visual stabilization should be information about the direction and speed of eye movement. Information about the amplitude of movement is also critical. Neuronal changes recorded from visual neurons during eye movements in various animals are transient bursts or transient decreases in spontaneous activity, usually related to the direction of the movement but essentially unrelated to its duration or its amplitude[2] (Jeannerod, Kennedy, & Magnin, 1979; Robinson & Wurtz, 1976). These features are basically in accord with the characteristics of the extraretinal signal as it could be postulated from the results of the psychophysical experiments.

---

[2] A valuable exception is the demonstration by Johnstone and Mark (1971) of "efference copy neurons" in the optic tectum of the fish coding the amplitude and the direction of saccadic eye movements.

The Paralyzed Eye Situation as a Cue
for Understanding Visuomotor Localization

Although the preservation of perceptual stability during eye movements could be explained, at least in part, by the action of signals
related to the displacement of the eyes, this cannot be the case for
visuomotor localization, which requires information about eye position
relative to the head. This difference raises the point of how eye-re-head
position can be encoded and monitored centrally, and of how this
monitoring can be demonstrated. The "paralyzed eye situation" has
long been considered critical for this demonstration. As first described
by Graefe (1870), people with paralysis of an extrinsic eye muscle display striking behavior when they attempt to reach a hand to objects
viewed in their peripheral visual field through their paralyzed eye alone.
Typically, they overreach in the direction of the attempted eye movement, which is prevented by the paralysis, and miss the target. This
phenomenon, called past-pointing, has now been fully confirmed by
further clinical cases (Adler, 1943; Jackson & Patton, 1909; Perenin,
Jeannerod, & Prablanc, 1977) and by experiments that use reversible
block of extraocular muscles in normal subjects (Brindley, Goodwin,
Kulikowski, & Leighton, 1976; Kornmüller, 1931; Siebeck, 1954;
Stevens, Emerson, Gerstein, Kallos, Neufeld, Nichols, & Rosenquist,
1976). In addition, attempts to move the eyes against the paralysis
produce an illusory displacement of the visual scene in the direction of
the attempted movement.

It is argued that establishing a comprehensive explanation for past-
pointing would represent an important step, since this explanation
would also hold for visuomotor localization. We think, however, that
this is not completely true. First, it has recently been shown that past-
pointing and illusory displacement occur only when the eye is partially
paralyzed and not if the paralysis is total (Brindley et al., 1976). Second, past-pointing is related to the attempt to make an eye movement
against the paralysis. If the subject is required to point at objects in his
peripheral visual field while keeping his gaze fixed ahead of him, no
past-pointing occurs (Perenin et al., 1977). These observations have to
be taken into account in the subsequent discussion.

The interpretation by Graefe of the phenomena related to paralysis
of extraocular muscles was that if one muscle is contracted more than
normally required for a given result, the increase in effort makes the
subject believe that his eye is displaced from its viridical position.
Through the perceived sensation of effort the subject overestimates the
rotation made by the eye and his visual field is displaced in the same
direction. Past-pointing would be a direct consequence of this displacement. According to what has been previously mentioned about percep-

tual stability, our explanation of the illusory displacement of the visual field produced by the attempted eye movement would be the following: a mismatch is created between the amplitude of the displacement of the retinal image and the corresponding extraretinal signal. Due to paralysis, the retinal motion signal is weak, while the extraretinal signal, reflecting the amplitude of the intended eye movement, is much stronger. As a corollary, the illusion of displacement disappears when the paralysis is complete and there is no retinal motion signal.

Now the question is, why should the position of pointing be displaced? In an attempt to answer this question we measured, in patients with partial paralysis of one eye, the movements of the normal, covered eye (Perenin et al., 1977). This study showed that the normal eye made movements about 35% larger than the actual position of a target would have required. This exaggerated movement could reflect either the increased effort needed to bring the paralyzed eye to target (according to Hering's law of equal innervation) or, more simply, the fact that the retinal error of the paralyzed eye remained uncorrected in spite of the attempted movement. In the latter alternative a situation of positive retinal feedback would be created that would drive the oculomotor system farther and farther (Young & Stark, 1963). Our interpretation of past-pointing was that this exaggerated oculomotor output would generate an overestimated spatial value for the target location. Kelso, Tuller, and Harris (1981) more correctly argued that, since eye and arm movements are coupled in a common reaching task, parametrization of the motor commands occurs over the total coupled system. As a consequence, the increase in force required to move the partially paralyzed eye is necessarily distributed to the system controlling the arm. Accordingly, if the subject is asked to fixate straight ahead while reaching for the target, thereby decoupling eye and arm movements, no past-pointing occurs. A similar reasoning applies to effects observed by Skavenski, Haddad, and Steinman (1972) when the subject had to maintain monocular fixation at a point source of light (in the dark) while his eye was loaded constantly by mechanical traction. In this situation, there was no change in the retinal locus of the target and no eye movement, since fixation was maintained. The only variable was the force exerted by the subject to oppose ocular displacement by the traction. The resulting effect, as experienced by the subject, was a displacement of the "straight ahead" direction as tested by a psychophysical method of subjective alignment.

This experiment indicates that the direction of the gaze axis relative to that of the head axis is not the only cue for determining the gaze position actually felt by the subject, or, more precisely, the gaze position actually used for spatially oriented behavior. In other terms, the mean position of the gaze can be felt as displaced, or can be centrally registered as displaced, in spite of its maintaining its position. We use the term *registered gaze position* to indicate the gaze position that is relevant for visuomotor coordination.

The Role of Registered vs. Actual Gaze Position in Guiding Behavior

In 1866, Helmholtz first reported the effects on visuomotor coordination produced by adaptation to prismatic displacement of the visual field. These effects appear to provide critical information concerning the problem of the encoding of gaze position. Briefly summarized, the experiment was as follows: a subject looking monocularly through a laterally displacing prism first misreaches for objects placed in front of him, but which he sees as laterally displaced. With repeated attempts to reach for the objects, he adapts rapidly to the new situation. On removal of the prism, misreaching reappears, now in the opposite direction. Helmholtz's interpretation of the phenomenon was that it is not the muscular feeling of the hand and arm which is at fault, nor the judgment of their position, but the "judgment of the direction of the gaze." Knowledge as to the direction of gaze axis, he thought, does not depend upon the actual position of the eyeball or the tension of the extraocular muscles, but simply upon "the effort of will involved in trying to alter the adjustment of the eyes."[3] This notion of a judgment of the direction of gaze resembles that of a registered gaze position.

More recent prism experiments have partly confirmed Helmholtz's hypothesis. If, for instance, an object placed directly in front of the subject appears displaced to the right when viewed through the prism, the gaze axis also has to deviate to the right in order to keep foveal fixation of that object. At the end of the adaptation period and after the prism has been removed, the gaze axis, examined in the absence of a fixation point, tends to keep a deviated posture, to the right in our example (Craske, 1967; Kalil & Freedman, 1966). It is hard to determine whether this new gaze position, produced by adaptation to the visuomotor conflict, in fact corresponds to a new central registration of gaze position. Lackner (1973) argued that the position of the eyes with

---

[3] Arguments for a central monitoring of extraretinal signals related to eye movements, and for their contribution to a central representation of extrapersonal visual space, can be traced in the literature as early as Charles Bell. This author (quoted by Wade, 1978) had noticed that visual afterimages seem to move with the eyes in spite of being stationary on the retina. Hence, Bell thought that ". . . vision in its extended sense is a compound operation, the idea of position of an object having relation to the activity of the muscles . . . . If we move the eyes by the voluntary muscles . . . we shall have the notion of place or relation raised in the mind."

respect to the head should not be dissociated from the head position with respect to the body. In Lackner's study, as well as in the above-mentioned experiments in which the position of the gaze after prism adaptation was measured, the subject's head was maintained in fixed alignment with the body axis. However, when the subject was asked in which direction he felt his head was pointing, he indicated a direction different from that of the actual head axis (i.e., he felt his head turned to the left). In other words, the gaze deviation to the right following adaptation to a right-displacing prism would be a consequence of a deviated registered head position to the left. This explanation would account for other effects observed after removal of the prism: the straight-ahead direction, the position of auditory targets, and the direction of reaching arm movements all appear to be deviated in the same direction as the registered head position (Lackner, 1973; McLaughlin & Rifkin, 1965).

The effects resulting from experimental deviation of one eye can be interpreted similarly. Olson (1980) surgically induced monocular strabismus in cats. Visuomotor localization in these animals was tested by measuring the accuracy of jumping to a target board, guided by either the normal or the deviated eye. Although jumps guided by the normal eye were accurate, jumps guided by the operated eye were systematically in error in the direction opposite to the eye deviation. According to Olson, "This error is in the direction one would predict from the cat's being unaware that its eye is deviated." In other words, the localization error might correspond to the difference between the registered and the actual gaze direction. Whether the discrepancy would in fact be between the registered and actual head, or eye, position is unclear. Indeed, Olson reported a torticollis in his strabismic cats whether they were to jump under control of the deviated eye or the normal eye.

Although the effects of surgically induced strabismus on visuomotor localization may be compensated for in time (within a few months according to Olson, 1980), the case may be different in some humans with strabismus lasting from early childhood. In such subjects, only one eye (the dominant) seems to be used for visuomotor localization, and vision with the other, deviated, eye is constantly suppressed. In testing the accuracy of pointing with the hand at targets under control of either eye, Mann, Hein, and Diamond (1979) found that the direction of pointing always reflected the momentary orbital position of the dominant eye, even when targets were seen by the suppressed eye. The question is, why should pointings to targets presented to the suppressed eye be influenced by the posture of the other eye? The answer given by Mann et al. is based on a conception of visuomotor development whereby the correspondence among the direction of objects in space, the retinal loci of their images, and the orbital posture of the eye is progressively incorporated in a representation of visual space that supports visually guided behavior. Consequently, "information about retinal locus of an object's image and information about the posture of the eye combine to permit movements to be directed toward that object" (Mann et al., 1979). In strabismic subjects this process would occur during development only for the dominant eye, so that information about the posture of that eye would be the only way to disambiguate retinal locus information when targets are presented to the suppressed eye.

Many of the abovementioned effects indicate that the actual gaze position may be ignored for the guidance of visuomotor behavior, although another position is monitored centrally and can be effectively used for directing actions in space. A logical conclusion that can be drawn from this fact is that afference resulting from either movements or static positions of the eyes should not be the only cue involved in this central monitoring. It has often been claimed that eyes lack position sense (Brindley & Merton, 1960). We have shown that subjects trained to fixate targets at various positions in their visual field could not replicate these positions by moving their eyes in the dark, or with the lids closed; instead, they made saccades about three times larger than actually required (Jeannerod, Gerin, & Mouret, 1965; see also Koerner, 1975).

This result, however, cannot be interpreted merely as evidence against ocular position sense. To our understanding, it means that the extraretinal signal (of either afferent or efferent origin) cannot be used alone for locating the absolute position of gaze axis in space (see also Skavenski, 1976); rather, the extraretinal signal must combine with retinal information to produce a behaviorally effective reference that takes into account, although is not necessarily superimposed upon, gaze position. The resulting effect would be the generation of an internal representation of a body-centered visual space, used for directing the movement of the hand toward the target. This hypothesis does not hold only for the encoding of gaze position. Prablanc, Echallier, Jeannerod, and Komilis (1979) showed that proprioceptive information about hand position with respect to the body (which is certainly much sharper than the extraretinal signal) is not sufficient in itself to ensure an accurate target-directed movement. A brief visual exposure of the hand is needed at the time of movement onset to calibrate hand position with respect to the internal representation of visual space. Hence the model of a bimodal reconstruction of the space for goal-directed actions can be generalized. Both the extraretinal signal for the eye and the position sense for the head or arm must be calibrated by vision in order to match the moving segment to the body reference and to the target location.

References

Adler, F. M. Pathologic physiology of convergent strabismus. *Archives of Ophthalmology*, 1943, *33*, 362-377.

Biguer, B., Jeannerod, M., & Prablanc, C. The coordination of eye, head and arm movements during reaching at a single visual target. *Experimental Brain Research*, 1982, *46*, 301-304.

Brindley, G. S., Goodwin, G. M., Kulikowski, J. J., & Leighton, D. Stability of vision with a paralyzed eye. *Journal of Physiology* (London), 1976, *258*, 65-66.

Brindley, G., & Merton, P. A. The absence of position sense in the human eye. *Journal of Physiology* (London), 1960, *153*, 127-130.

Campbell, F. W., & Wurtz, R. H. Saccadic omission: Why we do not see a grey-out during a saccadic eye movement. *Vision Research*, 1978, *18*, 1297-1303.

Craske, B. Adaptation to prisms: Change in internally registered eye position. *British Journal of Psychology*, 1967, *58*, 329-335.

Graefe, A., von. *Les paralysies des muscles moteurs de l'oeil* (A. Sichel, trans. from German). Paris: Delahaye, 1870.

Gregory, R. L. Eye movements and the stability of the visual world. *Nature*, 1958, *182*, 1214-1216.

Hallett, P. E., & Lightstone, A. D. Saccadic eye movement towards stimuli triggered by prior saccades. *Vision Research*, 1976, *16*, 99-106.

Helmholtz, H. von. *Handbuch der physiologischen Optik*. Leipzig: Vos, 1866, (English trans., Optical Society of America, 1925.)

Holtz, E. von, & Mittelstaedt, H. Das Reafferenzprinzip. Wechselwirkungen zwischen Zentralnervensystem und Peripherie. *Naturwissenschaften*, 1950, *37*, 464-476. (English trans. in *The behavioural physiology of animals and man*. London: Methuen, 1973, pp. 139-173.)

Jackson, J. H., & Paton, L. On some abnormalities of ocular movements. *Lancet*, 1909, 900-905.

Jeannerod, M., Gerin, P., & Mouret, J. Influence de l'obscurité et de l'occlusion des paupières sur le contrôle des mouvements oculaires. *Année Psychologique*, 1965, *65*, 309-324.

Jeannerod, M., Kennedy, H., & Magnin, M. Corollary discharge: Its possible implications in visual and oculomotor interactions. *Neuropsychologia*, 1979, *17*, 241-258.

Johnstone, J. R., & Mark, R. F. The efference copy neurone. *Journal of Experimental Biology*, 1971, *54*, 403-414.

Kalil, R. E., & Freedman, S. J. Persistence of ocular rotation following compensation for displaced vision. *Perceptual and Motor Skills*, 1966, *22*, 135-139.

Kelso, J. A. S., Tuller, B., & Harris, K. S. A "dynamic pattern" perspective on the control and coordination of movement. In P. MacNeilage (Ed.), *The production of speech*. New York: Springer, 1981.

Koerner, F. H. Non-visual control of human saccadic eye movements. In G. Lennerstrand & P. Bach-Y-Rita (Eds.), *Basic mechanisms of ocular mobility and their clinical implications*. Oxford: Pergamon Press, 1975, pp. 565-569.

Kornmüller, A. E. Eine Experimentelle Anästhesie der aüsseren Angenmuskeln am Menschen und ihre Auswirkungen. *J. f. Psychol. Neurol.* 1931, *41*, 351-366.

Lackner, J. R. The role of posture in adaptation to visual rearrangement. *Neuropsychologia*, 1973, *11*, 33-44.

Mann, V. A., Hein, A., & Diamond, R. Localization of targets by strabismic subjects: Contrasting patterns in constant and alternating suppressors. *Perception and Psychophysic*, 1979, *25*, 29-34.

Matin, L. Eye movements and perceived visual direction. In D. Jameson & L. Hurvich (Eds.), *Handbook of sensory physiology* (Vol. 7). Berlin: Springer, 1972, pp. 331-380.

Matin, L., & Pearce, D. G. Visual perception of direction for stimuli flashed during voluntary saccadic eye movements. *Science*, 1965, *148*, 1485-1488.

Mays, L. E., & Sparks, D. L. Saccades are spatially, not retinocentrically, coded. *Science*, 1980, *208*, 1163-1165.

McLaughlin, S. C., & Rifkin, K. I. Change in straight ahead during adaptation to prism. *Psychonomic Science*, 1965, *2*, 107-108.

Miles, F. A., & Evarts, E. V. Concepts of motor organization. *Annual Review of Psychology*, 1979, *30*, 327-362.

Olson, C. R. Spatial localization in cats reared with strabismus. *Journal of Neurophysiology*, 1980, *43*, 792-806.

Palka, J. Discrimination between movements of eye and object by visual interneurones of crickets. *Journal of Experimental Biology*, 1969, *50*, 723-732.

Perenin, M. T., & Jeannerod, M. Visual function within the hemianopic field following early cerebral hemidecortication in man. I. Spatial localization. *Neuropsychologia*, 1978, *16*, 1-13.

Perenin, M. T., & Jeannerod, M. Subcortical vision in man. *Trends in Neuroscience*, 1979, *2*, 204-207.

Perenin, M. T., Jeannerod, M., & Prablanc, C. Spatial localization with paralyzed eye muscles. *Ophtalmologica*, 1977, *175*, 206-214.

Poeppel, E., Held, R., & Frost, D. Residual visual function after brain wounds involving the central visual pathways in man. *Nature*, 1973, *243*, 295-296.

Prablanc, C., Echallier, J. F., Jeannerod, M., & Komilis, E. Optimal response of eye and hand motor systems in pointing at a visual target. II. Static and dynamic visual cues in the control of hand movement. *Biological Cybernetics*, 1979, *35*, 183-187.

Prablanc, C., Massé, D., & Echallier, F. Corrective mechanisms in visually goal-directed large saccades. *Vision Research*, 1978, *18*, 557-560.

Robinson, D. A. Oculomotor control signals. In G. Lennerstrand & P. Bach-Y-Rita (Eds.), *Basic mechanisms of ocular mobility and their clinical implications*. Oxford: Pergamon Press, 1975, pp. 337-374.

Robinson, D. L., & Wurtz, R. H. Use of an extra-retinal signal by monkey superior colliculus neurons to distinguish real from self-induced stimulus movement. *Journal of Neurophysiology*, 1976, *39*, 852-870.

Siebeck, R. Wahrnehmungsstörung und Störungswahrnehmung bei Augenmuskellähmungen. *Graefes Archiv fur Ophthalmologie*, 1954, *155*, 26-34.

Skavenski, A. A. The nature and role of extraretinal eye-position information in visual localization. In R. A., Monty & J. W. Senders (Eds.), *Eye movements and psychological processes*. Hillsdale, N. J.: Erlbaum, 1976, pp. 277-287.

Skavenski, A. A., & Hansen, R. M. Role of eye position information in visual space perception. In J. Senders, D. Fisher & R. Monty (Eds.), *Eye movements and the higher psychological functions*. New York: Erlbaum, 1978, pp. 15-34.

Skavenski, A. A., Haddad, G., & Steinman, R. M. The extraretinal signal for the visual perception of direction. *Perception and Psychophysic*, 1972, *11*, 287–290.

Sperry, R. W. Effect of 180° rotation of the retinal field in visuomotor coordination. *Journal of Experimental Zoology*, 1943, *92*, 263–279.

Sperry, R. W. Neural basis of the spontaneous optokinetic response produced by visual inversion. *Journal of Comparative and Physiological Psychology*, 1950, *43*, 482–489.

Steinman, R. M. Oculomotor effects on vision. In G. Lennerstrand & P. Bach-Y-Rita (Eds.), *Basic mechanisms of ocular mobility and their clinical implications.* Oxford: Pergamon Press, 1975, pp. 395–415.

Stevens, J. K., Emerson, R. C., Gerstein, G. L., Kallos, T., Neufeld, G. R., Nichols, C. W. & Rosenquist, A. C. Paralysis of the awake human: Visual perceptions. *Vision Research*, 1976, *16*, 93–98.

Teuber, H. L. Perception, in Field, J., Magoun, H. W., & Hall, V. E. (Eds.), *Handbook of Physiology*, Section I, Neurophysiology, American Physiological Society, Washington, 1960, 89–121.

Wade, N. J. Sir Charles Bell on visual direction. *Perception*, 1978, 7, 359–362.

Weiskrantz, L., Warrington, E. R., Sanders, M. D. & Marshall, J. Visual capacity in the hemianopic field following a restricted occipital ablation. *Brain*, 1974, *97*, 709–728.

Wiermsa, C. A. G., & Yamaguchi, T. Integration of visual stimuli by the crayfish central nervous system. *Journal of Experimental Biology*, 1967, *47*, 409–431.

Young, L. R. & Stark, L. A Discrete model of eye nacking movements. *IEEE Transactions*, 1963, *MIL 7*, 113–115.

Chapter 2

Maintenance of Equilibrium During Movement

M. Dufossé, J.M. Macpherson, J. Massion, and A. Polit

The mechanisms involved in the active displacement of a body segment from an initial position toward a final one have been the focus of many investigations. Most attention in the recent past has been devoted to the analysis of the movement per se, the central control of the dynamic phase, and the final joint position. Movement, however, represents only the tip of an iceberg, the concealed part being the body posture that supports the performance of movement. The interaction of movement and posture, which has been investigated very little, is the focal point of this chapter.

How Is Body Posture Established?

Body posture at any given moment results from the interaction of several mechanisms that maintain the position of the different joints against the force of gravity. Among the factors that prevent the collapse of the body under the influence of gravity, the mechanical properties of the skeleton are not of major importance except for some parts concerned with the erect posture, such as the vertebrae or the leg bones. Muscle tone, or more specifically the postural tone that is distributed among the muscles opposing gravity, is the means by which the joints resist the forces (mainly the force of gravity) exerted on the segments that are linked. Distribution of postural tone can be modified by several inputs, proprioceptive as well as exteroceptive. The influence of head and neck position on limb posture, for example, is important for adapting body posture to the surrounding environment (Magnus, 1924; Roberts, 1967).

    The position of a body segment also depends on the central command that adjusts the co-contraction of muscles surrounding the joint (Polit & Bizzi, 1979). This position can be maintained even against external forces such as gravity. Thus, we may speak of the *postural fixation* of a joint. Postural fixation usually serves to support the body segment distal to the joint. Postural fixation of the shoulder, for example, is used for the support of the arm, as postural fixation of the head on the trunk serves to support the head (Martin, 1967). However, postural fixation can also be used for the support of proximal and axial segments, for example, when fixation of the limb joints serves to support the body weight (supporting reaction).

While maintaining postural tone and postural fixation of the joints the conformation of the body segments must also obey a very fundamental rule, maintenance of equilibrium. The general posture of the body is always such that the resultant of internal forces (the center of gravity) projects inside the area delimited by the two feet for bipedal species and the four feet for quadrupeds, at least under static conditions. Several sensory cues, including visual, labyrinthine, and somesthetic, signal the displacement of the center of gravity and lead to a reflex readjustment of posture. They are thus involved in maintaining the center of gravity in an appropriate position.

How Do Movement and Posture Interact?

In his theoretical note concerning interaction between body posture and movement, Hess (1943; see also Jung & Hassler, 1960) subdivided any motor act into two components. One component, called *teleokinetic*, is the movement toward a target. The other, called *ereismatic*, is the component that "supports" the movement. The need for the supporting component is easily understood in view of the fact that the force that provokes the movement is accompanied by an equal force of opposite direction exerted on the nonmoving part of the body. In order to resist that force, joints of the supporting part need increased fixation. However, increased fixation may not be sufficient to resist the forces exerted by the moving segment; these forces may tend to displace the center of gravity and provoke falling. Therefore, there is usually an adjustment of the whole body posture that tends to relocate the center of gravity to a position appropriate for equilibrium maintenance.

There is another reason body posture is changed during movement. In contrast to rigid objects, living creatures are capable of changing their general shape. When a human bends forward, his center of gravity is displaced forward, which could provoke falling in the absence of appropriate postural changes in other body segments.

Quite a few interesting examples are mentioned in the literature that deal with the postural adjustments associated with movement in man. Babinski (1899) mentioned that arching of the head and back in standing man is accompanied by a flexion of the knees. This "synergy" permits the maintenance of the center of gravity in the same position. In patients with asynergia, arching of the back is not accompanied by a flexion of the knees and falling backward occurs (Figure 2-1). Another interesting example of synergy was given by Gurfinkel and Elner (1973), who observed that respiratory movements of the chest in standing man are accompanied by movements of the hips. These compensate for the displacement of the center of gravity that the chest movement would otherwise have provoked. Leg and arm movements are accompanied by marked postural changes in standing man (Alexeiev & Naydell, 1972; Belenkii, Gurfinkel, & Paltsev, 1967; Bouisset & Zattara, 1980; Nashner & Cordo, 1980). Postural changes also accompany movements when the arms are used for support of the body (Marsden, Merton, & Morton, 1978; Nashner & Cordo, 1980).

Several features are common to the various postural adjustments associated with movement. First, the myographic changes associated with the adjustment of posture precede by some 50 msec those of the

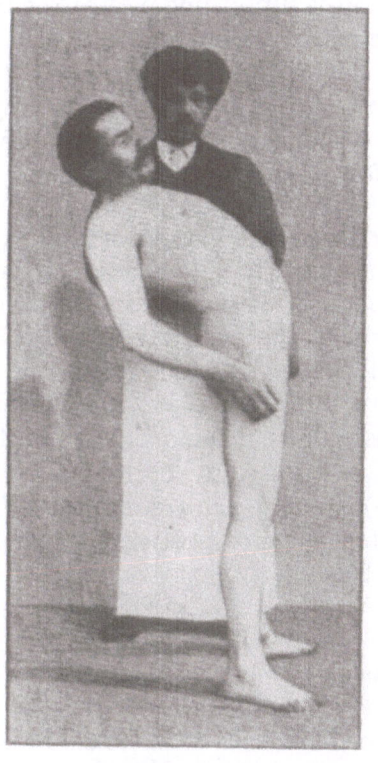 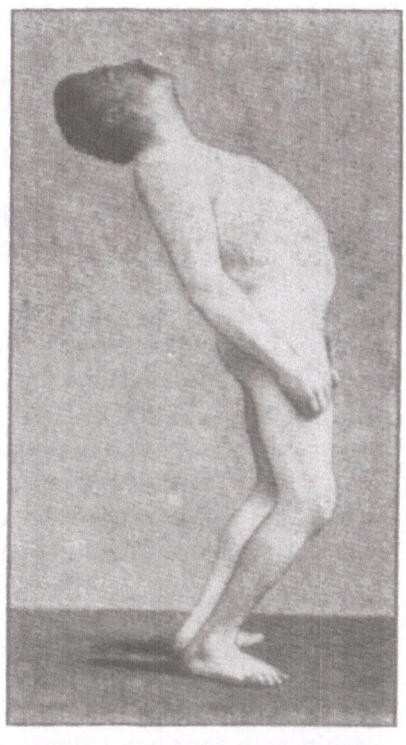

(a)   (b)

Fig. 2-1. Asynergia in a patient with cerebellar lesion. Compare the lack of knee flexion in the cerebellar patient (a) with the knee flexion of the normal subject (b). (From Babinski, J. De l'asynergie cérébelleuse. *Revue Neurologique*, (Paris), 1899, 7, 806–816.)

moving segment. When an arm is being raised, for example, myographic changes are seen in leg muscles before any change is observed in the deltoid muscle of the moving arm (Belenkii et al., 1967). Thus, postural changes anticipate the movement and compensate in advance for the forces that the movement exerts on the supporting part of the body. This phenomenon is also seen with load changes in an arm segment maintained in a fixed position (Marsden et al., 1978; Nashner & Cordo, 1980; Traub, Rothwell, & Marsden, 1980). Short-latency (50-msec) myographic changes are seen at the level of the supporting limbs before the load changes would have provoked any mechanical disturbance at that level. This "reflex" response thus anticipates the mechanical disturbances.

Second, postural adjustment is gated on or off, or modulated according to the body weight supported by a given limb. Sensory cues are probably involved in this mechanism. As noted above, myographic changes in leg muscles precede an arm movement in the standing subject. However, this postural adjustment is reduced or disappears when the subject is bending forward and is supported on a bar at chest level (Nashner & Cordo, 1980). When an arm is supporting part of the body weight, muscles of this arm show anticipatory postural adjustments that are otherwise lacking.

Finally, as shown by Gurfinkel and Elner (1973), postural adjustment shows long-term adaptation. It disappears in the normal healthy subject who has been lying in bed for several days and recovers very rapidly after training.

Very little is known about the central organization of postural adjustment in man. The short latency of the myographic changes observed when a load change is provoked at the level of a fixated arm joint led Nashner and Cordo (1980) to suggest that the adjustment is organized at a low hierarchical level. According to Babinski (1899), asynergia may result from cerebellar lesions, although this has been questioned by other authors (Thomas, 1940; Holmes, 1939). The contribution of the basal ganglia to postural adjustment was stressed by Martin (1967) and more recently by Traub et al. (1980). Finally, the frontal area of the cortex may be important in the performance of postural adjustments according to Gurfinkel and Elner (1973).

Coordination Between Posture and Movement in the Quadruped

The quadruped has been schematically represented by Gray (1944) as a table with four legs (Figure 2-2). In contrast to a rigid table, joints are inserted between the table and the legs and also between the different segments of the "table top"; this indicates that the rigidity of the quadruped can be modified by the action of the central nervous system on the muscles.

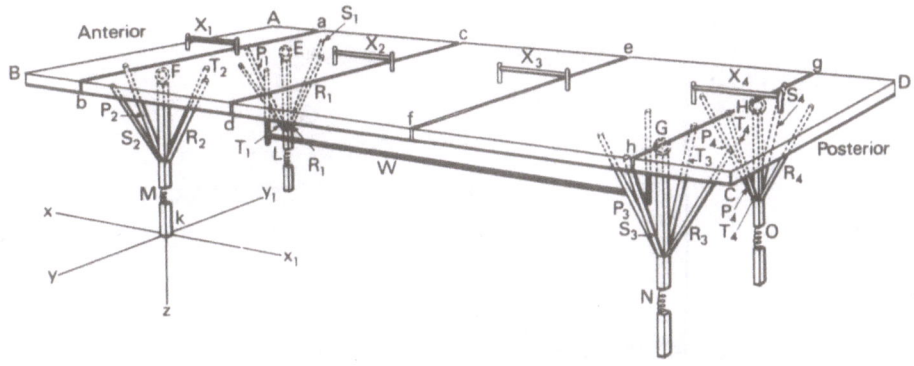

Fig. 2-2. Model of quadruped as a table with four legs. Note the joints in the back of the table and the ball joints with supports at the top of each leg. (From Gray, J. Studies in the mechanisms of the tetrapod skeleton. *Journal of Experimental Biology*, 1944, *20*, 88–116.

The weight of a quadruped is supported on four thrust points, the paws, and each limb displacement is of course accompanied by a new distribution of the weight on the three remaining limbs. The quadruped is an interesting model for the study of postural adjustment because, by using force platforms to measure the vertical forces exerted by each limb, it is possible to have a reasonable approximation of the postural changes that are taking place. This model has been used to analyze which type of body support the standing quadruped uses during single limb movement, and to investigate the central control of both movement and associated postural changes.

Two Patterns of Postural Adjustment

We can put a rigid table with four legs on the four force platforms and observe the force changes that take place when one platform is removed, that is, when one of the legs becomes unsupported. In this case, the weight of the table is supported by one diagonally opposite pair of feet, whereas the other pair is unloaded (Figure 2-3). The center of gravity remains at the same central position, as can be shown by calculating the center of pressure (resultant of forces recorded from the four platforms) before and after the drop of the one platform.

How closely does the behavior of a standing quadruped approach that of a rigid object? Ioffé and Andreyev (1969) were the first to notice that a diagonal bipedal stance accompanies a limb movement in the dog. If we put a cat on four force platforms and remove one of the supports unexpectedly (Figure 2-4), after a short time a typical diagonal bipedal stance develops, resembling that of a rigid table, with one pair of diagonally opposite limbs loaded and the other pair unloaded. The same type of diagonal pattern is observed when a horizontally directed perturbation is applied to a limb, forcing it to move. The diagonal pat-

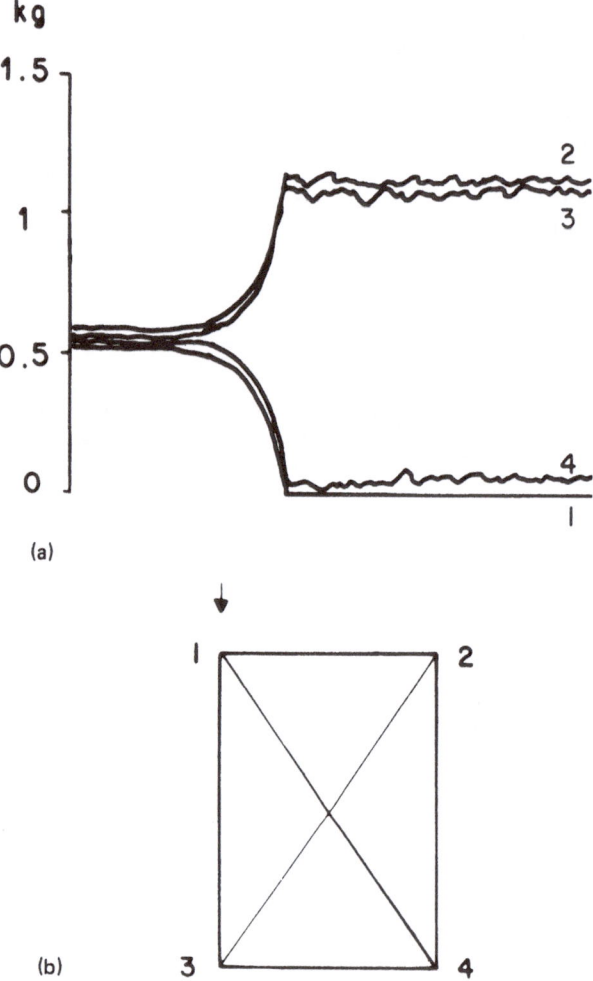

Fig. 2-3.(a) The rectangle represents a table with four feet, each corner corresponding to a foot. It stands with each foot on a force platform. When one platform
is dropped (*arrow*), the weight of the table is supported by one pair of diagonally
opposite feet, whereas the two other feet are unloaded. (b) Vertical forces recorded
from each platform before, during, and after the drop.

tern is also seen when an unexpected stimulation is applied to a central
command pathway for movement, such as a motor cortical site, provoking a contralateral flexion movement (Gahéry & Nieoullon, 1978). The
degree of diagonality seen with cortical stimulation is more or less
pronounced depending on the stimulated site, but for a given site, the
highest values are observed in the neutral situation (conditions of quiet
standing with the animal receiving a continuous stream of milk). Thus,
in this so-called neutral situation, the postural circuits that produce the
diagonal pattern are permanently ready to function. In general, the

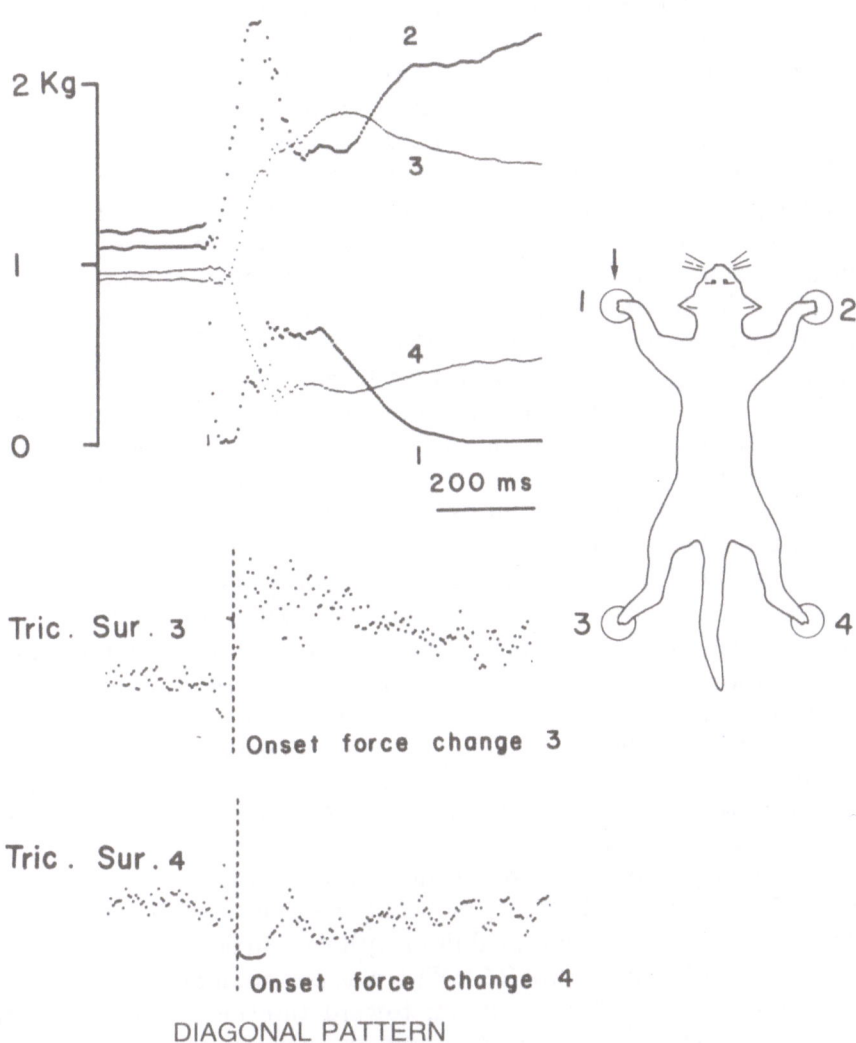

Fig. 2-4. Dropping of one force platform (*arrow*) provokes a diagonal postural adjustment in the standing cat. Mean value of 10 trials. After an initial phase during which a tendency to fall toward the unsupported limb is observed, a support with a diagonal pattern develops. The triceps suralis in the loaded hind limb (*3*) is activated and the triceps suralis in the unloaded hind limb (*4*) is inhibited. These changes are preceded by a brief phasic change of opposite direction that could result from stretch reflexes associated with the initial falling of the body toward the unsupported limb.

diagonal pattern is observed when an unforeseen event, external or internal (electrical stimulation), forces a limb to move.

With the diagonal pattern the center of gravity shows only a moderate or no displacement and the axis of the vertebral column remains rigid. Two torsion couples of opposite direction are exerted at the level of fore and hind limbs, as indicated by the force changes recorded during the diagonal pattern (Figure 2-4). These torsion couples should provoke rotation of the vertebral column, however, the rigidity of the back muscles is sufficient to oppose these couples and to prevent torsion movements. Macpherson, Dufossé, and Massion, (1980) demonstrated this rigidity by inserting rods in given vertebrae (T1, T12, L5) and measuring their lateral displacement and their rotation in the frontal plane. No significant lateral displacement or torsion of one segment with respect to the others was observed during the diagonal postural adjustment when one limb was forced to move by an external horizontal force. The diagonal pattern is not, however, a purely mechanical effect resulting from unloading of one limb, as in the case of the rigid table. An active neural contribution can be shown in the case of forelimb unloading caused by a horizontally directed perturbation. Inhibition of the triceps in the contralateral hind limb and activation of the triceps in the ipsilateral hind limb, are observed to precede force changes by 5–30 msec (Macpherson et al., 1980). In addition, in the case of stimulation of a cortical site provoking a hind limb movement, postural force changes of the fore limbs precede the force changes of the moving hind limb (Gahéry & Nieoullon, 1978).

In contrast to the diagonal pattern, a nondiagonal pattern is seen with the placing movement of a trained, free-standing cat when each movement is reinforced by milk reward. It is also seen with a conditioning paradigm in which the cat is trained to raise one limb when a conditioned stimulus (discontinuous sound) is delivered. This nondiagonal pattern is characterized by the fact that the weight of the moving limb is transferred to the symmetrical limb with very little contribution from the other two limbs (Figure 2-5). With this pattern a marked lateral displacement of the vertebral column toward the side of the supporting limb is seen. The back is not displaced as a whole. A lateral flexion is observed at the level of the fore half of the body in the case of fore limb movement.

The functional significance of each of these patterns, nondiagonal and diagonal, seems to be quite different (Figure 2-6). The diagonal pattern appears to be always available and ready to function. An unexpected event that forces one limb to move will make this pattern appear. It maintains the back rigid and the center of gravity at about the same position. Thus the general orientation of the body is maintained. It has, however, the disadvantage of being unstable due to the fact that it uses a bipedal stance. The nondiagonal pattern has been observed in moderate degree with a reinforced placing movement (Coulmance, Gahéry,

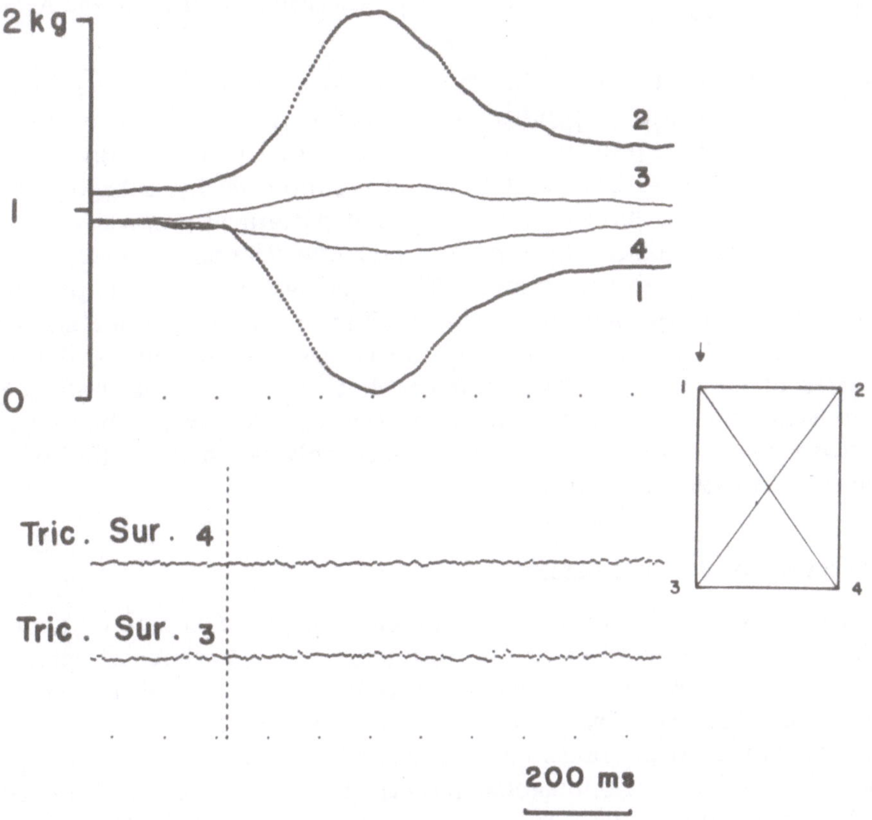

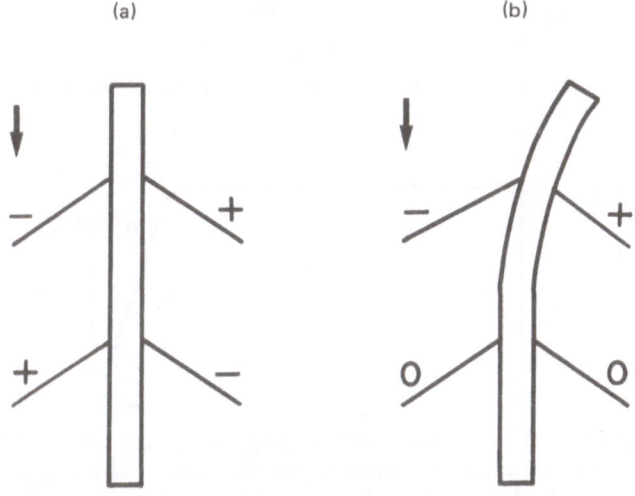

2 kg

2
3

I

4
I

0

Tric . Sur . 4

Tric . Sur . 3

1

2

3

4

200 ms

Fig. 2-5. Postural adjustment in conditioned fore limb movement is nondiagonal, that is, force changes are observed almost exclusively at the level of the fore limbs with the fore limb lift-off. Mean value of 15 trials. No myographic changes are observed in the triceps suralis of either hind limb.

(a)

(b)

−

+

+

−

−

+

0

0

Fig. 2-6. Two types of postural adjustment accompanying single limb movement in quadruped. (a) Diagonal pattern. (b) Nondiagonal pattern.

Massion, & Swett, 1979; Gahéry, Ioffé, Massion, & Polit, 1980) and very markedly with a conditioned lift-off movement. Usually, during the training procedure for the conditioned movement, the movement is first elicited by tactile stimulation and is supported by a diagonal pattern. As training proceeds, the diagonal pattern is progressively and spontaneously replaced by a nondiagonal one. The same process can be seen with the placing movement. The nondiagonal pattern is probably not a new one, formed during the training procedure, but already part of the animal's repertoire. It has the advantage of being more stable, the center of gravity being displaced nearer to the center of the triangle formed by the three remaining supporting legs. It may also be that this pattern is more energy efficient because only one-half of the body is concerned with the postural changes.

Presetting of Postural Pattern

Let us now examine in detail the preparation and initiation of a conditioned lift-off movement of one fore limb. The experimental paradigm consists of a preparatory signal (continuous tone of 0.6–1.2 seconds duration) followed by a conditioned stimulus (discontinuous tone) during which lift-off takes place (Figure 2-7). The lift-off movement is accompanied by a nondiagonal pattern. Do the central processes activated during the preparatory phase modify the state of the local circuits responsible for the postural adjustment in such a way that a nondiagonal pattern is prepared? In other words, does movement preparation also involve preparation of the appropriate postural pattern? If postural preparation does take place, is it restricted to the adjustments associ-

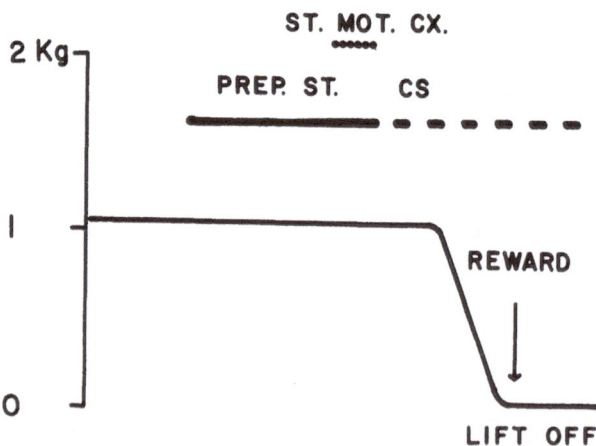

Fig. 2-7. Experimental paradigm with a preparatory period (*PREP*, continuous tone) followed by a conditioned stimulus (*CS*, discontinuous tone), the last stimulus being followed by limb lift-off. At the end of the preparatory period a perturbing stimulus, for example, stimulation of the motor cortex (*ST MOT CX*) is delivered.

ated with movement of the prepared limb or is it also apparent for the other limbs?

One answer to this problem is provided by the situation in which a horizontal force is applied to the toes of a hind limb during the period preparatory to a movement of a fore limb. As can be seen in Figure 2-8, the hind limb imposed movement is accompanied by a diagonal pattern, whereas the fore limb conditioned movement that immediately follows is accompanied by the expected nondiagonal pattern. This indicates

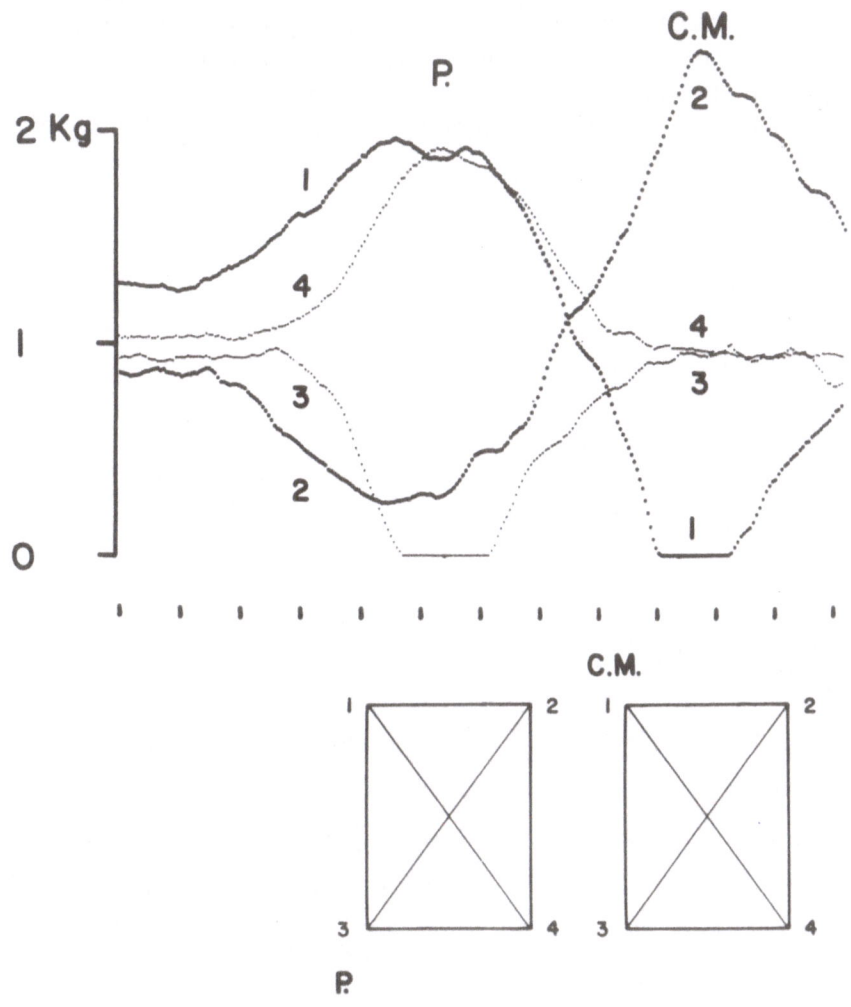

Fig. 2-8. Single trial with the same paradigm as Figure 2-7. During the preparatory period, a horizontal force makes the left hind limb move (perturbation, *P*). This movement is accompanied by a postural adjustment with a diagonal pattern. The succeeding conditioned stimulus is followed by lift-off of the left fore limb. This conditioned movement (*CM*) is accompanied by a nondiagonal pattern.

that the two postural patterns can be observed depending on the site of the movement and the way it is elicited.

Is there nevertheless preparation of the nondiagonal pattern at the level of the limb that is preparing the movement? In order to examine this question cortical stimulation was delivered to sites on the left and right motor cortex, both of which produce a contralateral fore limb flexion. When stimulation was delivered at the end of the preparatory period to the motor cortex ipsilateral to the moving limb (i.e., provoking a movement of the nonprepared side), the postural adjustment was comparable to that observed during the neutral sessions (Figure 2-9). When the cortical stimulation was delivered contralateral to the prepared limb (i.e., provoking a movement of the same limb that performs the lift-off movement), the pattern of the postural adjustment had a tendency to change and to be less diagonal and closer to the nondiago-

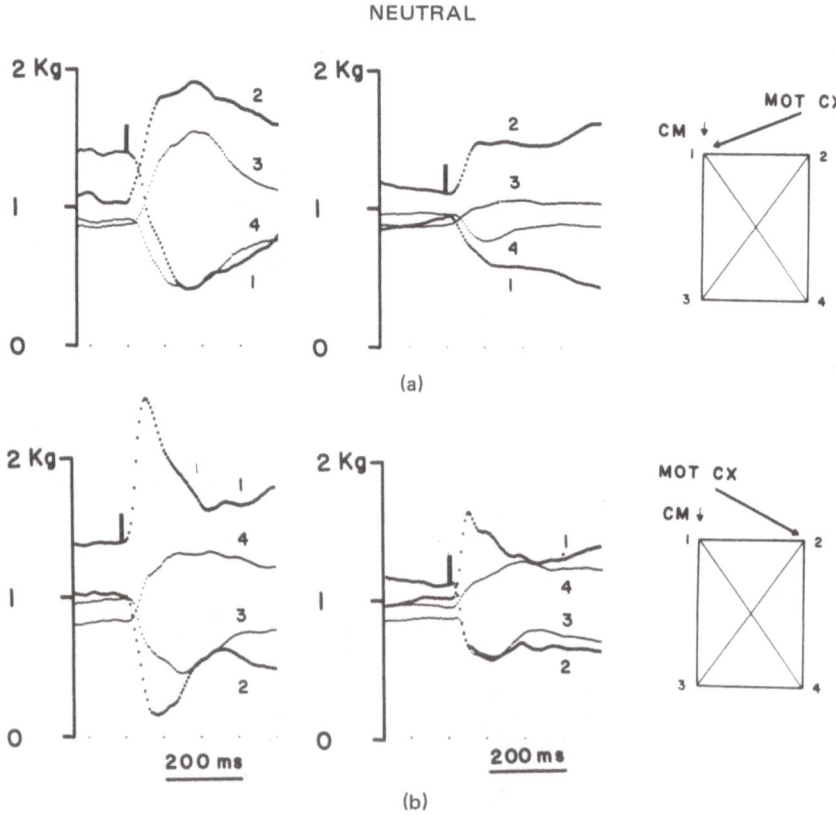

Fig. 2-9.(a) Effect of cortical stimulation in a neutral situation. (b) Effect of cortical stimulation during the preparatory phase to a conditioned lift-off movement of left fore limb (CM). Note that the postural pattern seen with lateral cortical stimulation (MOT CX) ipsilateral to the conditioned movement is unmodified, whereas that seen with contralateral stimulation is less diagonal than in the neutral stimulation.

nal pattern. The change in pattern was usually better observed during the second or third session, when both cortical stimulation and conditioned signal were associated, than during the first.

This experimental finding indicates that the postural pattern provoked by cortical stimulation can be modified to be less diagonal when delivered during the sessions of conditioned movement. This change may be lateralized, that is, occur only for stimulation of the side of the cortex related to the limb performing the conditioned movement. This result can be interpreted in two ways. First, cortical stimulation may reveal a local preparatory setup for the postural circuits that are associated with the conditioned lift-off movement. Second, cortical stimulation may act as a conditioned stimulus itself because of its occurrence, even if only rarely, during the conditioning sessions and has a tendency to elicit a postural pattern that resembles that of the conditioned movement. Both mechanisms are probably involved; however, at the present stage of the experiments, the question remains open.

Central Organization of Postural Changes

Several models have been proposed in order to explain how movement is organized centrally. One of the best known is that proposed by Allen and Tsukahara (1974), in which two categories of central brain structures are involved in the task of movement execution. In the first class are structures involved in movement preparation and initiation: associative cortical areas, motor cortex, neocerebellum, and basal ganglia. The second group of structures is concerned with movement execution: intermediate cerebellum, red nucleus, and several types of feedback loop, internal as well as external. Are the postural changes that accompany movement performance organized at the level of brain structures involved in the preparation and initiation of movement, or within the circuits responsible for movement performance?

Evidence for a Circuit Organized at a Low Hierarchical Level

Experiments with central stimulation give some evidence that circuits providing for postural support to movement might be organized at some low level. As shown by Gahéry and Nieoullon (1978), electrical stimulation of a cortical site provoking a contralateral movement in the standing cat does not result in falling. Apparently postural adjustment maintains equilibrium. Latency measurements of force changes of the moving and supporting limbs show that, at least for a cortically induced hind limb movement, the force change of each of the supporting limbs precedes that of the moving limb. Thus, the postural adjustment, usually diagonal, appears to be provoked in a feedforward manner. Postural

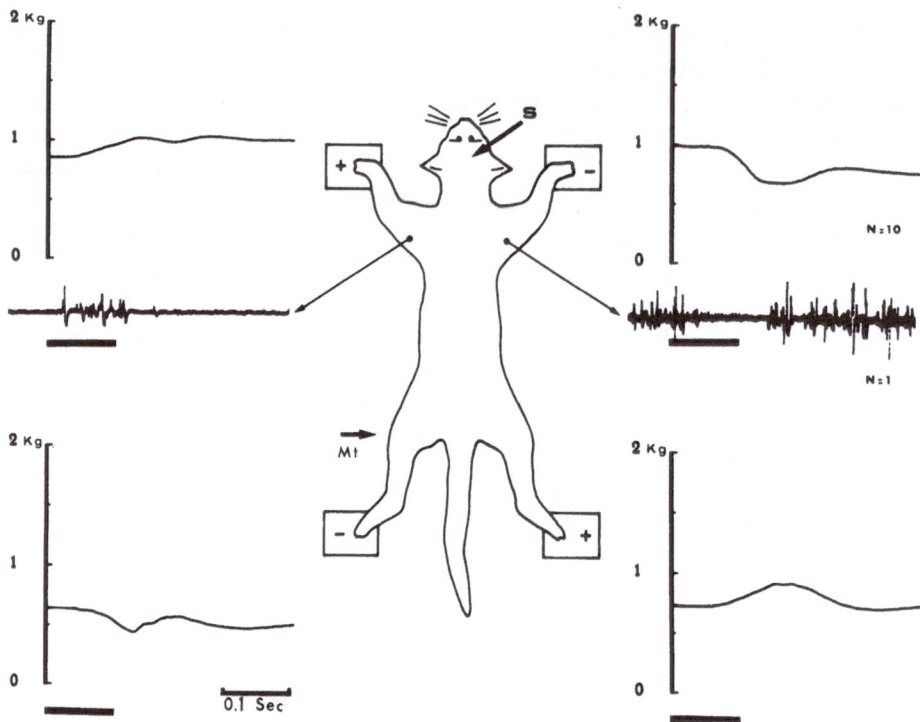

Fig. 2-10. Stimulation (0.5 msec, 100 Hz, 0.1 second) within the red nucleus provokes contralateral hind limb flexion (*MVT*). Intensity of stimulation is adjusted in
order to be subthreshold for movement. Note the diagonal pattern of the postural
adjustment (unloading of right fore limb and left hind limb, loading of left fore
limb and right hind limb) and the early myographic changes in the fore limb triceps.

adjustment is simultaneous with movement when cortical stimulation is
delivered after section of the pyramidal tract (Nieoullon & Gahéry,
1978). Appropriate postural adjustment accompanies movement provoked by stimulation of another central command pathway for movement, the rubrospinal tract. As can be seen in Figure 2-10, stimulation
of a site within the red nucleus that provokes a contralateral hind limb
movement is accompanied by a postural adjustment of the diagonal
type.

As a whole, these results could be interpreted as indicating that each
site producing a contralateral movement may also contain the entire
organization that provides for the associated postural adjustment. It is
more likely, however, that a single circuit for the postural adjustment
exists which is activated by the same impulses that elicit the movement;
this circuit could be located at a site where the pyramidal and rubrospinal pathways converge. Convergence of pyramidal and rubrospinal
fibers is seen at the bulbar level (lateral reticular nucleus) and spinal
cord. Sherrington (1906) showed in the decerebrate cat that there is
an organization linking flexors and extensors of diagonally opposite

limbs and this organization might be appropriate for postural adjustment with a diagonal pattern.

Role of Motor Cortex, Red Nucleus, and Cerebellum

The role of the motor cortex, red nucleus, and cerebellum in the postural adjustments associated with limb movement has been investigated mainly with lesion and single-unit recording experiments. After total cerebellectomy, the postural adjustments are deeply depressed, but the pattern is not abolished. Thus circuits outside the cerebellum can be responsible for the postural adjustments. Nevertheless, the cerebellum plays an important role in the control of these adjustments (Regis, Trouche, & Massion, 1976).

The red nucleus, which is one of the main output pathways by which cerebellar influence can reach the spinal cord, is involved dynamically during postural adjustment (Ioffé, 1973a). However, unitary discharges are mainly related to the change in the state of contraction of the contralateral musculature, whether the muscles are involved in movement or in postural adjustment (Padel & Steinberg, 1978).

The ventrolateral thalamic nucleus is located on another important pathway by which the cerebellum can influence the spinal cord, that is the cerebellocorticospinal pathway. Unit activity in this nucleus is modified when contralateral limbs are involved in the postural adjustment (Smith, Massion, Gahéry, & Roumieu, 1978). However, in contrast with red nucleus cells, the unit discharges are not concerned with specific muscle activity and, for many cells, are seen throughout the motor sequence.

Complete lesion of the motor cortex depresses the contralateral movement and its associated adjustment as a whole, including ipsilateral muscles involved in the adjustment. In contrast, ipsilateral placing and its associated postural adjustment (involving muscles of both sides) is unmodified (Massion, 1979). This observation is consistent with the notion that the motor cortex could exert tonic control on a subcortical circuit responsible for postural adjustment (Figure 2-11). Alternatively, the postural adjustment could be depressed due to the lack of phasic impulses associated with movement control which normally originate in the motor cortex or pass through it.

What is the specific action of the central motor pathways involved in the control of postural adjustment? One possibility is that they are associated with the regulation and coordination of a given postural pattern. That is, the diagonal pattern. The cerebellorubral and cerebellocortical pathways would thus be included in regulating loops comparable to those that intervene during locomotion (Grillner, 1975). A second and not exclusive possibility is that part of the posture-related activity is associated with a central command involved in the selection of the appropriate pattern, as a function of the evolving motor act. This selection would take place either during the preparatory phase preceding movement execution or with movement initiation.

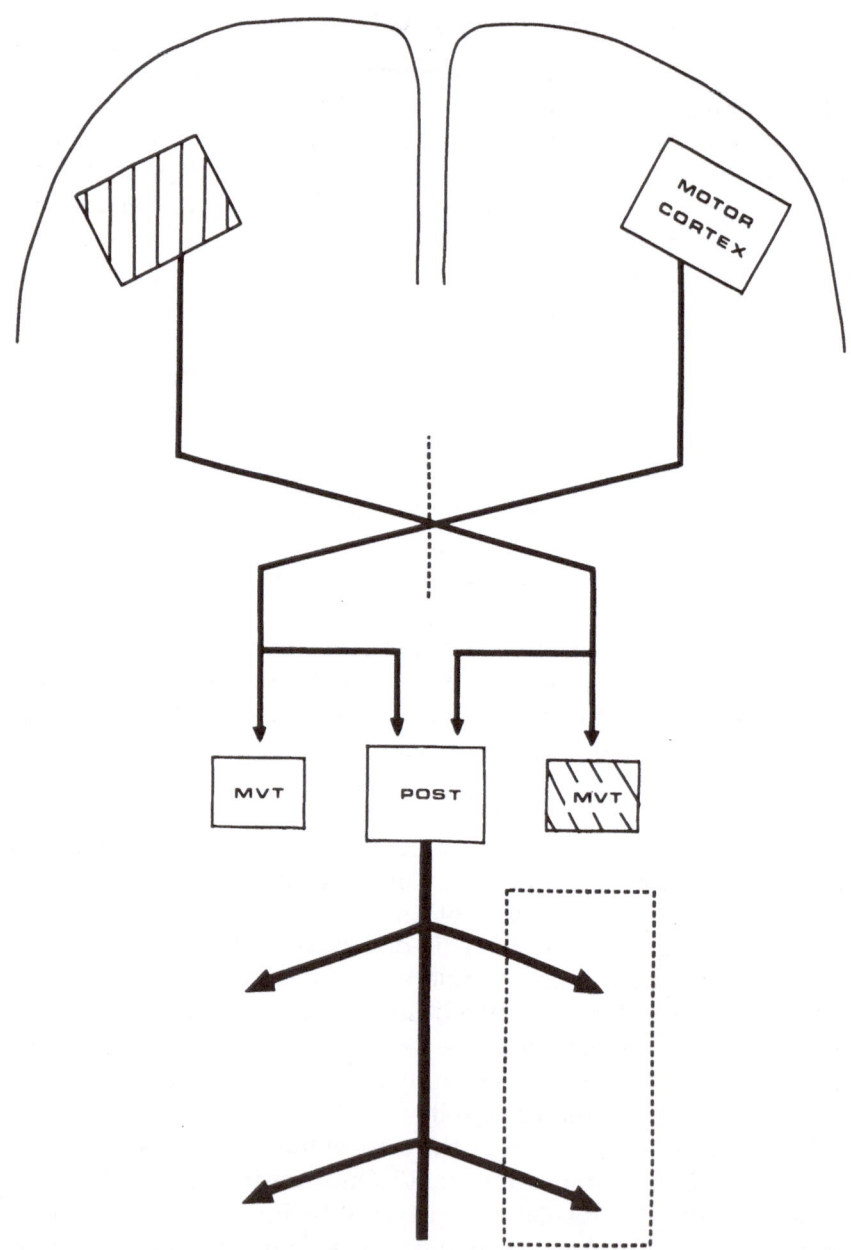

Fig. 2-11. Feedforward command of movement and posture from one motor cortex—one of the hypotheses proposed to explain the link between postural adjustment and limb movement. *MVT*, movement; *POST*, posture; *hatched areas*, side of motor cortical lesion and impaired contralateral movement.

Conclusions

The central nervous system appears to provide a postural change automatically when a movement is carried out; the main consequence of this change is to permit the maintenance of equilibrium during movement execution. In the standing quadruped, a diagonal pattern of postural adjustment is observed with any limb movement provoked by an unexpected event. This pattern is probably organized at the bulbospinal level and seems to be the most appropriate way to provide a temporary support (even if this support if unstable) for the body weight when the support of one limb is lost. A nondiagonal pattern is seen mainly with conditioned movements of a limb. It provides more stable support and may be more energy efficient.

As two patterns have been observed, one may ask how the central nervous system is able to select the appropriate pattern. The diagonal pattern appears to be permanently ready to function for any unexpected perturbation forcing one limb to move. Thus, one may suppose that one of the main functions of the central structures associated with movement preparation and initiation is to select the appropriate postural pattern and suppress inappropriate patterns. This might be achieved by pyramidal tract fiber collaterals (Shinoda, Zarzecki, & Asanuma, 1979) or by those that have been demonstrated in the rubrospinal (Shinoda, Ghez, & Arnold, 1977), reticulospinal (Peterson, Maunz, Pitts, & Mackel, 1975), and vestibulospinal (Abzug, Maeda, Peterson, & Wilson, 1974) tracts as well. These collaterals distribute along the spinal cord and would act on the propriospinal circuits involved in the postural support of movement, as suggested by Ioffé (1973a & b). The basic diagonal postural pattern would then be gated off or modulated and another pattern more appropriate for each specific intentional movement would appear.

Acknowledgments

This work was supported by ATP INSERM A 650 5174.

References

Abzug, C., Maeda, M., Peterson, B. W., & Wilson, V. J. Cervical branching of lumbar vestibulospinal axons. *Journal of Physiology* (London), 1974, *243*, 499-522.

Alexeiev, M. A., & Naydel, A. V. The mechanisms of interrelationship between human muscle activity in complex motor tasks. *Zhurnal Physiol Fiziologicheskii*, 1972, *58*, 1721-1730 (in Russian).

Allen, G. I., & Tsukahara, N. Cerebro-cerebellar communication systems. *Physiological Reviews*, 1974, *54*, 957-1006.

Babinski, J. De l'asynergie cérébelleuse. *Revue Neurologique* (Paris), 1899, *7*, 806-816.

Belenkii, V. E., Gurfinkel, V. S., & Paltsev, E. I. On elements of control of voluntary movements. *Biofizica*, 1967, *12*, 135-141 (in Russian).

Bouisset, S., & Zattara, M. Anticipatory muscular activities prior to a movement. *Neuroscience Letters, Suppl.* 1980, *5*, S114.

Coulmance, M., Gahéry, Y., Massion, J., & Swett, J. E. The placing reaction in the standing cat: A model for the study of posture and movement. *Experimental Brain Research*, 1979, *37*, 265-281.

Gahéry, Y., Ioffé, M., Massion, J., & Polit, A. The postural support of movement in cat and dog. *Acta Neurobiologial Experimentalis*, 1980, *40*, 741-756.

Gahéry, Y., & Nieoullon, A. Postural and kinetic coordination following cortical stimuli which induce flexion movements in the cat's limbs. *Brain Research*, 1978, *155*, 25-37.

Gray, J. Studies in the mechanisms of the tetrapod skeleton. *Journal of Experimental Biology*, 1944, *20*, 88-116.

Grillner, S. Locomotion in vertebrates: Central mechanisms and reflex interaction. *Physiological Reviews*, 1975, *55*, 247-304.

Gurfinkel, V. S., & Elner, A. M. On two types of static disturbances in patients with local lesions of the brain. *Agressologie*, 1973, *14D*, 64-72.

Hess, W. R. Teleokinetisches und ereismatisches Kräftesystem in der Biomotorik. *Helvetica Physiologica et Pharmocologica Acta*, 1943, *1*, C62-C63.

Holmes, G. The cerebellum of man. *Brain*, 1939, *62*, 1-30.

Ioffé, M. *Cortico-spinal mechanisms of instrumental motor reactions.* Moscow: Nauka, 1973 (in Russian). (a)

Ioffé, M. Pyramidal influences in establishment of new motor coordinations in dogs. *Physiology and Behavior*, 1973, *11*, 145-153. (b)

Ioffé, M. E., & Andreyev, A. E. Inter-extremities coordination in local motor conditioned reactions of dogs. *Zhurnal Vysshei Nervnoi Deiatel'nosti imeni I.P. Pavlova*, 1969, *19*, 557-565 (in Russian).

Jung, R., & Hassler, R. The extrapyramidal motor system. In J. Field, H. W. Magoun, & V. E. Hall (Eds.), *Handbook of Physiology.* Section I: *Neurophysiology* (Vol. 2). American Physiological Society, 1960, chap. 35.

Macpherson, J. M., Dufossé, M., & Massion, J. Two types of postural adjustment to limb flexion in the cat. *Society of Neurosciences Abstracts*, 1980, *6*, 464.

Magnus, R. *Körperstellung.* Berlin: Springer, 1924.

Marsden, C. D., Merton, P. A., & Morton, H. B. Anticipatory postural responses in the human subject. *Journal of Physiology* (London), 1978, *275*, 47P.

Martin, J. P. *The basal ganglia and posture.* London: Pitman, 1967.

Massion, J. Role of motor cortex in postural adjustments associated with movement. In H. Asanuma & V. J. Wilson (Eds.), *Integration in the nervous system.* Tokyo: Igaku-Shoin, 1979, pp. 239-260.

Nashner, L. M., & Cordo, P. J. Coordination of arm movements and associated postural adjustments in standing subjects. *Society of Neurosciences Abstracts*, 1980, *6*, 394.

Nieoullon, A., & Gahéry, Y. Influence of pyramidotomy on limb flexion movements induced by cortical stimulation and on associated postural adjustment in the cat. *Brain Research*, 1978, *155*, 39-52.

Padel, Y., & Steinberg, R. Red nucleus cell activity in awake cats during a placing reaction. *Journal de Physiologie* (Paris), 1978, *74*, 265-282.

Peterson, B. W., Maunz, R. A., Pitts, N. G., & Mackel, R. G. Patterns of projection and branching of reticulospinal neurons. *Experimental Brain Research*, 1975, *23*, 333-351.

Polit, A., & Bizzi, E. Characteristics of motor programs underlying arm movements in monkeys. *Journal of Neurophysiology*, 1979, *42*, 183-194.

Regis, H., Trouche, E., & Massion, J. Effet de l'ablation du cortex moteur ou du cervelet sur la coordination posturocinétique chez le chat. *Electroencephalography and Clinical Neurophysiology*, 1976, *41*, 348-356.

Roberts, T. D. M. *Neurophysiology of postural mechanisms.* London: Butterworths, 1967.

Sherrington, C. S. *The integrative action of the nervous system.* London: Constable, 1906.

Shinoda, Y., Ghez, C., & Arnold, A. Spinal branching of rubrospinal axons in the cat. *Experimental Brain Research*, 1977, *30*, 203-218.

Shinoda, Y., Zarzecki, P., Asanuma, H. Spinal branching of pyramidal tract neurons in the monkey. *Experimental Brain Research*, 1979, *34*, 59-72.

Smith, A. M., Massion, J., Gahéry, Y., & Roumieu, J. Unitary activity of ventrolateral nucleus during placing movement and associated postural adjustment. *Brain Research*, 1978, *149*, 329-346.

Thomas, André. *Equilibre et équilibration.* Paris: Masson, 1940.

Traub, M. M., Rothwell, J. C., & Marsden, C. D. Anticipatory postural reflexes in Parkinson's disease and other akinetic-rigid syndromes and in cerebellar ataxia. *Brain*, 1980, *103*, 393-412.

Chapter 3

Neural Mechanisms of Visual Orientation in Rodents: Targets Versus Places

Melvin A. Goodale

In many vertebrates, including mammals, the primary sensory system for the organization of behavior in space is vision. For example, the complex and exquisite movements that are made by predators as they pursue their prey are in many cases clearly under visual control, as are the equally impressive maneuvers made by the prey as they attempt to elude their would-be captors. But both predator and prey are not simply changing their direction and rate of locomotion in response to the movements of one another; they are also avoiding obstacles and negotiating barriers in the terrain through which they are moving. Moreover, the prey may be locomoting toward a particular and visible target such as a tree into which the predator cannot follow, or toward a particular but invisible place such as a burrow in which it can take refuge. Finally, during the performance of these patterns of behavior, the information arriving through the eyes is constantly being integrated with information provided by the other receptor systems.

In the visual neurosciences there has been a tendency to lump all of these visuomotor behaviors together under the general heading of orientation behavior, to distinguish it from the identification or recognition functions of the visual system. In terms of the predator–prey example discussed above, those parts of the visual system involved in recognizing or identifying a prey or predator have traditionally been thought to be distinct from those parts of the visual system controlling orientation toward a prey object or away from a predator. This *two visual systems* hypothesis was made anatomically explicit by Schneider (1967, 1969). He suggested that the phylogenetically older pathway from retina to superior colliculus is involved in the control of orientation behavior enabling an animal to localize a stimulus in visual space, whereas the more recently evolved geniculostriate system participates in the identification of the visual stimulus. This distinction implies that the visual pathways are not a unitary system, but are instead a network of relatively independent visuomotor channels. However useful the two visual systems hypothesis might have been in the past, it now appears to be an oversimplification. The term *orientation behavior*, for example, has been used in different contexts to refer to a wide variety of behaviors, including saccadic eye movements, head movements, tracking, reaching with the forelimbs, locomotion toward a particular place within a frame of reference, and the maintenance of upright posture. While it is clear that these patterns of behavior are oriented with respect either to specific visual stimuli or to some sort of spatial schema, it is equally clear that very different control mechanisms are likely to be involved in each. Certainly, it is a fair assumption that the superior colliculus is neither necessary nor sufficient for the control of all visuomotor skills observed in mammals.

In most vertebrates, there are at least seven or eight retinofugal targets, and each of these projections is further elaborated after the primary synapse (Ebbesson, 1970; Riss & Jakway, 1970). These separate inputs and their interconnections are reflected in the complex array of visually guided movements that most mammals are capable of. Two visual systems are not enough—a multichannel (and interactive) model is needed.

Two types of visuomotor problems that mammals must deal with in moving around their environments—making visually guided movements toward targets, and making visually guided movements toward places—are examined in this chapter. The viewpoint is presented that target-directed movements consist of several different visuomotor programs, and at least three separate retinofugal pathways participate in the control and integration of these programs. It is also suggested that the visual control of movements directed toward places in space that are not marked by local cues is even more complex, depending not only on sensory information, but also on memory of the environment. Much of the evidence presented is derived from my own work on the visually guided behavior of two rodent species, the laboratory rat and the Mongolian gerbil (*Meriones unguiculatus*).

Visually Guided Movements Toward Targets

When a rat or gerbil runs toward a visual target, a number of different patterns of behavior can be observed (Goodale, Foreman, & Milner, 1978; Goodale & Milner, 1982; Goodale & Murison, 1975; Ingle, 1982; Ingle, Cheal, & Dizio, 1979). After the target has been detected, a series of discrete head movements are initiated that vary in amplitude as a function of the initial position of the target in the visual field. At the same time (or shortly thereafter), the animal begins to locomote toward the target, typically accelerating its rate of locomotion to a peak ve-

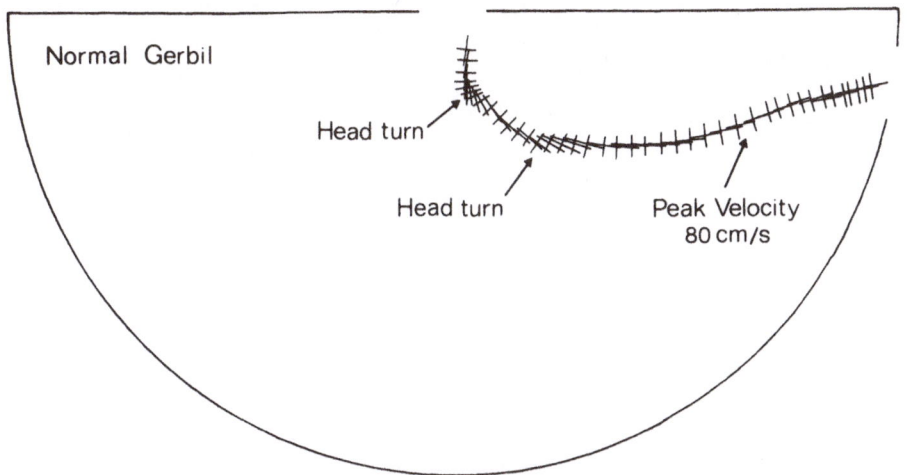

Fig. 3-1. Orientation movements made by a normal gerbil toward a discrete visual target. (This is a schematic drawing of the arena illustrated in Figure 3-3.) The gerbil has entered the arena (through opening *A*) and has initiated a series of head turns toward the target (opening *B*). The individual line segments that describe the path the gerbil followed were drawn from successive frames of film (64 frames/sec) and indicate the longitudinal axis of the gerbil's head on each frame. The line segment bisecting this axis and perpendicular to the path marks a vertical plane passing through the center of the two eyes. From this reconstruction a number of features of the gerbil's behavior can be directly observed, including successive head turns and changes in velocity.

locity and then rapidly decelerating as it makes the final approach. This sequence of events is illustrated in Figure 3-1.

All these patterns of visuomotor behavior are modifiable if the stimulus conditions vary. Thus, if a moving target is used, the amplitude of the orienting head movements and the direction of the locomotion show compensatory changes. For example, if a target is first introduced at a point 20° from the vertical midline and moves rapidly into the gerbil's temporal visual field, the amplitude of the initial head turn is larger than 20° (Ingle et al., 1979). The amplitude of later turns also shows this overshoot, allowing the animal to achieve a "collision" course with the moving target. The ecological utility of this behavior is obvious: only by making a predictive orientation movement would a gerbil be able to pursue and catch fast-moving prey.

In the laboratory, rodents typically are required to make their visually guided movements directly toward the target. A hungry gerbil does not have to avoid an obstacle or barrier as it runs toward the baited disk or the small opening in an arena wall that contains a food reward. The normal environment of these animals is rarely so accommodating, however. Often an animal has to run around a clump of vegetation or negotiate some other obstacle in order to obtain a food item it has spotted.

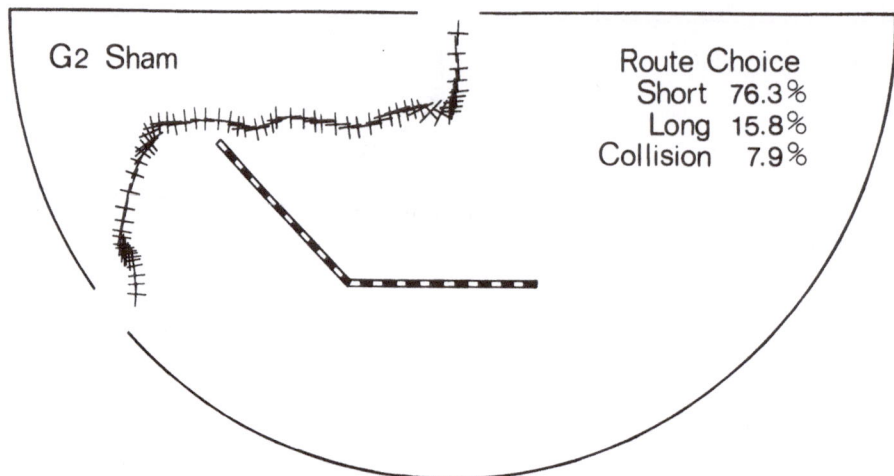

Fig. 3-2. Orientation movements made by a normal (sham-operated) gerbil approaching a target located behind a transparent grid barrier. (This is a schematic drawing of the arena illustrated in Figure 3-3.) The head and optic axes have been drawn from successive frames of film. This plot shows the gerbil moving into the arena (from opening *A*), making a head turn toward the target hold (*B*), avoiding the barrier, and locomoting toward the opening in the arena wall. Notice that the gerbil took the shorter of the two routes around the barrier. The performance of this gerbil on 52 barrier trials is summarized on the right.

I recently looked at such barrier avoidance in a laboratory setting (Goodale & Milner, 1982). Gerbils trained to run toward a small target located in any one of a number of different positions in the wall of a semicircular arena would almost immediately avoid a transparent grid suddenly placed between them and the target. As illustrated in Figure 3-2, they usually took the shorter of the two routes around the barrier. The latter observation suggests that relatively complex computations are involved in the control of visually guided locomotion toward a discrete target when an indirect route must be followed to acquire the target.

Orientation toward a visual stimulus, then, consists of a sequence of separate visuomotor acts, each of which can vary as a function of the characteristics of the stimulus and the situation in which the stimulus occurs. In the following discussion, possible neural substrates of these patterns of behavior are described.

Role of the Superior Colliculus

In rodents, a large fraction of the retinofugal fibers terminates in the superior colliculus (Lashley, 1934; Lund, 1972; Nauta & Straaten, 1947), where they distribute across the superficial and intermediate laminae in an orderly and retinotopic fashion (Lund, 1972). Early stimulation studies of the colliculus, principally in cats (Apter, 1946;

Hess, Bürgi, & Bucher, 1946; Hyde & Eason, 1959; Hyde & Eliasson, 1957), led to the hypothesis that this midbrain structure is important in the mediation of what Hess et al. (1946) termed "visual grasp reflexes," whereby objects in the peripheral visual fields are relocated on the central field by movements of the head and eyes. A similar orientation function for the colliculus was emphasized by Schneider (1967, 1969). Lesions of the superior colliculus result in orientation deficits in rats (Goodale et al., 1978; Goodale, Milner, & Rose, 1975; Goodale & Murison, 1975), gerbils (Goodale & Milner, 1982; Ingle, 1982; Mlinar & Goodale, 1980), and hamsters (Mort, Finlay, & Cairns, 1980; Schneider, 1967, 1969). Not all classes of visually oriented behavior, however, are equally affected, and the severity of the deficit is field dependent.

I recently examined the behavior of gerbils when they were required to detect and locomote toward visual targets presented in different parts of their visual field. An example of an apparatus used in these observations is shown in Figure 3-3. The training procedure was quite straightforward. Gerbils were required to run to a hole, 5 cm in diameter, which could be located in any one of several different positions in the wall of the semicircular arena. Upon entering the hole, they were presented with a sunflower seed. By filming the behavior of the gerbils

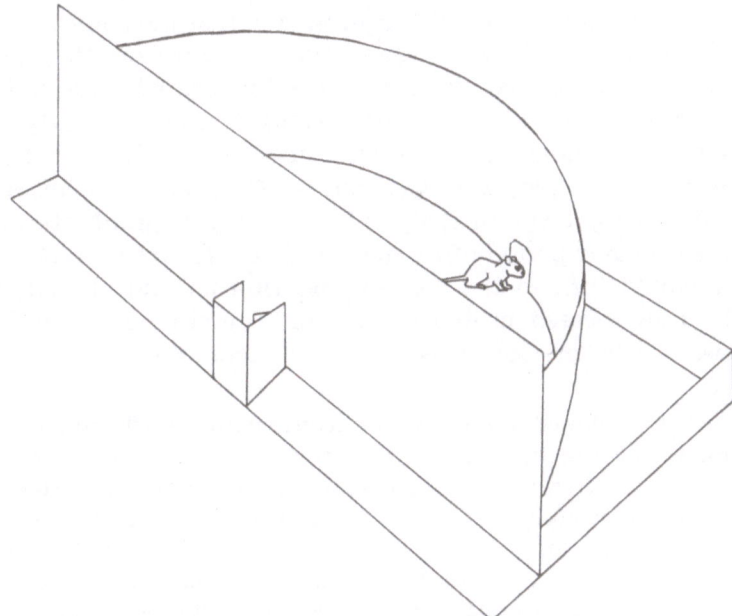

Fig. 3-3. Arena used to record orientation toward discrete visual targets in the gerbil. The radius of the arena was 35 cm and the opening in the arena wall was 5 cm wide. After the gerbil passed through the opening it received a sunflower seed.

through a 45° mirror suspended over the arena, I was able to plot the route each animal followed from the arena entrance to the hole. With this technique, it was also possible to record any head movements that an animal made as it oriented toward a hole located in its peripheral visual field.

Normal gerbils rapidly learned to detect and run toward the hole wherever it was located. The same was true for gerbils with large lesions of the posterior neocortex (including area 17 and extending into areas 18a and 18b, and sometimes areas 19 and 7). In other words, elimination of the entire geniculostriate system in adult animals did not interfere with their ability to detect a small dark hole against the white background of the wall and to run toward it with great accuracy. The same was not true for adult gerbils that sustained large bilateral lesions of the superior colliculus. In the early stages of training, these animals often ran toward another part of the arena before entering the hole, and the more peripheral the hole, the more often they did this. Although initially this result appeared to support the division of labor implied by the two visual systems hypothesis, the eventual outcome of these experiments required a more complicated account.

When the film records were carefully analyzed it became apparent that the gerbils with collicular lesions learned to run toward the hole as efficiently as normal animals when the hole was located within about 40° of the frontal midline. As shown in Figure 3-4, it was only when the hole was located outside this central area that the colliculectomized gerbils had difficulty. With extensive training, however, they were able to locomote toward even these peripherally located targets. This was true even for those with very deep lesions extending slightly into the dorsal tegmentum and periaqueductal gray matter. Nevertheless, their behavior on these occasions was far from normal. As illustrated in Figure 3-4, a colliculectomized gerbil would run forward into the arena and pause for some period of time, sometimes well over 1 second. Eventually it would turn in the correct direction and run directly toward the hole, now located in its rostral field. The visuomotor deficit does not appear to be one of visually guided locomotion per se; rather, it

---

Fig. 3-4. Routes followed in the arena by representative gerbils from the four different groups: sham operated gerbils (*Sham*), those with lesions of the posterior neocortex (*VC*), those with collicular lesions (*SC*), and those with pretectal lesions (*PT*). Four different positions of the opening in the arena wall are shown: in the top row the opening is located in the central position of the gerbil's field as it enters the arena; in each succeeding row, the opening is located more peripherally. Each column represents the performance of a single gerbil with the opening in the four positions. Each small dot represents the center of the gerbil's head on successive frames of film (64 frames/sec). The only animal to show inefficient running is the gerbil with lesions of the superior colliculus, and only on openings located more than 40° from the midline.

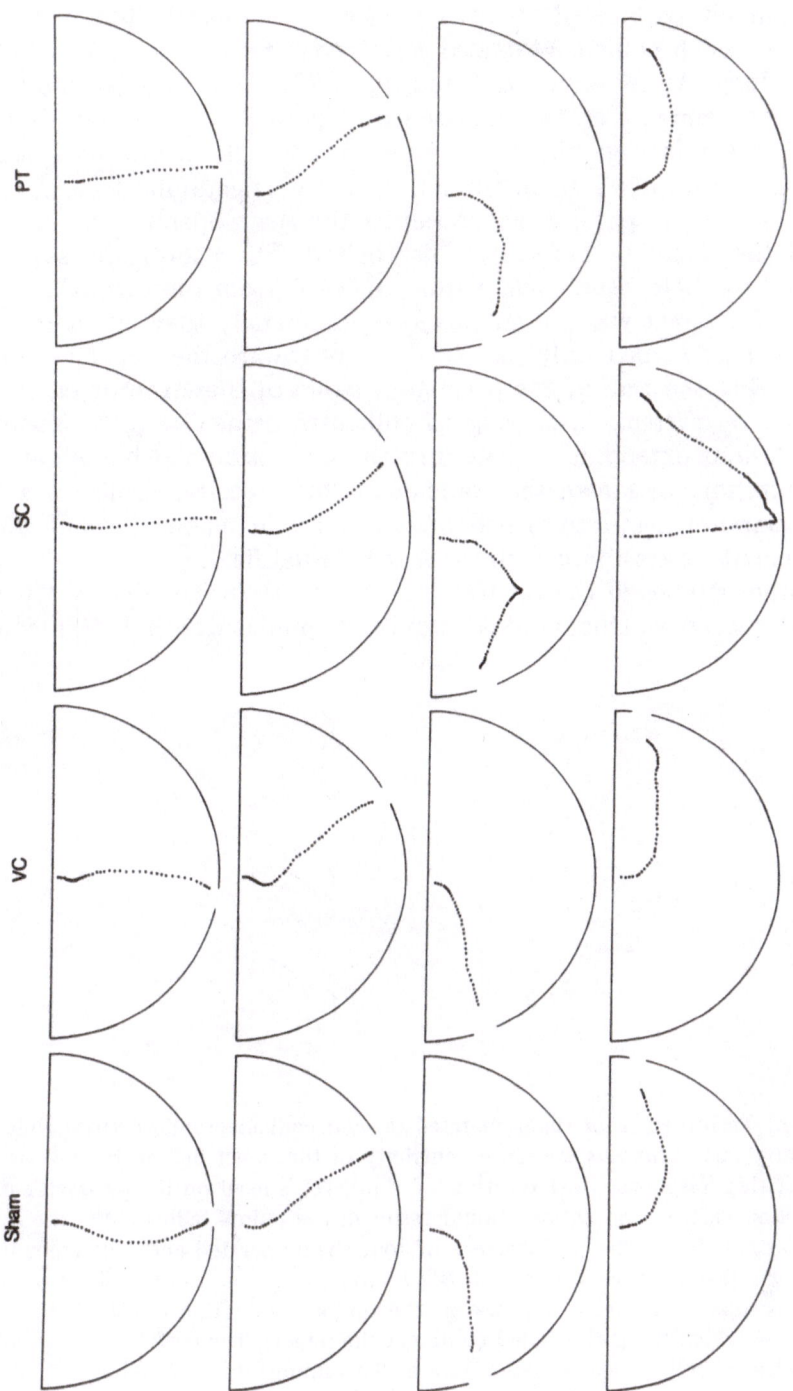

reflects a failure to initiate a rapid orientation movement of the head toward targets located beyond 40° from the midline.

A similar pattern of results emerged from several other perimetry tasks in which colliculectomized gerbils were tested (Goodale & Milner, 1982; Ingle, 1982; Mlinar & Goodale, 1980). The findings from one of these experiments are summarized in Figure 3-5. Sham-operated and colliculectomized gerbils were trained to enter the center of a circular perimeter through a small tunnel. The colliculectomized gerbils were again able to respond as accurately as the sham-operated animals provided the target (a 2-cm black disk baited with a sunflower seed) was located in their visual field within 35°–40° from the vertical midline. When the target was presented more peripherally they either failed to detect it or turned only part of the way toward the target before encountering the wall of the perimeter. None of these results can be explained by systematic sparing of collicular tissue. Even those animals with lesions extending slightly into the pretectum and including all of the anterior portion of the colliculus (which receives projections from the temporal hemiretina) had no difficulty in making visually guided movements toward targets in the rostral visual field.

Earlier studies in the rat (Goodale et al., 1978; Goodale & Murison, 1975) had shown that colliculectomized animals were able to locomote

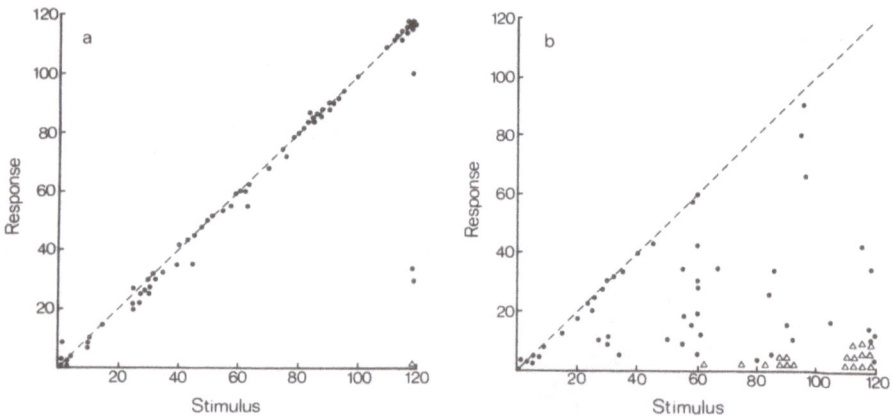

Fig. 3-5. Performance of sham-operated (a) and colliculectomized (b) gerbils on a perimetry task. *Stimulus*, retinal eccentricity of the target disk in the left or right visual field; *Response*, final position of the gerbil's head on the perimeter of the apparatus. Perfect orientation should result in the points falling along the $X = Y$ axis (*broken line*). The performance of four sham-operated and four colliculectomized gerbils on 18 trials is shown. *Black circle*, a response in the direction of the stimulus; *open triangle*, a response in the opposite direction (most of these were misses, in which the gerbil failed to detect the target). The performance of colliculectomized gerbils began to deteriorate as the stimulus was moved beyond 30°–40° from the rostral visual axis; these animals failed to detect many of the targets beyond 60° from the midline.

as efficiently and as accurately as normal animals to visual targets located within 20° of their visual midline. Tests with targets located more peripherally were not attempted. More recently, hamsters with undercuts of the superior colliculi have also been shown to be capable of accurately locomoting toward visual targets located within 30° of their visual midline (Mort et al., 1980). Locomoting toward the target in all these experiments was always rewarded with food. When novel visual stimuli were used and orientation was not rewarded, colliculectomized rats never responded to stimuli presented in their visual periphery, even though the same stimuli proved to be quite distracting for sham-operated and posterior decorticate animals (Goodale et al., 1978; Goodale & Murison, 1975). Hamsters with collicular lesions showed similar deficits when novel visual stimuli were presented (Mort et al., 1980). The latter observations suggest that variables such as previous training, reward, and/or novelty may affect the performance of colliculectomized animals.

All of these results suggest that the superior colliculus is not an essential component in the neural circuitry mediating visually guided locomotion toward discrete visible landmarks, provided those landmarks are located in the rostral visual field. However, collicular mechanisms appear to contribute significantly to the capacity for oriented head movements toward targets located in the peripheral visual field. Animals with lesions of the superior colliculus either fail to detect targets located beyond 40° from their midline or make slow and incomplete head turns in the correct direction. Thus, although the retinotectal pathway does not mediate all classes of orientation behavior, it is essential to the mediation of rapid head turns toward stimuli appearing in the visual periphery. Even this collicular function, however, may be modulated by input from other retinofugal pathways.

Role of the Geniculostriate Pathway

The projections from the lateral geniculate nucleus to area 17 have been regarded for many years as essential to pattern or form vision. Since Lashley's pioneering work on the rat's visual system, a large number of investigators (e.g., Bauer & Cooper, 1964; Goodale & Cooper, 1965; Horel, 1968; Thompson & Rich, 1963) have shown that lesions of the posterior neocortex in this species result in profound disturbances in the performance of visual pattern discriminations. More recent studies, in which the lesions have been confined to area 17, have revealed only a modest reduction in visual acuity postoperatively (Dean, 1978). A similar pattern of results has emerged from studies on the hamster (Emerson, 1980; Schneider, 1967, 1969). Schneider also observed that, although hamsters with lesions of the posterior neocortex were unable to perform horizontal-vertical discriminations, they could orient as well as normal animals toward sunflower seeds introduced anywhere in their visual field. It was this dissociation that led Schneider to suggest that the geniculostriate component of his two visual systems model was involved in stimulus identification but not in stimulus localization.

As already noted, gerbils with large bilateral lesions of the posterior neocortex (including area 17 and extending into surrounding peristriate areas) could detect and locomote toward visual targets as rapidly and accurately as normal animals. Furthermore, they could do this no matter where the target was located in their visual field. While this result could be interpreted as support for Schneider's (1967, 1969) model, it stands in sharp contrast to the results of our earlier experiments with the rat (Goodale et al., 1978; Goodale & Murison, 1975). These experiments had shown that rats with large aspiration lesions of the posterior neocortex were far less efficient than normal rats in locomoting toward a target located in their rostral visual field. There were, however, some important differences in the two experimental situations. For the rats, the target was a rear-illuminated plexiglas door, 5 cm in diameter, located in the wall of a large open arena measuring 90 X 90 cm in which the level of ambient illumination was relatively high. For the gerbils, a dark hole 5 cm in diameter provided a high-contrast target against the homogeneous white wall of a much smaller arena. Thus, the acuity demands of the two situations were very different, and the deficit we observed in the posterior decorticate rats may have reflected the impairment in visual acuity that accompanies area 17 lesions (Dean, 1978). This interpretation receives some support from the results of an experiment that I conducted in collaboration with David Ingle.

Large aspiration lesions of the posterior neocortex were performed in 3-day-old gerbils. The gerbils were later trained to run toward a target hole located in any one of a number of different positions in the wall of a semicircular arena. The dimensions of this apparatus were identical to those of the arena shown in Figure 3-3. In contrast to the homogeneous white background used in the other studies, the walls of this arena were covered with a vertical black and white grating. Although normal gerbils had no trouble guiding their head turns and locomotion toward the small opening in the striped wall, the posterior decorticate animals were unable to detect the target hole visually even when it was located in the center of their visual field. Typically, they ran forward from the arena entrance until they encountered the wall opposite, where they appeared to search for the opening using their vibrissae. The runs of posterior decorticate gerbils in this and in the white-walled arena are compared in Figure 3-6. The poor performance of the posterior decorticate gerbils when the target was embedded within a grating pattern is reminiscent of the deficits observed in rats with similar lesions when they were required to run toward a low-contrast target.

All these results suggest that whatever visual mechanisms are neces-

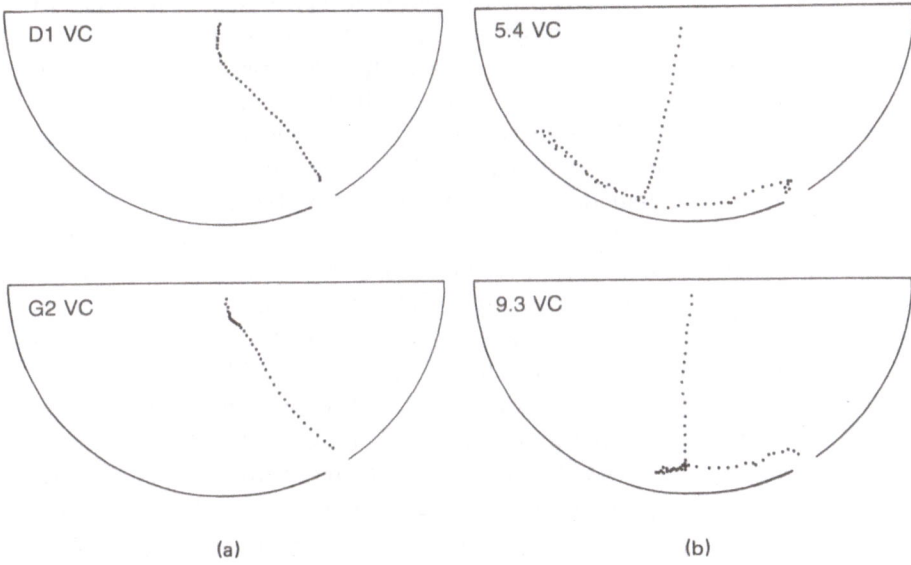

Fig. 3-6. Routes followed by posterior decorticate gerbils in an arena with homogenous white walls (a) and in one with the walls covered with a vertical black–white
grating (b). Each dot represents the center of the gerbil's head on successive frames
of film.

sary for disembedding the critical features of a target from its surround,
some networks within the cortical elaboration of the geniculostriate
pathway play an essential role in the analysis. While such a conclusion is
neither surprising nor novel, it does imply that relatively complex feature detection sometimes precedes orienting movements mediated by
the superior colliculus. In other words, cortical input to the superior
colliculus may play an important role in modulating collicular output.

   This hypothesis has received some support from an experiment with
moving targets reported by Ingle (1977b). As discussed earlier, normal
gerbils can predict the path of a moving target in such a way that they
turn and run toward it on a "collision" course (Ingle et al., 1979). If a
moving target is introduced in the frontal field and moves temporally
into the visual periphery, then the head turns that the gerbil makes will
overshoot the initial position of the target by a considerable amount.
Stationary targets or ones moving in a nasal direction normally elicit
undershoots. When gerbils with lesions of the posterior neocortex (centered on area 17 and extending into area 18) were tested in this apparatus, they continued to make brisk head turns toward targets presented
anywhere in their visual field. Nevertheless, an important component of
the orienting response was absent in these animals; they failed to make
any overshoots toward targets moving from the central field to the
periphery. In other words, posterior decorticate gerbils, while able to

initiate rapid head turns toward the point in space where a target first appeared, were unable to adjust the amplitude of these movements to achieve an efficient "collision" course with a temporally moving target.

This finding suggests that some part of the primary visual cortex provides the visuomotor channels mediating orienting behavior with information about the direction in which a target is moving and perhaps its velocity. It is not clear how this information is transmitted. One possible route is via corticotectal projections to the intermediate and deeper laminae of the superior colliculus. Projections from primary visual areas of the cortex have been shown to terminate retinotopically in the superior colliculus of at least three rodent species: the rat (Lund, 1964, 1966), the hamster (Rhoades & Chalupa, 1978b), and the gerbil (Caviness & Sherman, 1979). Moreover, like the cat (Berman & Cynader, 1975; Rosenquist & Palmer, 1971; Wickelgren & Sterling, 1969), at least one rodent, the hamster, shows a massive reduction in the directional selectivity of collicular units following lesions of the posterior neocortex (Chalupa & Rhoades, 1977; Rhoades & Chalupa, 1978a). While these results implicate the geniculostriate pathway in the modulation of orientation behavior, there is no evidence to suggest which, if any, of the cortical areas beyond area 17 in the rodent might be critical for predictive overshoots to moving targets. In this regard, Ingle (personal communication 1982) argued that outflow from area 18 may prove to be important since in the cat, at least, this areas has a significantly greater proportion of direction-selective cells than area 17 (Orban, Kennedy, & Maes, 1979).

Corticofugal control of orientation behavior may also involve pathways to areas outside the superior colliculus. As emphasized in the previous section, animals with large bilateral lesions of the superior colliculus continue to demonstrate reliable and accurately oriented turns toward baited targets located up to 40° on either side of the frontal midline, as well as occasional abnormal responses to stimuli located more peripherally. Ingle (1982) showed that gerbils receiving similar lesions as neonates will turn toward targets up to 70° from the midline, but that this residual ability disappears entirely after bilateral lesions of posterior neocortex in adulthood. This result suggests that projections from the visual area of the neocortex to visuomotor nuclei outside the collicular laminae can, in the absence of the superior colliculus, mediate head turns and other changes in posture constituting an orientation movement toward a visual target. Thus, the geniculostriate pathway might be capable of mediating orientation movements without collicular participation.

While this conclusion seems reasonable, it must be accepted with caution. Since the colliculi were aspirated when these animals were only a few days old, it is possible that incoming corticotectal fibers could have continued to grow, forming aberrant pathways to portions of the deeper collicular laminae or other brain stem areas spared by these early lesions. The effect of subsequent cortical lesions would have been to deafferent these aberrant pathways. Alternatively, the large cortical lesions could have depressed the activity of retinofugal targets in the posterior thalamus and pretectal complex that might have been responsible for the residual visuomotor behavior. Retinal projections to these structures may have been normal, but then again the early collicular ablations could have resulted in aberrant retinal projections to nuclei in these areas (or to spared portions of the colliculi themselves). In either case, their activity could have been affected by the later cortical lesions.

Whatever the contribution of geniculostriate projections to orientation behavior might be in animals with neonatal collicular lesions, their role in the mediation of the visually guided movements that survive collicular ablations performed in adulthood remains to be examined. One way to examine the problem would be to make discrete lesions in the posterior neocortex of gerbils that have already received bilateral ablations of the superior colliculus as adults. If the geniculostriate projections do mediate oriented turns in the absence of the colliculus, then small lesions in specific locations within area 17, and perhaps other visual areas in the peristriate belt, should produce "behavioral scotomata" in the residual field that correspond to the retinotopy of the damaged area in the cortex.

The experiments demonstrating residual visuomotor behavior in colliculectomized animals show some interesting parallels with the experiments demonstrating predictive orientation movements in normal animals. The $40°$ on either side of the frontal midline in which stimuli continue to elicit rapid and accurate head turns after collicular lesions corresponds to the portion of the field in which temporally moving targets elicit compensating overshoots from intact gerbils (Ingle et al., 1979). Just as stimuli presented outside this area often fail to elicit orientation movements from colliculectomized animals, moving targets that are initially presented to normal animals more than $40°$ from the midline and continue to move into the peripheral field fail to elicit overshoots and instead result in the same sort of undershoots that stationary targets elicit. The fact that the residual visuomotor behavior that survives collicular lesions is elicited by stimuli presented within the frontal field provides evidence that in normal animals the geniculostriate system may play a role in controlling orienting behavior to stimuli located in this portion of the field. This region includes the entire binocular segment in rats and gerbils. Nevertheless, it should not be forgotten that gerbils with complete bilateral lesions of area 17 and surrounding peristriate areas can run directly toward high-contrast targets presented anywhere within a homogeneous field. Thus, while the geniculostriate pathway modulates orientation movements (and might even mediate such movements in the rostral field without collicular participation), it is not necessary when targets are stationary and highly salient.

Role of the Pretectal Complex

In the experiments discussed thus far the targets were in plain view, unobstructed by barriers, and the animals were required to run directly toward them. What happens when the target is placed behind a barrier? As shown in Figure 3-2, normal gerbils had no trouble avoiding a transparent grid barrier located in front of the target hole in the semicircular arena. Indeed, they most often took the shorter route around the barrier when the target was closer to one edge than the other. In this experiment, the barrier subtended about 80° of the gerbil's visual field at the entrance to the arena and consisted of 16 vertical black stripes each occupying 2.5° and spaced at equal intervals across a piece of transparent plexiglass. Gerbils with complete bilateral ablations of the superior colliculus could negotiate this barrier as efficiently as normals. In contrast, the performance of gerbils with large bilateral lesions of the posterior neocortex was impaired on this test (Figure 3-7). Not only would they run into the barrier when the target was located behind it, they would also run toward it on about one-half of the trials in which the target was not obscured by the barrier but was instead in plain view on one side or the other. In other words, animals with lesions of the posterior neocortex appeared to be unable to disembed the target from the grid barrier, or even to discriminate between the barrier and the target on those trials in which they were separated.

The nature of their deficit, however, is not entirely clear. While loss in acuity accompanying damage to area 17 is probably very important, the poor performance may also have reflected a disturbance in stereopsis, motion parallax, or some other mechanism contributing to depth perception. Despite these uncertainties, the result of this experiment indicates that visual areas in the posterior neocortex contribute to the visuomotor channels that enable an animal to negotiate barriers.

The performance of another group of gerbils, however, suggests that the geniculostriate projection may not be the only pathway that plays a significant role in this situation. Gerbils with large lesions in the pretectal area were also impaired on the barrier task. They showed difficulty in avoiding collision only when the target was directly behind the barrier (Figure 3-7). This result suggests that their deficit, unlike that of the posterior decorticates, was not compounded by a reduction in acuity. The precise mechanisms that were disturbed by the lesions remain unknown.

In lower vertebrates, such as the frog (Ingle, 1977a), there is a loss of barrier avoidance following large posterior thalamic lesions in the area considered to be homologous with the pretectal projections of "higher" vertebrate classes (Ebbesson, 1970; Riss & Jakway, 1970). Frogs with such lesions continue to snap accurately at prey presented anywhere in their visual field. In contrast, frogs with lesions of the optic tectum, considered homologous with the mammalian superior colliculus, no longer respond to prey but avoid barriers with considerable accuracy (Ingle, 1973, 1977a). This double dissociation of deficits resembles to some extent the results of the gerbil experiments discussed above. Gerbils with pretectal lesions showed brisk head turns and other orienting movements toward targets located in their peripheral visual fields but had difficulty negotiating barriers; gerbils with collicular lesions had no trouble avoiding the barrier but failed to respond normally to targets located in their peripheral visual fields. Thus, in both vertebrate classes the pretectum and the optic tectum (or the superior colliculus) may play complementary roles in the control of orientation behavior. The pretectal complex in rodents, however, consists of several separate and noncolaminar retinofugal targets (Scalia & Arango, 1979). At present, there is no clear behavioral evidence to suggest what the function of the different pretectal nuclei might be.

It is likely that the pretectal area in rodents, like that of other mam-

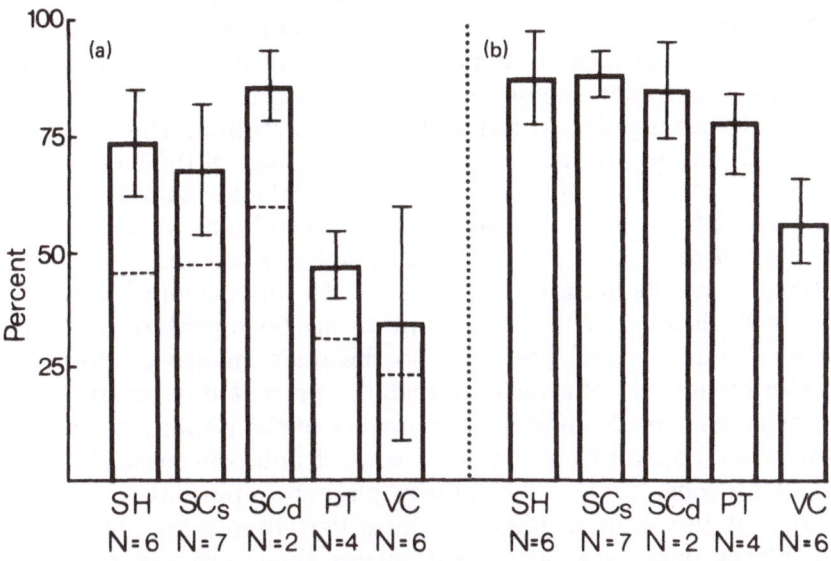

Fig. 3-7. Performance on the barrier-avoidance task by five different groups of gerbils: those sham-operated gerbils (*SH*), those with superficial collicular lesions (*SC_s*), those with deep collicular lesions (*SC_α*), those with pretectal lesions (*PT*), and those with lesions of the posterior neocortex (*VC*). (a) Mean (± SD) percentage of trials in which each group avoided the barrier. *Broken line*: mean percentage of trials in which each group took the shorter route around the barrier. On all these trials the target was located behind the barrier and the animal locomoted toward it without first touching the arena perimeter. (b) Mean (± SD) percentage of trials in which each group ran toward the opening when it was not obscured by the barrier.

mals (Berman & Payne, 1980; Berson & Graybiel, 1978; Giolli & Guthrie, 1967; Harting & Noback, 1971), receives input from, and projects (via related thalamic nuclei) to, some of those areas of the cortex that were ablated in our gerbils. This may explain why lesions of the posterior neocortex in the gerbil produced deficits on the barrier-avoidance task that resembled those produced by pretectal lesions. Nevertheless, since the terminal fields of a number of different thalamic nuclei, including the dorsal lateral geniculate nucleus, were destroyed when the neocortex was aspirated, the deficits in the posterior decorticate group could have arisen from disruption of the cortical components of several different visuomotor systems.

Visually Guided Movements Toward Places

It is a common observation that animals can locate places in space that are not marked by local cues. Such behavior is very different from moving toward a visible target under direct visual control. An animal may be using vision to guide its movements, but the place it is moving toward may provide no distinctive visual cues requiring the animal to organize its movements with respect to a spatial frame of reference. Even though landmarks located some distance from the goal might provide some important cues, many animals, including rodents, are able to approach the same point in space from any direction. This implies that the animal has an internal representation or map of the location of the goal with respect to visible landmarks. Celestial cues, as well as information derived from other sensory systems, may also be incorporated into this internal representation. Although much has been made of the "spatial map" and its putative relationship to hippocampal activity (see O'Keefe & Nadel, 1978), little attention has been paid to the contribution of different visual pathways to the maintenance of this type of spatially organized behavior. The evidence presented in this section suggests that areas in the posterior neocortex of the rat may be implicated in the visual control of locomotion toward points in space that are not marked by local cues, and thus may play an important role in spatial memory in this animal. Rats, like most vertebrates, are able to distinguish places they have visited from places they have not, even though such places might not be proximate cues. In the laboratory, this ability has been effectively demonstrated with the use of a radial maze, consisting of eight equally spaced arms that project radially from a central platform (Olton & Samuelson, 1976). The radial maze task differs from more traditional maze tasks in that the animal is not required to follow a particular route from start to finish; instead, it is released on the central platform and is allowed to obtain a small amount of food from the end of each of the arms (Figure 3-8). A rat's performance on this task is measured in terms of the number of arm entries it makes before obtaining all of the food. Thus, a rat must make at least eight choices, but may require many more. A choice score can be calculated by recording the number of different arms a rat enters in its first eight choices; the higher the score (to a maximum of 8.0), the better the performance.

Rats are remarkably efficient in this task, and yet their high level of performance is not dependent on either response strategies or intramaze cues. Olton and Samuelson (1976) found in their original study that rats did not rely on regular patterns, chains of responses, or any other algorithms to "solve" the maze, nor did they use cues initially present within the maze, or ones they may have deposited themselves, such as odor trails. Indeed, their ability to discriminate between arms that they had previously entered and arms that they had not seemed to be largely dependent on cues available outside of the maze itself. Several recent studies have confirmed these observations (Olton & Collison, 1979; Olton, Collison, & Werz, 1977) and have indicated that visual cues within the environment surrounding the radial maze seem to be particularly important to the maintenance of efficient performance (Dale & Innis, 1980; Olton, 1978; Zoladek & Roberts, 1978). Moreover, rats

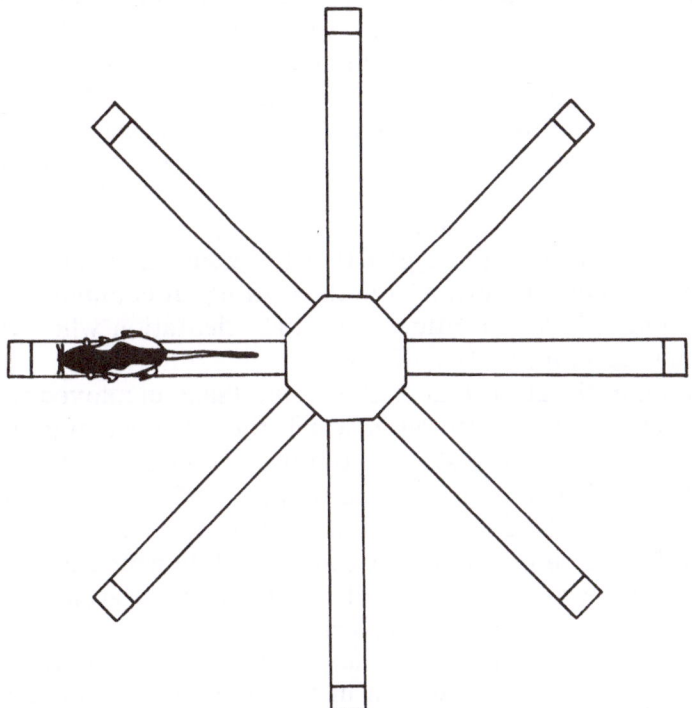

Fig. 3-8. Overhead view of the radial maze. Food is located in small rectangular containers at the end of each arm and cannot be seen by the rat until it has reached the container.

attend to the spatial arrangement of the different visual cues rather than using individual stimuli as markers for particular arms of the maze (Suzuki, Augerinos, & Black, 1980).

Since visual cues are so important, it is not surprising to find that blind rats take much longer than sighted ones to reach an efficient level of performance on the radial maze (Dale & Innis, 1980). Similarly, sighted rats trained in the dark take many more trials to achieve high choice scores than animals trained with visual cues available (Dale, 1980). Nevertheless, when sighted rats that are performing efficiently in the light are subsequently blinded by enucleation, they show only transient deficits in performance, or no deficits at all (Dale & Innis, 1980). At first these results may appear contradictory. The paradox is resolved, however, if vision is crucial for acquisition while other sensory information maintains performance in the absence of visual cues. If efficient performance depends on the construction of an internal representation of space, then nonvisual cues must be incorporated into this representation even though its basic structure is determined by the dominant visual cues. If this is the case, then one might ask what structures in the visual system contribute to the formation of such a spatial map, and are these structures still necessary even when vision is no longer available?

Lashley (1929, 1943) and Tsang (1934, 1936) had found earlier that lesions of the posterior neocortex (centered on area 17) in rats produced larger deficits in acquisition and retention of performance on a complex Lashley III maze than did either sectioning of the optic radiations or peripheral blinding by enucleation. Moreover, Lashley (1943) showed that combining enucleation and lesions of the posterior neocortex resulted in even greater deficits, and Tsang (1936) reported that cortical lesions could also produce deficits in previously enucleated rats. These results led them to suggest that the visual areas in the posterior neocortex are not only involved in the learning of complex mazes when vision is intact, but also influence spatial orientation when visual cues are not, and never have been, available.

The training situation that Lashley and Tsang employed was a traditional, if rather difficult, maze in which the rat was required to follow the "correct" route from start to finish in order to reach the food reward most efficiently. Foreman and Stevens (1979) showed that lesions of area 17 in the rat also interfere with acquisition and retention of accurate performance on the radial maze, a task in which spatial memory is strongly implicated. Their results are difficult to evaluate, however, since the performance of the decorticates was not compared with that of rats blinded peripherally. It is not clear whether the deficit after cortical lesions is due to a reduction in the amount of incoming visual information or to a disturbance of the spatial map. To answer this question, it was necessary to compare the effects of enucleation with those of cortical lesions in the same experiment.

In our first experiment (Goodale & Dale, 1981), rats were trained every day on an eight-arm radial maze (Figure 3-8) until they reached criterion performance (an average choice score of 7.2 over 5 successive training days). Subsequently, they received either sham operations, bilateral eye enucleations, lesions of the posterior neocortex, or combined enucleations and lesions of the posterior neocortex. After a 2-week recovery period, the four groups of rats were returned to the radial maze and were retrained for 50 daily trials or until they reattained criterion performance. Whereas the enucleated rats with intact brains showed only a slight, but significant, decrement in their performance, the two groups with lesions in the posterior neocortex each showed a massive deficit (Figure 3-9). When the posterior neocortex was aspirated in animals from the enucleated group, performance degraded significantly.

These results are consistent with the earlier observations of Foreman and Stevens (1979), who suggested that rats with ablations of the posterior neocortex lack the ability to discriminate between those arms

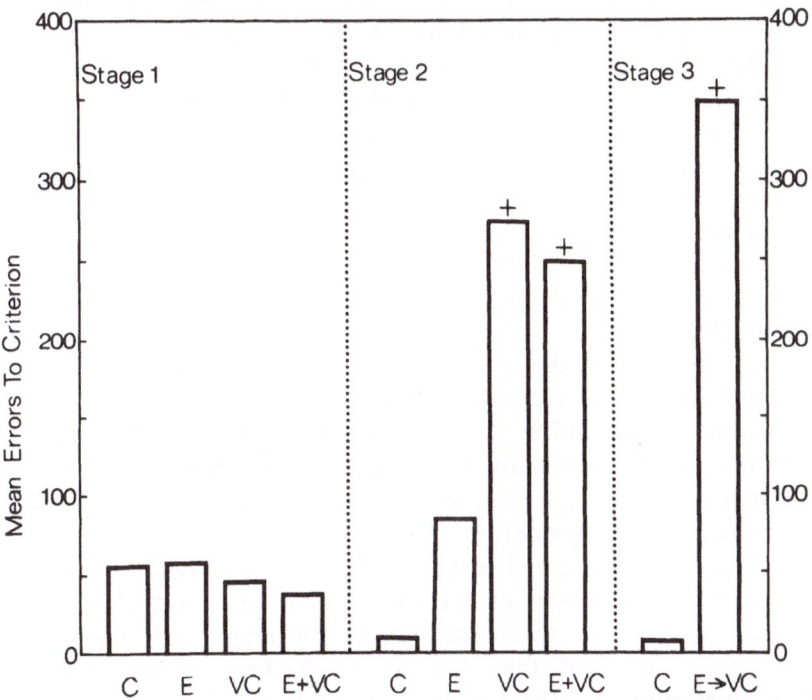

Fig. 3-9. Performance of four different groups of rats in the radial maze experiment. (a) In Stage 1, none of the groups had received surgery. (b) In Stage 2, Group C ($N = 5$) received sham operations, Group E ($N = 6$) enucleations, Group VC ($N = 5$) posterior neocortex lesions, and Group $E + VC$ ($N = 5$) both enucleations and posterior neocortex lesions. (c) In Stage 3 Group C received a second sham-operation, and Group E lesions of the posterior neocortex (becoming Group $E \rightarrow VC$).

that they had entered and those that they had not. Since none of the rats in their study had their eyes enucleated, it might be argued that the poor performance of their posterior decorticates was simply due to the reduction in acuity that follows damage to areas 17, 18, and 18a (Dean, 1978, 1981). However, the fact that the enucleated animals with intact brains in our experiment performed so much better than animals with lesions of the posterior neocortex, whether enucleated or not, suggests that such lesions disrupt more than the processing of incoming visual information. The enucleated animals had no vision at all, yet they showed only a modest performance decrement.

The lesions in our experiments were large: they included area 17 and extended into areas 18, 18a, and the posterior portion of area 7 in most animals. Thus, the cortical targets of both the lateral geniculate nucleus and the lateral posterior nucleus of the thalamus were affected. For this reason, it is unclear whether one or both of these pathways (or their cortical elaboration within area 7 or other areas of the neocortex) are critical for efficient performance on the radial maze. Foreman and Stevens (1979) reported that rats with lesions of the superior colliculus were not nearly so impaired as posterior decorticates. This result suggested that the cortical projections from the lateral posterior nucleus, which receives a major inflow from the superior colliculus, may not be as important to performance on the radial maze as the geniculostriate pathway. Thomas (1970) and Thomas and Weir (1975) argued that lesions of the "parietal" cortex, apparently including area 7, in the rat interfere with retention on the Lashley III maze. Whatever the critical areas might be, some portion of the terminal cortical fields of one or both of these "visual" thalamic nuclei is clearly implicated. In our study two animals with large lesions of the posterior neocortex that spared the largest portion of the cortical targets of both the lateral geniculate nuclei and the lateral posterior nuclei showed only minor and very transient deficits on the radial maze. Although other parts of the neocortex have not been systematically explored in this context, as far as the posterior neocortex is concerned, the critical area or areas seem to coincide with the visual projection areas.

The deficits we observed in the posterior decorticates were very different from the disturbances described in rats with extensive lesions of the hippocampus or interruptions of its major afferent pathways (Olton, 1977; Olton, Walker, & Gage, 1978). Our animals showed none of the perseverative behavior typically associated with hippocampal damage and rarely repeated sequences of choices within a trial. The fact that hippocampal lesions disrupt performance on the radial maze supports speculation that this structure may provide part of the substrate for spatial memories (Olton & Samuelson, 1976; Thompson, 1974) or cognitive maps (O'Keefe & Nadel, 1978). However, the poor postoperative performance of rats with lesions of visual areas in the posterior neocortex suggests that these areas may play an important role in the organization of sensory input before its transmission to higher order systems involving structures such as the hippocampus. Participation of these cortical visual areas may continue to be necessary even when visual information from the environment is no longer available. Thus, lesions of the posterior neocortex that include the visual projection systems are far more devastating than peripheral blinding, provided that visual cues played a significant role in the original learning of the radial maze. What if vision was never used at all to learn the maze? Would lesions of the posterior neocortex have the same effect? If visual cues provide the basic structure of the spatial map into which nonvisual cues are later incorporated, then lesions of the posterior neocortex might not be expected to affect the performance of rats that learned the maze as enucleates.

In a later experiment, Robert Dale, Nancy Innis, and I made large posterior neocortex lesions in three rats that had learned the radial maze as blinded animals. Three additional enucleated rats that had also learned to perform well in the maze received sham operations. Both groups of animals had been overtrained for many trials prior to surgery. As shown in Table 3-1, except for a modest performance decrement on the first day of postoperative training, the posterior decorticates were indistinguishable from the sham-operated animals, even though their lesions were at least as large as those carried out in the first experiment. Thus, if vision is not used to learn the maze, not only is acquisition of efficient performance much slower, but subsequent lesions of posterior neocortex have no effect on performance. As the first experiment showed, if vision is used, acquisition is much faster, and lesions of the posterior neocortex affect performance even more than peripheral blinding. Taken together, the results of these two experiments suggest that the areas in the posterior neocortex that are implicated in spatial memory function only when the animal learns the maze with visual cues available.

This interpretation of the results, however, must be regarded as tentative. The rats in our second study showed a great deal of response patterning preoperatively and tended to choose the eight arms in a particular order from trial to trial. It is possible that it was this response strategy that enabled the animals to perform so well after posterior decortication. Response patterning is certainly not an uncommon strategy for blind animals to use when learning the radial maze (Dale, 1980). Nevertheless, the large amount of overtraining that these animals received before undergoing surgery could also have contributed to the use of such a nonspatial solution to the problem. Whatever the reason, the results of the second experiment clearly indicate that under some conditions lesions of the posterior neocortex do not disrupt radial maze performance, and the deficits we observed in the first experiment could

Table 3-1. Postoperative Choice Scores and Error Scores for Enucleated Rats After Lesions of the Posterior Neocortex or Sham Operations

Rat no.	Day 1		Day 2		Day 3		Day 4		Day 5	
	Choices	Errors	Choices	Errors	Choices	Errors	Choices	Errors	Choices	Errors
Lesions										
1	6	4	7	5	8	0	8	0	8	0
2	5	9	7	2	8	0	8	0	8	0
4	7	0	8	0	8	0	8	0	8	0
Mean	5	4.3	7.3	2.3	8	0	8	0	8	0
Sham operations										
3	7	2	7	1	8	0	8	0	8	0
6	7	1	7	1	8	0	8	0	8	0
7	7	1	8	0	8	0	8	0	8	0
Mean	7	1.3	7.3	7	8	0	8	0	8	0

not have been due to some nonspecific disturbance in performance. Thus, Lashley's (1929, 1943) original result is confirmed, but the conclusions are sharpened: the integrity of the so-called visual areas of the neocortex is essential to the maintenance of spatial behavior if that behavior was originally organized under visual control, even if vision is no longer available.

Conclusions

Rodents, like most animals, must organize their behavior in space in order to survive and reproduce. Much of this organization depends on information derived from photoreceptors, but the movements that a rodent makes in visual space are controlled by a number of different stimulus parameters and are certainly not mediated by a single visuomotor channel. Even a behavior such as approaching a discrete visual target is not as simple as it might first appear to be. A sequence of movements is involved, the spatial and temporal characteristics of which vary as a function of the nature of the eliciting stimulus and the situation in which it occurs. Moreover, the simple division of labor implied by the two visual systems model is not sufficient to account for the neural organization of these different patterns of behavior. The neural substrate is highly complex and involves at least three separate retinofugal targets: the superior colliculus, the lateral geniculate nucleus, and the pretectal area. Although the different visuomotor channels in which each of these structures participates are interactive, it is clear that the superior colliculus is neither necessary nor sufficient for all orientation behaviors.

Visually guided movements directed toward places that are not marked by distinctive local cues involve another level of complexity. These movements are controlled not only by incoming visual information but also by a highly organized spatial memory. Whereas the hippocampus has long appeared to be an essential part of the substrate for this spatial memory or map, it is now apparent that visual projections to the posterior neocortex also play a major role in the organization of the map. Indeed, this network may provide the basic framework for an internal representation of space into which information from other sensory systems can be incorporated.

Orientation behavior cannot be regarded as a unitary functional class of behavior mediated exclusively by the superior colliculus. Many different patterns of behavior are oriented in visual space and many different visuomotor channels are involved. In the rodent, we are only beginning to understand these behaviors and their underlying neural mechanisms.

Acknowledgments

This work was supported in part by Natural Sciences and Engineering Research Council of Canada Grant A6313. I am grateful to Debra Stewart, who prepared the illustrations and graphic material.

References

Apter, J. T. Eye movements following strychninization of the superior colliculus in cats. *Journal of Neurophysiology*, 1946, *9*, 43-88.

Bauer, J. H., & Cooper, R. M. Effect of posterior cortical lesions on performance of a brightness discrimination task. *Journal of Comparative and Physiological Psychology*, 1964, *58*, 84-92.

Berman, N., & Cynader, M. Receptive fields in cat superior colliculus after visual cortex lesions. *Journal of Physiology*, 1975, *245*, 261-270.

Berman, N., & Payne, B. R. Cross corticofugal projections in the cat. *Society for Neuroscience Abstracts*, 1980, *6*, 482.

Berson, D. M., & Graybiel, A. M. Thalamocortical projections and histochemical identification of subdivision of the LP-pulvinar complex in the cat. *Society for Neuroscience Abstracts*, 1978, *4*, 620.

Caviness, V. S., Jr., & Sherman, H. B. *Retinal and neocortical afferents to the superior colliculus of the gerbil.* Paper presented at Satellite Symposium on Comparative Aspects of Vision in Rodents at the 9th annual meeting of the Society for Neuroscience, Atlanta, November 1979.

Chalupa, L. M., & Rhoades, R. W. Responses of visual, somatosensory and auditory neurons in the golden hamster's superior colliculus. *Journal of Physiology*, 1977, *270*, 595-626.

Dale, R. H. I. The role of vision in the rat's radial maze performance (Doctoral dissertation, University of Western Ontario, 1979). *Dissertation Abstracts International*, 1980, *40*, 5047B.

Dale, R. H. I., & Innis, N. Spatial memory without vision: Radial maze performance of blind rats. *University of Western Ontario Research Bulletin*, No. 523, 1980.

Dean, P. Visual acuity in hooded rats: Effects of superior collicular and posterior neocortical lesions. *Brain Research*, 1978, *156*, 17-31.

Dean, P. Visual pathways and acuity in hooded rats. *Behavioral Brain Research*, 1981, *3*, 239-272.

Ebbesson, S. On the organization of central visual pathways in vertebrates. *Brain, Behavior and Evolution*, 1970, *3*, 178-194.

Emerson, V. F. Grating acuity of the golden hamster: The effects of stimulus orientation and luminance. *Experimental Brain Research*, 1980, *38*, 43-52.

Foreman, N. P., & Stevens, R. G. *Visual lesions and radial arm maze performance in rats.* Paper presented at the annual meeting of the British Psychological Society, Nottingham, April 1979.

Giolli, R. A., & Guthrie, M. D. Organization of projections of visual areas I and II upon the superior colliculus and pretectal nuclei in the rabbit. *Brain Research*, 1967, *6*, 388-390.

Goodale, M. A., & Cooper, R. M. Cues utilized by normal and posterior-neodecorticate rats in the Yerkes brightness discrimination task. *Psychonomic Science*, 1965, *3*, 513-514.

Goodale, M. A., & Dale, R. H. I. Radial-maze performance in the rat following lesions of posterior neocortex. *Behavioral Brain Research*, 1981, *3*, 273-288.

Goodale, M. A., Foreman, N. P., & Milner, A. D. Visual orientation in the rat: A dissociation of deficits following cortical and collicular lesions. *Experimental Brain Research*, 1978, *31*, 445-457.

Goodale, M. A., & Milner, A. D. Fractionating orientation behavior in the rodent. In D. Ingle, M. Goodale, & R. Mansfield (Eds.), *Analysis of visual behavior*. Cambridge, Mass. M.I.T. Press, 1982, pp. 267-299.

Goodale, M. A., Milner, A. D., & Rose, J. Susceptibility to startle during ongoing behavior following collicular lesions in the rat. *Neuroscience Letters*, 1975, *1*, 333-337.

Goodale, M. A., & Murison, R. C. C. The effects of lesions of the superior colliculus on locomotor orientation and the orienting reflex in the rat. *Brain Research*, 1975, *88*, 243-255.

Harting, J. K., & Noback, C. R. Subcortical projections from the visual cortex in the tree shrew. *Brain Research*, 1971, *25*, 21-33.

Hess, S., Bürgi, S., & Bucher, V. Motor function of tectal and tegmental area. *Monatschrift für Psychiatrie und Neurologie*, 1946, *112*, 1-52.

Horel, J. A. Effects of subcortical lesions on brightness discrimination acquired by rats without visual cortex. *Journal of Comparative and Physiological Psychology*, 1968, *65*, 103-109.

Hyde, J. E., & Eason, R. G. Characteristics of ocular movements evoked by stimulation of brainstem in cats. *Journal of Neurophysiology*, 1959, *22*, 666-678.

Hyde, J. E., & Eliasson, S. G. Brainstem induced eye movements in cats. *Journal of Comparative Neurology*, 1957, *108*, 139-172.

Ingle, D. Two visual systems in the frog. *Science*, 1973, *181*, 1053-1055.

Ingle, D. Detection of stationary objects by frogs (*Rana pipiens*) after ablation of optic tectum. *Journal of Comparative and Physiological Psychology*, 1977, *91*, 1359-1364. (a)

Ingle, D. Role of visual cortex in anticipatory orientation toward moving targets by the gerbil. *Society for Neuroscience Abstracts*, 1977, *3*, 68. (b)

Ingle, D. Organization of visuomotor behaviors in vertebrates. In D. Ingle, M. Goodale, & R. Mansfield (Eds.), *Analysis of visual behavior*. Cambridge, Mass. M.I.T. Press, 1982, pp. 67-109.

Ingle, D. J., Cheal, M., & Dizio, P. Cine analysis of visual orientation and pursuit by the Mongolian gerbil. *Journal of Comparative and Physiological Psychology*, 1979, *93*, 919-928.

Lashley, K. S. *Brain mechanisms and intelligence: A quantitative study of injuries to the brain*. New York: Dover, 1963. (Originally published, 1929).

Lashley, K. S. The mechanism of vision. VII. The projection of the retina upon the primary optic centers of the rat. *Journal of Comparative Neurology*, 1934, *59*, 341-373.

Lashley, K. S. Studies of cerebral function in learning. XII. Loss of the maze habit after occipital lesions in blind rats. *Journal of Comparative Neurology*, 1943, *79*, 431-462.

Lund, R. D. Terminal distribution in the superior colliculus of fibers originating in the visual cortex. *Nature*, 1964, *264*, 1283-1285.

Lund, R. D. The occipitotectal pathway of the rat. *Journal of Anatomy*, 1966, *100*, 51-62.

Lund, R. D. Anatomic studies on the superior colliculus. *Investigative Ophthalmology*, 1972, *11*, 434–441.

Mlinar, E., & Goodale, M. A. *Visual search in Mongolian gerbils with lesions of the superior colliculus.* Paper presented at the annual meeting of the Canadian Psychological Association, Calgary, June 1980.

Mort, E., Finlay, B. L., & Cairns, S. J. The role of the superior colliculus in visually-guided locomotion and visual orienting in the hamster. *Physiological Psychology*, 1980, *8*, 20–28.

Nauta, W. J. H., & Straaten, J. J. van. The primary optic centers of the rat: An experimental study by the "bouton" method. *Journal of Anatomy*, 1947, *81*, 127–134.

O'Keefe, J., & Nadel, L. *The hippocampus as a cognitive map.* Oxford, England: Oxford University Press, 1978.

Olton, D. S. Spatial memory. *Scientific American*, 1977, *236*, 82–98.

Olton, D. S. Characteristics of spatial memory. In S. H. Hulse, H. Fowler, & W. K. Honig (Eds.), *Cognitive processes in animal behavior*. Hillsdale, N.J.: Erlbaum, 1978.

Olton, D. S., & Collison, C. Intramaze cues and "odor trails" fail to direct choice behavior on an elevated maze. *Animal Learning and Behavior*, 1979, 7, 221–223.

Olton, D. S., Collison, C., & Werz, M. A. Spatial memory and radial arm maze performance in rats. *Learning and Motivation*, 1977, *8*, 289–314.

Olton, D. S., & Samuelson, R. J. Remembrance of places passed: Spatial memory in rats. *Journal of Experimental Psychology: Animal Behavior Processes*, 1976, *2*, 97–116.

Olton, D. S., Walker, J. A., & Gage, F. H. Hippocampal connections and spatial discrimination. *Brain Research*, 1978, *139*, 295–308.

Orban, G. A., Kennedy, H., & Maes, H. Comparison of neuronal properties in areas 17 and 18 of the cat. *Society for Neuroscience Abstracts*, 1979, *5*, 801.

Rhoades, R. W., & Chalupa, L. M. Functional and anatomical consequences of neonatal visual cortical damage in the superior colliculus of the golden hamster. *Journal of Neurophysiology*, 1978, *41*, 1466–1494. (a)

Rhoades, R. W., & Chalupa, L. M. Functional properties of the corticotectal projection in the golden hamster. *Journal of Comparative Neurology*, 1978, *180*, 617–634. (b)

Riss, W., & Jakway, J. S. A perspective on the fundamental retinal projections of vertebrates. *Brain, Behavior, and Evolution*, 1970, *3*, 30–35.

Rosenquist, A. C., & Palmer, L. A. Visual receptive-field properties of cells in the superior colliculus after cortical lesions in the cat. *Experimental Neurology*, 1971, *33*, 629–652.

Scalia, F., & Arango, V. Topographic organization of the projections of the retina to the pretectal region in the rat. *Journal of Comparative Neurology*, 1979, *186*, 271–292.

Schneider, G. E. Contrasting visuomotor functions of tectum and cortex in the golden hamster. *Psychologische Forschung*, 1967, *31*, 52–62.

Schneider, G. E. Two visual systems: Brain mechanisms for localization and discrimination are dissociated by tectal and cortical lesions. *Science*, 1969, *163*, 895–902.

Suzuki, S., Augerinos, G., & Black, A. H. Stimulus control of spatial behavior on the eight-arm maze in rats. *Learning and Motivation*, 1980, *11*, 1–18.

Thomas, R. K. Mass function and equipotentiality: A reanalysis of Lashley's retention data. *Psychological Reports*, 1970, *27*, 899-902.

Thomas, R. K., & Weir, V. K. The effects of lesions in the frontal or posterior association cortex of rats on maze III. *Physiological Psychology*, 1975, *3*, 210-214.

Thompson, R. Localization of the "maze memory system" in the white rat. *Physiological Psychology*, 1974, *2*, 1-17.

Thompson, R., & Rich, I. Differential effects of posterior thalamic lesions on retention of various visual habits. *Journal of Comparative and Physiological Psychology*, 1963, *56*, 60-65.

Tsang, Y. -C. The functions of the visual areas of the cerebral cortex of the rat in the learning and retention of the maze. I. *Comparative Psychology Monographs*, 1934, *10* (4, Serial No. 50).

Tsang, Y. -C. The functions of the visual areas of the cerebral cortex of the rat in the learning and retention of the maze, II. *Comparative Psychology Monographs*, 1936, *12*, (2, Serial No. 57).

Wickelgren, B., & Sterling, P. Influence of visual cortex on receptive-fields in the superior colliculus of the cat. *Journal of Neurophysiology*, 1969, *32*, 16-23.

Zoladek, L., & Roberts, W. A. The sensory basis of spatial memory in the rat. *Animal Learning and Behavior*, 1978, *6*, 77-81.

Chapter 4

Role of the Monkey Superior Colliculus in the Spatial Localization of Saccade Targets

David L. Sparks and Lawrence E. Mays

It has long been recognized that information about the position of the eyes in the orbit plays an important role in the perception of visual direction. With the eyes, head, and body stationary, there is a one-to-one correspondence between the direction of a visual stimulus and the location of its image on the retina. However, since the eyes do not remain stationary, the perception of visual direction must be based upon a combination of information about the location of the retinal image and the position of the eyes in the orbit (Helmholtz, 1866; Holst & Mittelstaedt, 1950; Matin, 1972; Shebilske, 1977; Skavenski, 1976).

Earlier models of the saccadic eye movement system (Robinson, 1973; Young & Stark, 1963) assumed that saccade targets were localized using retinal information alone. Recent experiments have shown, however, that saccade targets are localized using a combination of retinal and eye position signals (Hallet & Lightstone, 1976; Mays & Sparks, 1980b, 1981); spatial models of the saccadic system, consistent with these and other data, have been developed (Robinson, 1975; Zee, Optican, Cook, Robinson, & Engel, 1976). We have suggested a role for the superior colliculus in the initiation of saccades to targets localized in a spatial frame of reference (Mays & Sparks, 1980a; Sparks, Mays, & Pollack, 1977).

In this chapter we present a brief review of the evidence implicating the superior colliculus in the control of saccades. We summarize the data and models suggesting that saccade targets are localized in a spatial, rather than a retinocentric, frame of reference and discuss the neural basis for the spatial localization of saccade targets.

Role of the Superior Colliculus in Saccadic Eye Movements

Converging lines of evidence derived from anatomical, microstimulation, chronic unit recording, and lesion experiments indicate that the primate superior colliculus is an important structure in the neural control of saccadic eye movements (Goldberg & Robinson, 1978; Sparks & Mays, 1981; Sparks & Pollack, 1977; Sprague, 1975; Wurtz & Albano, 1980).

Anatomical Studies

Projections to the Superior Colliculus

The superior colliculus receives inputs from sensory systems that direct eye movements, and projects to brain stem areas known to be important in generating oculomotor commands. Visual signals reach the superior colliculus directly from the retina and indirectly from the visual cortex (see Lund, 1972a, 1972b, for references). Other areas containing visually responsive neurons (including the pretectal area, parietal cortex, frontal eye fields, and prefrontal cortex) also project extensively to the superior colliculus (Edwards, Ginsburgh, Henkel, & Stein, 1979; Goldman & Nauta, 1976; Kunzle & Akert, 1974; Kunzle, Akert, & Wurtz, 1976; Kuypers & Lawrence, 1967). Both the direct and indirect visual inputs to the superior colliculus are arranged retinotopically (Cynader & Berman, 1972; Graybiel, 1976; Hendrickson, Wilson, & Toyne, 1970; Hubel, LeVay, & Wiesel, 1975; Pollack & Hickey, 1979) and are maintained in topographical register with each other (McIlwain, 1977; Schiller, Stryker, Cynader, & Berman, 1974). In the monkey, each colliculus contains a representation of the contralateral visual field, although neurons with receptive fields near the vertical meridian may discharge to stimuli presented in the ipsilateral visual field. Central regions of the contralateral visual field project to the anterolateral superior colliculus, while the most eccentric parts of the visual field are represented in caudal regions of the colliculus. Upper regions of the visual field are represented in the medial superior colliculus and lower regions are represented laterally.

The superior colliculus receives auditory inputs from the inferior colliculus, dorsomedial periolivary nucleus, nuclei of the trapezoid body, ventral nucleus of the lateral lemniscus, and temporal cortex (Barnes, Magoun, & Ranson, 1943; Edwards et al., 1979; Whitlock & Nauta, 1956). There are inputs to the superior colliculus from the following somatosensory areas: the spinotectal tract, sensory trigeminal complex, dorsal column nuclei, lateral cervical nucleus (Edwards et al., 1979; Poirier & Bertrand, 1955), and parietal lobe (Petras, 1971). In the cat and mouse, auditory and somatosensory inputs are also arranged topographically (Drager & Hubel, 1975, 1976; Gordon, 1973; Stein, Magalhaes-Castro, & Kruger, 1976), but the pattern of inputs from these sensory systems has not been studied adequately in the monkey.

Projections from the Superior Colliculus

The superior colliculus has both ascending and descending efferent projections (Wurtz & Albano, 1980); however, it is through the descending efferents that the superior colliculus exerts its most direct influence upon eye movements (Edwards & Henkel, 1978; Graham, 1977; Harting, 1977; Harting, Huerta, Frankfurter, Strominger, & Royce, 1980; Weber & Harting, 1978). The paramedian pontine reticular formation, inferior olivary complex (especially subnucleus b of the medial accessory nucleus), dorsal lateral pontine nuclei, nucleus reticularis tegmenti pontis, and rostral interstitial nucleus of the medial longitudinal fasciculus receive input from the superior colliculus. The paramedian pontine reticular formation is thought to be responsible for generating the horizontal components of all quick movements—saccades and the quick phases of vestibular and optokinetic nystagmus (Henn & Cohen, 1976; Keller, 1974; Robinson, 1975). The rostral interstitial nucleus of the medial longitudinal faciculus contains neurons that generate a burst of spike activity before vertical saccades (Buttner, Buttner-Ennever, & Henn, 1977; Graybiel, 1977; King & Fuchs, 1977). The superior colliculus projects to the posterior cerebellar vermis via the inferior olivary complex, dorsal lateral pontine nuclei, and nucleus reticularis tegmenti pontis (Frankfurter, Weber, & Harting, 1977; Frankfurter, Weber, Royce, Strominger, & Harting, 1976; Weber, Partlow, & Harting, 1978). The posterior vermis is the region of the cerebellum most closely involved in saccadic eye movements (Kase, Miller, & Noda, 1980; Optican & Robinson, 1980; Ritchie, 1976; Ron & Robinson, 1973).

Chronic Unit-Recording Experiments

Neurons in the intermediate layers of the superior colliculus have saccade-related discharges that precede saccade onset (Schiller & Koerner, 1971; Sparks, Holland, & Guthrie, 1976; Wurtz & Goldberg, 1972a). Each of these neurons discharges before saccades that are within a specific range of directions and amplitudes (the movement field), and there is a topographical arrangement of neurons with movement fields in the intermediate layers of the superior colliculus. Neurons that discharge prior to small saccades are located in the anterior colliculus; neurons firing before large saccades are found caudally. Neurons near the midline discharge before saccades with up components and neurons located laterally fire before saccades with down components.

One class of superior colliculus neurons, the saccade-related burst units, generates a discrete pulse of spike activity beginning approximately 20 msec before saccades into the center of their movement field (Sparks, 1978). The saccade-related burst units meet two criteria for participation in the initiation of saccades; (a) for visually elicited saccades that vary in latency, the pulse of spike activity of these units is tightly coupled to saccade onset; and (b) in a behavioral task in which a visual target does not always elicit a saccade, the occurrence of the pulse of spike activity is highly correlated with the occurrence of the saccade.

Although the pulse of spike activity of the saccade-related burst units is tightly coupled to saccade onset, the pattern of spike activity originating from a single unit does not encode saccade direction or amplitude. For these neurons, identical discharges precede saccades with a wide range of directions and amplitudes (Sparks & Mays, 1980). Saccade direction and amplitude are encoded by the location of the active population of neurons in the superior colliculus. The problem of how information about saccade direction and amplitude is extracted from the spatial distribution of superior colliculus activity has not been resolved (see Sparks & Mays, 1981).

Microstimulation Studies

Microstimulation of the superior colliculus produces a short-latency (20–30 msec), contralateral saccade. The direction and amplitude of the stimulation-induced saccade are a function of the site of stimulation and are relatively independent of both the position of the eye in the orbit and stimulation parameters (Robinson, 1972; Schiller & Stryker, 1972; Schiller, True, & Conway, 1979b). The motor map of saccade direction and amplitude, obtained by stimulation of the superior colliculus, corresponds closely to the map derived from chronic unit recordings.

Lesion Experiments

In the monkey, lesions of the superior colliculus do not eliminate visually elicited saccades. Following superior colliculus lesions, fewer spontaneous saccades occur, visually elicited saccades have longer latencies, and there is a slight deficit in saccadic accuracy (Albano & Wurtz, 1978; Mohler & Wurtz, 1977; Pasik, Pasik, & Bender, 1966; Schiller, True, & Conway, 1979a, 1980; Wurtz & Goldberg, 1972b). The failure of superior colliculus lesions to abolish visually elicited saccades should not be interpreted as evidence that the superior colliculus does not participate in the initiation of saccades. Lesions of the frontal eye fields or the visual cortex do not abolish visually guided saccades either. However, combined lesions of the superior colliculus and visual cortex (Mohler & Wurtz, 1977) or combined lesions of the superior colliculus and frontal eye fields (Schiller et al. 1979a, 1980) do eliminate visually guided saccades. A reasonable interpretation of the lesion data is that, in the normal animal, the superior colliculus is involved in saccade initiation, but in the absence of the superior colliculus an alternative pathway is available.

Models of the Saccadic Eye Movement System

Retinocentric Models

Early models of the saccadic system assumed that the computation of a saccade vector was based upon *retinal error*, the distance and direction of the target image from the fovea (Robinson, 1973; Young & Stark, 1963). The visual system was thought to compute retinal error and to pass this information directly to the oculomotor system, which, in turn, generated a command to correct for the retinal error. In these models, saccades were assumed to be preprogramed or ballistic since the vector of the saccade was determined at saccade onset.

The major features of retinocentric models of the saccadic system are illustrated in Figure 4-1. The retinal error (RE) signal created by the appearance of a visual target depends upon the position of the target

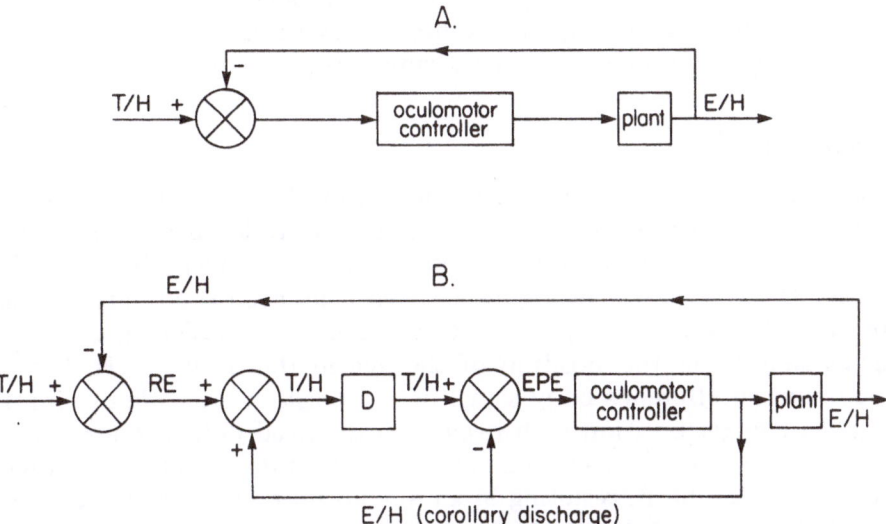

Fig. 4-1. Simplified versions of retinocentric (a) and spatial (b) models of the saccadic system. *T/H*, target position with respect to the head; *E/H*, position of the eyes in the head; *RE*, retinal error (the distance and direction of the retinal image of the target from the fovea); *D*, delay; *EPE*, eye position error (the difference between current and desired eye position).

with respect to the head (T/H) and upon the position of the eyes in the orbit (E/H). The oculomotor controller uses the retinal error signal to form command signals that rotate the globe a specific distance and direction.

The best argument in support of retinocentric models is provided by the functional organization of the superior colliculus. Schiller and Koerner (1971) hypothesized that the superior colliculus codes the location of a visual target relative to the fovea and triggers a saccade that produces foveal acquisition of that target. For example, the appearance of a target 20° to the right of fixation would elicit a discharge of superficial layer visual neurons in a particular region of the left superior colliculus. In this manner, retinal error is represented by the site of activity in the retinotopic map of the superficial layers. The discharge of visual neurons in the superficial layers coding retinal error is assumed to activate underlying regions of the superior colliculus containing neurons that discharge before saccades. Since the movement field map of the deeper neurons corresponds to the retinotopic map of the overlying superficial neurons, a 20° rightward saccade would occur, directing the fovea toward the region of the visual field containing the target.

The foveation hypothesis of superior colliculus function implies a retinocentric model of the saccadic system and the alignment of the visual, and motor maps in the superior colliculus appears to support a retinocentric model. However, there are many problems with the foveation hypothesis and retinocentric models (Sparks & Mays, 1981; see also Goodale, Chapter 3, this volume), and there is evidence that an alternative, spatial model of the saccadic system is correct.

Spatial Models

Other models of the saccadic system assume that visual targets for saccadic eye movements are localized with respect to the head or body and not with respect to the retina. A simplification of one such model (Zee et al., 1976) is shown in Figure 4-1. With the head stationary, the appearance of a visual target creates a retinal error (RE) signal that is dependent upon the position of the eye in the orbit (E/H). Retinal error and an efference copy signal of eye position are added to form a signal of target position with respect to the head (T/H). After a delay (D), the current eye position is subtracted from the target position with respect to the head, resulting in an eye position error (EPE), which is the vector of the saccade currently required to bring the image of the target onto the fovea. If eye position changes in the interval between the computation of retinal error and the command to move the eye, eye position error will be different from retinal error; otherwise, retinal error and eye position error are the same. In this model, the command signals of the oculomotor controller are based upon a continuous comparison of current and desired eye position.

Spatial models differ from retinocentric models in three important ways: First, in spatial models, the vector of a saccade is not determined by retinal information alone, rather, retinal signals are continuously combined with information about eye position to localize the target in a head or body frame of reference. Second, in spatial models the command is "go to a certain position in the orbit," not, as in the retinocentric model, "move a certain distance and direction." Third, in some spatial models, the saccade is not preprogramed or ballistic, but guided to its destination by a neural circuit that continuously compares actual and desired eye position.

Evidence for Spatial Models

Saccades can be made on the basis of cues other than retinal error. Auditory cues that are localized in a head frame of reference (Zahn, Abel, & Dell'Osso, 1978) and somatosensory cues can initiate saccades. Hallet and Lightstone (1976) found that subjects could make a saccade to the location of a visual stimulus that was flashed, briefly, during a prior saccade. Since the position of the eyes changed after the flash, retinal error information alone could not be used to compute the vector of the subsequent saccade to the position of the flashed target. The computation of the distance and direction of the second saccade must take into account information about the direction and amplitude of the first saccade.

We recently conducted an experiment that also strongly supports the view that saccade targets are localized in a nonretinocentric frame of reference (Mays & Sparks, 1980b, 1981). The experimental paradigm for this study is shown in Figure 4-2. Macaque monkeys were trained to look at small light-emitting diode targets in an otherwise darkened room. The search coil technique (Fuchs & Robinson, 1966) was used to measure horizontal and vertical eye position. On a typical trial, the fixation target (represented by the intersection of the axes) was extinguished and an eccentric target (T) was illuminated for 100 msec. The monkey usually looked to the position of target T with a latency of 160–200 msec. Randomly, on 30% of the trials, after target T was extinguished, but before the animal could begin a saccade, the eyes were driven to another position (S) in the orbit by electrical stimulation of the superior colliculus. Under these circumstances, if the monkey attempts to look to the position where target T appeared, where will the animal look?

Retinocentric models require that the vector of a saccade be based

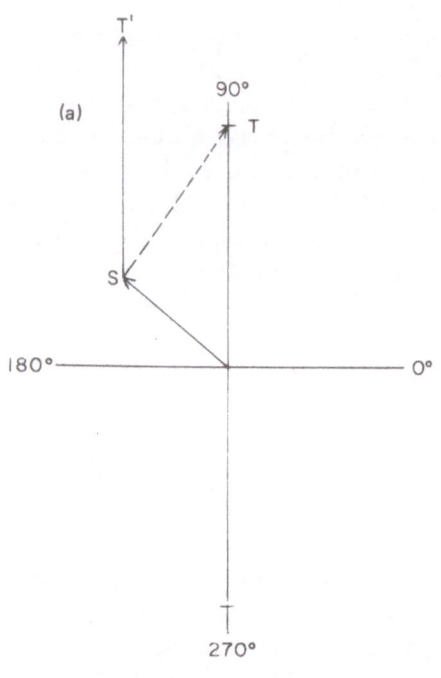

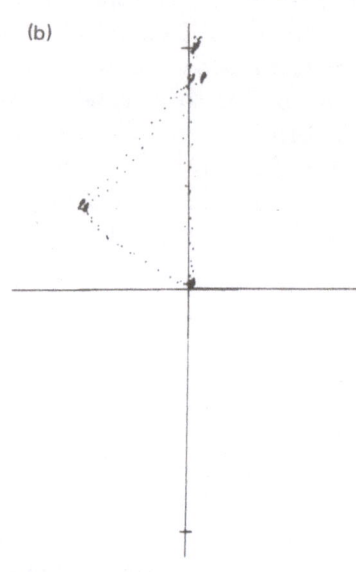

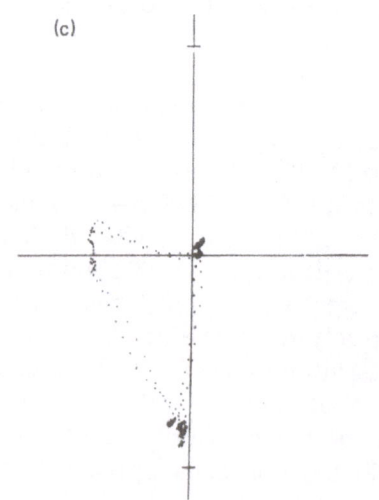

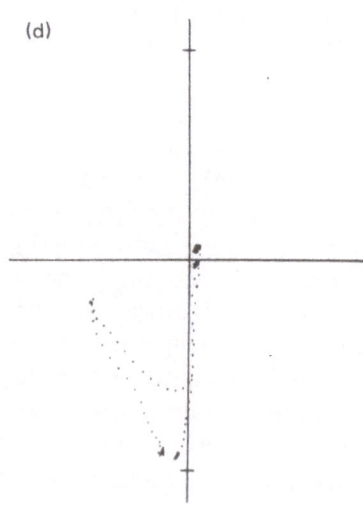

entirely upon a retinal error signal, and predict that the animal should produce a saccade with a predetermined distance and direction. Thus, the animal should look from S to T^1. Spatial models assume that the retinal error signal will be combined with information about the change in eye position produced by superior colliculus stimulation and predict that the animal will look to the position of the target in space, that is, to position T.

Two important features of the experimental design should be emphasized. First, except for the fixation target and the briefly flashed target, the task was performed in total darkness. Thus, the targets could not be localized using visual background cues as an external frame of reference. Second, the target at position T was extinguished before the stimulation-induced saccade. If the animal made a saccade to the position of the target, it could not be based upon a visual updata of target position.

Representative data obtained in this experiment are shown in Figure 4-2B. The trajectories of four eye movements sampled at 2-msec intervals are shown as X-Y plots. On two trials, the superior colliculus was not stimulated, and so the monkey made saccades directly to the location of the briefly flashed target. On the other trials, electrical stimulation drove the eyes away from the original fixation point (up and left). Nevertheless, the monkey made a compensatory saccade to within a few degrees of the actual position in space of the now darkened target.

Fig. 4-2. Interaction of visually and stimulation-induced saccades. (a) Experimental paradigm. In complete darkness, while monkeys are fixating a center target (*intersection of axes*), a peripheral target T is flashed for 100 msec. After the offset of target T, but before the animal initiates a saccade to the target, the superior colliculus is electrically stimulated, driving the eyes to position S. (b) Trajectories of saccades made to the visual target, with and without superior colliculus stimulation. The X-Y plots of the eye movements occurring on four trials are superimposed. The target was flashed $20°$ above the fixation point. *Dots*: horizontal and vertical position of the eye sampled at 2-msec intervals. On two trials, no superior colliculus stimulation occurred and the animal made accurate saccades to the position of the target. On the other trials, superior colliculus stimulation drove the eyes upward and leftward, but after a brief delay, the animal made saccades to the approximate position of the target in space. (c) same as (b), except the target was flashed $20°$ below the fixation point. (d) Trajectory of eye movements when the stimulation-induced saccade interrupted the visually triggered saccade. The target was flashed for 100 msec $20°$ below the fixation point. On one trial, the superior colliculus was not stimulated and the animal looked to the remembered location of the target. On the other trial, the animal began an eye movement to the target, but the stimulation-induced saccade interrupted this movement, driving the eyes to a different position. After a delay of approximately 12 msec, the monkey initiated a new saccade to the approximate position of target T.

Comparable data are shown in Figure 4-2c for when the flashed target was in a different location. In general, regardless of the position of target T and regardless of the vector of the saccade required to compensate for the stimulation-induced movement, the monkey was able to make a saccade to the approximate position of the target in space.

This finding supports a spatial, rather than a retinocentric model since on stimulation trials saccades to the actual target locations could not be directed by retinal error alone. Furthermore, since the occurrence of the stimulation trials was completely unpredictable, compensation for the stimulation-induced perturbation could not have been predetermined; rather, the target must have been localized using both retinal information and information about the stimulation-induced change in eye position.

Two trials in which the target was presented $20°$ below the fixation point are shown in Figure 4-2d. On one trial, without superior colliculus stimulation, the monkey looked directly to the position of the target. On the other trial, superior colliculus stimulation was timed so that the stimulation-induced movement interrupted the visually elicited saccade. Nevertheless, after a delay of 10–12 msec, the animal initiated a saccade to the approximate location of the target. Since the animal could look to the target position on trials in which the visually initiated movement was interrupted by stimulation, the eye position signal used to program the compensatory saccade must be based upon the accomplished, rather than the intended, movement of the eye.

In general, the results of this experiment support the hypothesis that an accurate eye position signal is continuously combined with retinal signals to provide a representation of the target position in space. The data also indicate that, following a saccade, there is little or no delay in the availability of a precise eye position signal.

Neural Representations of Saccade Targets

Identification of Neurons Encoding Saccade Targets in Different Coordinate Systems

Spatial models of the saccadic system imply that there are at least three neural representations of a visual target to which a saccade is made (Figure 4-3a); the first is a representation of the target as a retinal error signal; the second represents the target location in a head or body frame of reference and is based upon a combination of retinal and eye position information; and the third is a representation of eye position error, the difference between current and desired eye position. What are the response properties of neurons that encode the position of a target in these three different ways and under what conditions could such neurons be identified?

Activation of neurons signaling retinal error is dependent upon excitation of receptors in a particular region of the retina. Accordingly, the receptive field moves with each change in gaze (Figure 4-3b, left). It should be noted that neurons that appear to be responsive to activation of a particular region of the retina may actually be signaling eye position error. The distinction between neurons encoding retinal error and eye position error depends upon a critical test, described below.

Neurons representing a saccade target in a head or body frame of reference respond to targets in a specific region of the visual environment, regardless of the position of the eye in the orbit. Thus, as shown in Figure 4-3b (right), the plot of the receptive field of such a neuron while the animal is fixating point 1 will be the same as the plot of the receptive field while the animal is fixating point 2. These neurons are responsive to stimuli occupying a particular region of visual space, regardless of the retinal locus of the target image. The spatial properties of these neurons would not be detected in an acute recording study nor in a chronic recording experiment in which the receptive field was plotted with a single fixation point.

Neurons coding an eye position error signal alter their discharge rate when there is a certain difference between current and desired eye position. Presumably, different populations of neurons are responsive to different directions and magnitudes of eye position error. An eye position error can be produced in two ways: by the appearance of a visual target located a certain distance and direction from the fixation point; or by a change in eye position after the appearance of a visual target. These situations are shown in Figure 4-3c. On the left, the response field of a hypothetical eye position error neuron is represented as a circle (RF). With fixation straight ahead, the neuron discharges in response to saccade targets presented in the circular response field. The neuron discharges not because a particular region of the retina was activated, but because, after the appearance of the saccade target, the difference between current and desired eye position requires a saccade with a particular vector. Note that if this is the only test applied, the discharge of the cell appears simply to reflect retinal error.

A saccade target presented at position A outside the response field will not activate the neuron as long as fixation of the center point is maintained. Suppose, however, after target A disappears, gaze is shifted to a new fixation point but a saccade to the remembered location of target A is still required (Figure 4-3c, right). Because the difference between current and desired eye position is in the response field of the neuron, a change in discharge rate will signal this new eye position error. This signal is created, not by the appearance of a new target, but by a change in eye position occurring after the original target disappeared.

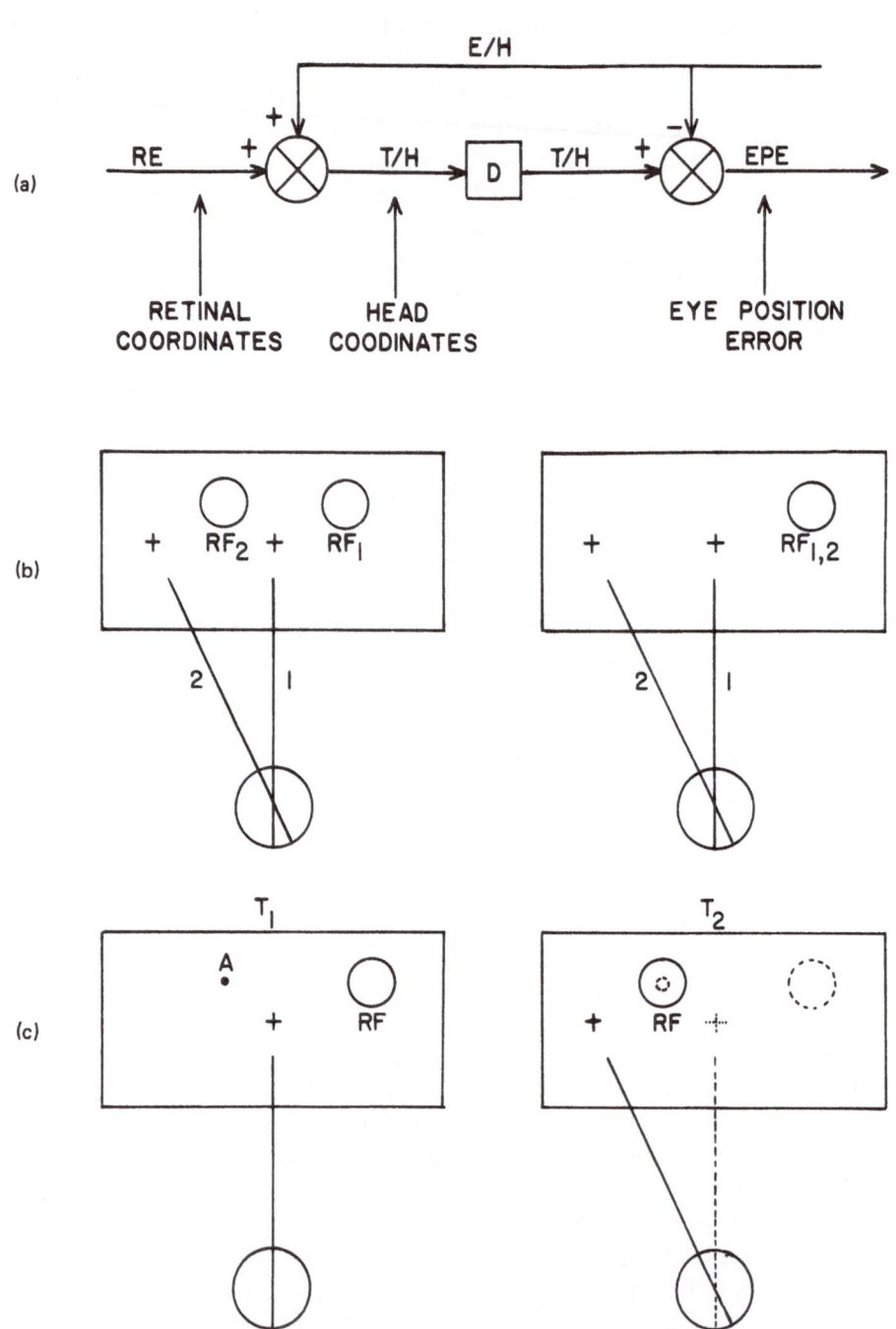

Experimental Data

Recordings from the Superior Colliculus

Recently, we obtained evidence that the superior colliculus contains a retinal error signal, a representation of eye position error, and a motor command to correct eye position error (Mays & Sparks, 1980a). Monkeys were trained to perform two types of tasks, shown in Figure 4-4. On single-saccade trials (Figure 4-4a), while the animal was fixating a center target 0, the fixation target was extinguished and target A appeared at an eccentric position. If the animal made a saccade to target A within 500 msec and maintained fixation for 1 sec, a water reinforcement was delivered. By systematically varying the position of target A, it was possible to plot the receptive fields of visually responsive neurons and the movement fields of neurons producing saccade-related discharges.

On double-saccade trials (Figure 4-4b), after fixation of the center target 0 for a variable period, the offset of the fixation target was followed by a brief presentation of target B. The duration of target B was less than the reaction time of the monkey, and thus target B was extinguished before a saccade occurred. The monkey learned that on trials with a brief presentation of target B, reward was contingent upon a saccade to B (0–B) and a second saccade (B–0) back to the position of the fixation target.

Fig. 4-3. Response properties of three types of hypothetical neurons. (a) The spatial model implies that there will be at least three neural representations of a saccade target: a representation in retinal coordinates, a representation in head coordinates, and a representation of eye position error (*EPE*). *RE*, retinal error; *E/H*, position of the eyes in the head; *T/H*, target position with respect to the head. (b) Response properties of neurons encoding the position of the target in retinal and head coordinates. *Left*: Neurons signaling the position of the target in retinal coordinates have a retinal receptive field. With fixation of point *1*, the neuronal discharge changes when stimuli activate a specific region of the retina (receptive field 1, RF_1). If the position of gaze changes (fixation of point 2), the receptive field moves (RF_2) since a specific region of the retina still must be activated in order to produce a neuronal response. *Right*: Neurons encoding the position of a target in head coordinates discharge whenever a stimulus appears in a specific region of the visual environment ($RF_{1,2}$), regardless of gaze direction (fixation 1 or fixation 2). The response of these neurons depends upon both retinal and eye position information. (c) Response properties of neurons encoding eye position error. Neurons conveying an eye position error signal alter their discharge rate when there is a specific difference between current and desired eye position. This difference may occur because of the appearance of a visual target that produces an eye position error (*left*) or because of a change in eye position after the brief appearance of a visual target (*right*). *RF*, response field of eye position error neuron; *A*, Target position.

Typically, on double saccade trials, the vector of the B–0 saccade was the same as the vector of the 0–A saccade. If the 0–A saccade was in the movement field of the neuron being studied, then the B–0 saccade, having the same vector, would also be in the movement field. However, the computation of the vector of the B–0 saccade must be based upon both retinal information and information about the change in eye position that occurred as a result of the 0–B saccade. Therefore, neurons discharging before the B–0 saccade must be part of the neural circuitry after the combination of retinal and eye position signals. Note, also, that the double-saccade trial type is the paradigm required to identify neurons encoding eye position error; that is, a change in eye position, the 0–B saccade, creates an eye position error.

Several classes of neurons could be identified using these behavioral tasks. Visually responsive neurons isolated in the superficial layers of the superior colliculus were activated by stimulation of a particular region of the retina. They responded only to the appearance of a visual stimulus in their receptive field, and the location of the receptive field shifted with each change in gaze. Thus, because of the retinotopic map,

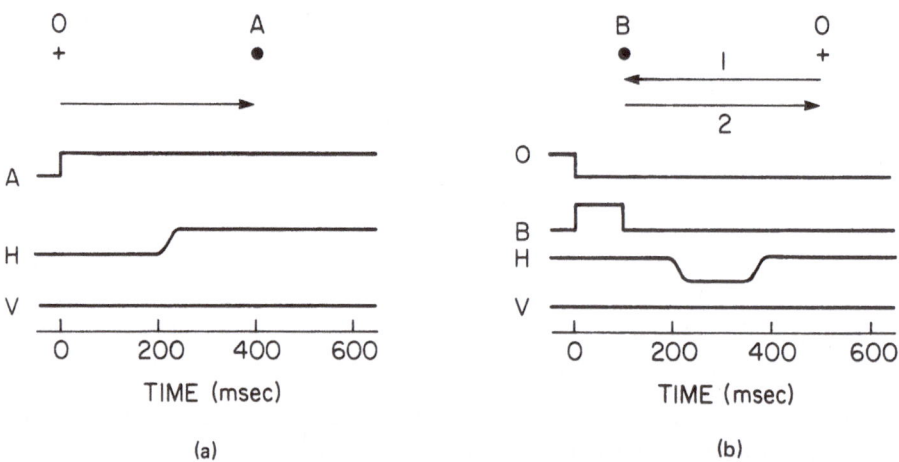

(a)   (b)

Fig. 4-4(a). Single-saccade trial. After fixation of target 0, the fixation target is turned off and target A appears at an eccentric location. Typically, the animal makes a saccade to the target with a latency varying between 160 and 200 msec. *A*, onset (upward deflection) of target A; *H*, horizontal eye position (upward deflection represents a rightward movement); *V*, vertical eye position. (b) Double-saccade trial. After fixation of target 0, the fixation target is turned off and target B is presented for 50–100 msec. The brief presentation of target B is a signal to the monkey to look to the remembered position of target B and then back to the position of target 0. *1*, the vector of the 0–B saccade to the position of target B; *2*, the vector of the B–0 saccade to the position of target 0; *0*, onset (upward deflection) and offset (downward deflection) of target 0; *B*, onset and offset of target B; *H*, horizontal eye position; *V*, vertical eye position.

the site of neuronal activity in the superficial layers of the superior colliculus represents a map of retinal error.

We isolated one type of neuron in the intermediate layers (called a quasi-visual, or QV cell) that may signal eye position error. Figure 4-5 shows the discharge of one such cell. On single-saccade trials, the neuron appeared to be visually responsive and to have a receptive field in the lower left quadrant of the visual field (Figure 4-5c). On these trials, the onset of the discharge was tightly coupled to target onset, not sac-

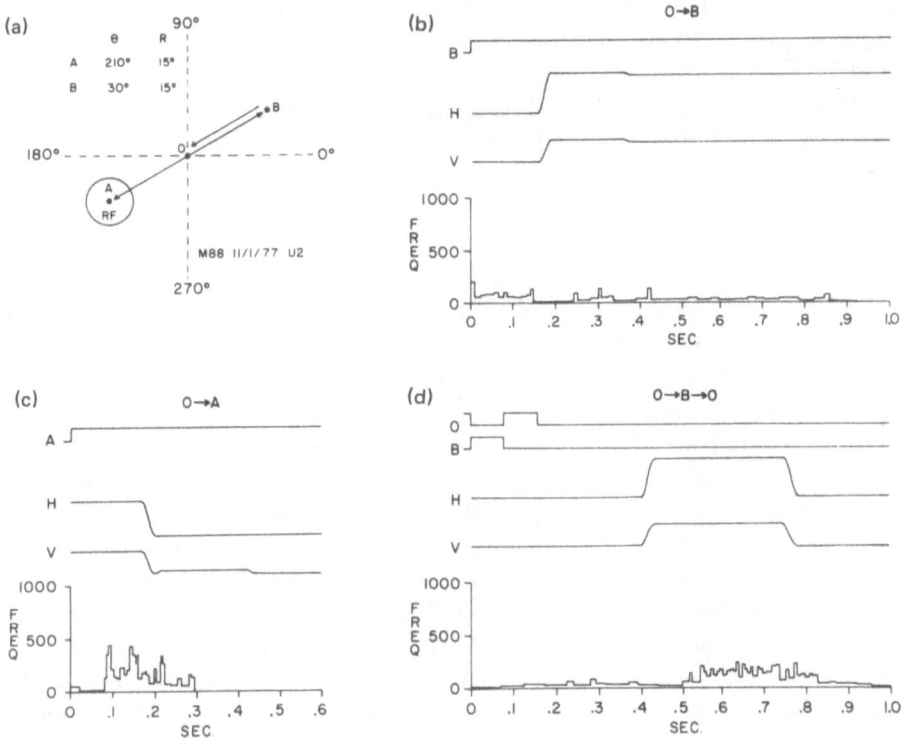

Fig. 4-5. Discharge pattern of a QV cell. (a) Vectors of the saccades occurring on single- (0–A) and double- (0–B and B–0) saccade trials. (b) Lack of neuronal response on a control trial (0–B saccade). *B*, onset of target B; *H*, horizontal eye position; *V*, vertical eye position; *bottom trace*, instantaneous spike frequency of the QV cell during this trial. (c) Typical response of a QV cell on single-saccade trials. A vigorous discharge, tightly coupled to target *A* onset, is observed. Typically, the increase in spike frequency continues until after the 0–A saccade. (d) Typical response of a QV cell on double-saccade trials. No observable alteration in discharge rate is related to the onset of targets B or 0, but after the eyes reach position B, an increase in discharge frequency begins and continues until after the return (B–0) saccade. (Adapted from Mays, L. E., & Sparks, D. L. Dissociation of visual and saccade-related responses in superior colliculus neurons. *Journal of Neurophysiology*, 1980, *43*, 207–232.)

to target duration. However, QV cells also gave a vigorous discharge on double-saccade trials (Figure 4-5d). On these trials, target B was flashed in the right visual field, the field opposite to the cell's receptive field. However, after the eyes reached position B, the cell began to discharge and continued to do so until after the saccade back to position B. This increase in discharge frequency was not associated with the appearance of target B or the 0–B saccade (Figure 4-5b).

Most intermediate-layer neurons with saccade-related discharges (saccade-related neurons) also discharged on double-saccade trials. A burst of activity preceded saccades in their movement field even though the position of the eyes changed after the appearance of the saccade targets. Thus, the discharge of saccade-related neurons occurs after retinal and eye position signals have been combined.

The spatial and temporal patterns of activity of visual, QV, and saccade-related neurons during a typical double-saccade trial are reconstructed in Figure 4-6. Traces on the left show the activity patterns of these cells and the corresponding filled circles (1–5) show the locations of cells plotted on a top view of the superior colliculus. With the eyes directed at the fixation target 0, the onset of target B (20° to the right) elicits a response from visual cells in the superficial layer of the left superior colliculus (represented by site 1). This response (shown in trace 1) is phasic, which is characteristic of visual cells in the superficial layer. Target B also evokes a sustained response (trace 2) from a neighboring population of QV cells (site 2). Although the duration of target B is brief, the activity of QV cells holds the eye position error signal in spatial register until a saccade is made. The QV cells at site 2 will fire as long as a 20° rightward saccade is needed. A saccade trigger could be combined with this QV eye position error signal to activate a population of saccade-related cells at site 4 (trace 4), thereby eliciting a saccade to B. When the animal looks to B, the eye position error signal is no longer "20° right," and so the activity of QV cells at site 2 will cease after a brief delay. When the eyes are directed at B, a saccade back to 0 is required. The current eye position (i.e., the eyes directed at B) is then subtracted from the remembered spatial location of 0 in a head frame of reference. The result is an eye position error signal of "20° left," which is signaled by activation of QV cells at site 3. These cade onset, and the duration of the neuronal response was proportional cells will remain active (trace 3) until a saccade trigger arrives, which can activate nearby saccade-related cells (trace 5) at site 5, thereby producing a saccade back to 0. This second saccade will cause the eye position error signal to be changed from "20° left," and QV cell activity at site 3 will cease shortly afterward. Thus, the discharge of QV cells appears to signal eye position error and to hold this information in spatial register until a saccade occurs or is cancelled. If a target is displaced or if the eyes move after a brief target disappears, the site of QV cell activity shifts to a location that represents the new eye position error.

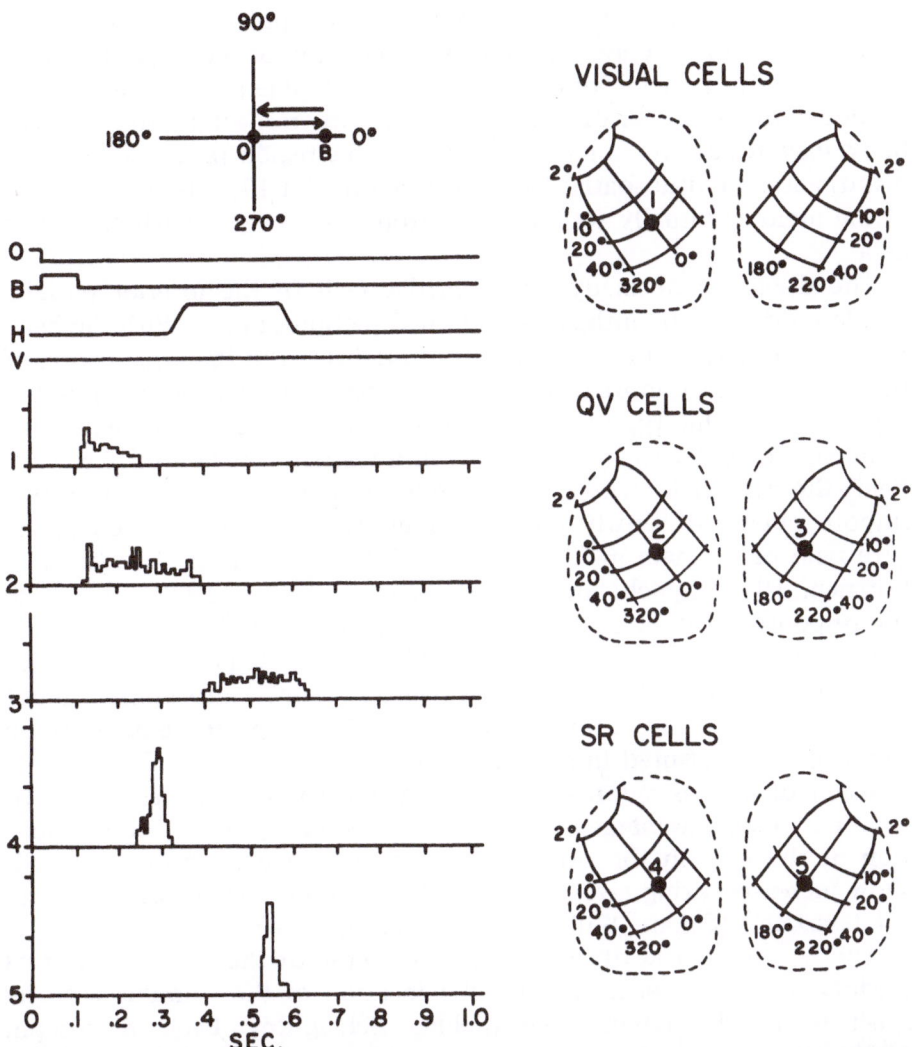

Fig. 4-6. Visual, QV, and saccade-related (*SR*) cell activity on a double-saccade trial. Vectors of double saccade (0–B and B–0) and neuronal responses. *0*, onset of target 0; *B*, onset of target B; *H*, horizontal eye position; *V*, vertical eye position; *traces 1–5*, activity patterns of cells *1–5*. Activity patterns of cells *1–5*. Location of cells *1–5* plotted on a top view of the superior colliculus. (From Mays, L. E., & Sparks, D. L. Dissociation of visual and saccade-related responses in superior colliculus neurons. *Journal of Neurophysiology*, 1980, *43*, 207–232.)

Representations of Target Position in Space

Our experiments suggest that neuronal activity in the primate superior colliculus may encode both retinal error and eye position error. We did not isolate neurons encoding the position of a target in head or body coordinates in the monkey superior colliculus. Nevertheless, spatial models of the saccadic system require a neural representation of the target in a head frame of reference and, presumably, the properties of QV cells in the superior colliculus are based upon a subtraction of eye position from a stored representation of the position of the target in space.

The inferior parietal lobule represents a possible site where neurons encoding the position of a visual target with respect to the head or body may reside. In humans, lesions of this region produce dramatic disturbances in visuospatial behavior (see Chapter 17, this volume), and in the monkey visually responsive neurons have been identified in this area.

The discharge of many light-sensitive cells of the inferior parietal lobule appears to be influenced by the direction of gaze. With the head fixed, physically identical visual stimuli delivered to the same region of the retina produce more or less vigorous responses, depending upon the angle of gaze. One type of parietal neuron displays an increase in spike frequency during fixation of targets in a restricted region of the visual field, the gaze field. The same neurons show increased activity during smooth pursuit of slowly moving targets when the targets are in the gaze field of the neuron (Anderson & Mountcastle, 1980; Hyvarinen & Poranen, 1974; Lynch, Mountcastle, Talbot, & Yin, 1977). Although the responses of inferior parietal lobule neurons are described as being jointly dependent upon retinal stimulation and gaze position, it cannot be stated unequivocally that these neurons convey information about the position of a target with respect to the head. Experiments in which the target is presented in specific regions of the visual field, while the position of gaze is sytematically varied, have not been described. To what extent the discharge of these neurons is merely related to the position of the eyes in the orbit is also unclear since some of these cells show increased firing rates during particular directions of gaze, even in total darkness (Sakata, Shibutani, & Kawano, 1980).

The responsiveness of neurons in the region of the thalamic internal medullary lamina depends upon the location of the stimulus with respect to the head–body axis (Schlag, Schlag-Rey, Peck, & Joseph, 1980). The properties of these neurons are described in detail by Schlag and Schlag-Rey (Chapter 5, this volume).

The smooth pursuit eye movement system, which attempts to match the velocity of the eyes to target velocity, also requires a combination of retinal and eye position information. With the eyes and head free to move, the computation of target velocity is derived by combining signals of retinal image velocity, eye velocity, and head velocity. Kase, Noda, Suzuki, and Miller (1979) found that Purkinje cells and mossy fibers in the posterior cerebellar vermis produced discharge rates proportional to absolute target velocity. When the monkey accurately tracked a sinusoidally oscillating target (so that retinal image velocity was negligible), neuronal activity was proportional to the velocity of the eye movement. When the monkey fixated a stationary target (so that eye position signals were constant) and a second target was oscillated, spike activity was proportional to the retinal image velocity of the second target. Thus, eye velocity and retinal image velocity signals were combined to provide a signal of absolute target velocity and the activity of neurons in the cerebellar vermis was an accurate representation of this signal.

Few experiments capable of detecting neurons encoding the location of a visual target in other than retinal coordinates have been conducted. Consequently, much remains to be learned about the neural mechanisms underlying the localization of objects in visual space.

Acknowledgments

This work was supported by National Eye Institute Research Grants EY-01189 and EY-02293 and by CORE Facility Grant EY-03039.

References

Albano, J. E., & Wurtz, R. H. Modification of the pattern of saccadic eye movements following ablation of monkey superior colliculus. *Neuroscience Abstracts*, 1978, *8*, 161.

Anderson, R. A., & Mountcastle, V. B. The direction of gaze influences the response of many light sensitive neurons of the inferior parietal lobule (area 7) in waking monkeys. *Neuroscience Abstracts*, 1980, *6*, 673.

Barnes, W. T., Magoun, H. W., & Ranson, S. W. The ascending auditory pathway in the brainstem of the monkey. *Journal of Comparative Neurology*, 1943, *79*, 129-152.

Buttner, U., Buttner-Ennever, J. A., Henn, V. Vertical eye movement related unit activity in the rostral mesencephalic reticular formation of the alert monkey. *Brain Research*, 1977, *130*, 234-252.

Cynader, M., & Berman, N. Receptive field organization of monkey superior colliculus. *Journal of Neurophysiology*, 1972, *41*, 1394-1417.

Drager, U. C., & Hubel, D. H. Responses to visual stimulation and relationship between visual, auditory, and somatosensory inputs in mouse superior colliculus. *Journal of Neurophysiology*, 1975, *38*, 690-713.

Drager, U. C., & Hubel, D. H. Topography of visual and somatosensory projections to mouse superior colliculus. *Journal of Neurophysiology*, 1976, *39*, 91-101.

Edwards, S. B., & Henkel, C. K. Superior colliculus connections with the extraocular motor nuclei in the cat. *Journal of Comparative Neurology*, 1978, *179*, 451-468.

Edwards, S. B., Ginsburgh, C. L., Henkel, C. K., & Stein, B. E. Sources of subcortical projections to the superior colliculus in the cat. *Journal of Comparative Neurology*, 1979, *184*, 309-330.

Frankfurter, A., Weber, J. T., & Harting, J. K. Brain stem projections to lobule VII of the posterior vermis in the squirrel monkey: As demonstrated by the retrograde axonal transport of tritiated horseradish peroxidase. *Brain Research*, 1977, *124*, 135-139.

Frankfurter, A., Weber, J. T., Royce, G. J., Strominger, N. L., & Harting, J. K. An autoradiographic analysis of the tecto-olivary projection in primates. *Brain Research*, 1976, *118*, 245-257.

Fuchs, A. F., & Robinson, D. A. A method for measuring horizontal and vertical eye movement chronically in the monkey. *Journal of Applied Physiology*, 1966, *21*, 1068-1070.

Goldberg, M. E., & Robinson, D. L. Visual system. Superior colliculus. In R. Masterton (Ed.), *Handbook of behavioral neurobiology* (Vol. 1). New York: Plenum Press, 1978, pp. 119-164.

Goldman, P. S., & Nauta, W. J. H. Autoradiographic demonstration of a projection from prefrontal association cortex to the superior colliculus in the rhesus monkey. *Brain Research*, 1976, *116*, 145-149.

Gordon, B. Receptive fields in deep layers of cat superior colliculus. *Journal of Neurophysiology*, 1973, *38*, 157-178.

Graham, J. An autoradiographic study of the efferent connections of the superior colliculus in the cat. *Journal of Comparative Neurology*, 1977, *173*, 629-654.

Graybiel, A. M. Evidence for banding of the cat's ipsilateral retinotectal connection. *Brain Research*, 1976, *114*, 318-327.

Graybiel, A. M. Organization of oculomotor pathways in the cat and rhesus monkey. In R. Baker & A. Berthoz (Eds.), *Control of gaze by brain stem neurons*. Amsterdam: Elsevier/North-Holland, 1977.

Hallet, P. E., & Lightstone, A. D. Saccadic eye movement towards stimuli triggered by prior saccades. *Vision Research*, 1976, *16*, 99-106.

Harting, J. K. Descending pathways from the superior colliculus: An autoradiographic analysis in the rhesus monkey (*Macaca mulatta*). *Journal of Comparative Neurology*, 1977, *173*, 583-612.

Harting, J. K., Huerta, M. F., Frankfurter, A. J., Strominger, N. L., & Royce, G. J. Ascending pathways from the monkey superior colliculus: An autoradiographic analysis. *Journal of Comparative Neurology*, 1980, *192*, 853-882.

Helmholtz, H. von [*A treatise on physiological optics* (Vol. 3)] (J. P. C. Southall, Ed. and Trans.). New York: Dover, 1962. (Originally published, 1866.)

Hendrickson, A., Wilson, M. E., & Toyne, M. J. The distribution of optic nerve fibers in *Macaca mulatta*. *Brain Research*, 1970, *23*, 425-427.

Henn, V., & Cohen, B. Coding of information about rapid eye movements in the pontine reticular formation of alert monkeys. *Brain Research*, 1976, *108*, 307-325.

Holst, E. von, & Mittelstaedt, H. Das Reafferenzprinzip (Wechselwirkungen zwischen Zentralnervensystem und Peripherie). *Naturwissenschaften*, 1950, *37*, 464-476. (Reprinted and translated in P. C. Dodwell (Ed.), *Perceptual processing: Stimulus equivalence and pattern recognition*. New York: Appleton, 1971.)

Hubel, D. H., LeVay, S., Wiesel, T. N. Mode of termination of retinotectal fibers in Macaque monkey: An autoradiographic study. *Brain Research*, 1975, *96*, 25-40.

Hyvarinen, J., & Poranen, A. Function of the parietal associative area 7 as revealed from cellular discharges in alert monkeys. *Brain*, 1974, *97*, 673-692.

Kase, M., Miller, D. C., & Noda, H. Discharges of Purkinje cells and mossy fibers in the cerebellar vermis of the monkey during saccadic eye movements and fixation. *Journal of Physiology* (London), 1980, *300*, 539-555.

Kase, M., Noda, H., Suzuki, D. A., & Miller, D. C. Target velocity signals of visual tracking in vermal Purkinje cells of the monkey. *Science*, 1979, *205*, 717-720.

Keller, E. L. Participation of medial pontine reticular formation in eye movement generation in monkey. *Journal of Neurophysiology*, 1974, *37*, 316-320.

King, W. M., & Fuchs, A. F. Neuronal activity in the mesencephalon related to vertical eye movements. In R. Baker & A. Berthoz (Eds.), *Control of gaze by brain stem neurons*. Amsterdam: Elsevier/North-Holland, 1977.

Kunzle, H., & Akert, K. Efferent connections of cortical area 8 (frontal eye field) in *Macaca fascicularis*. A reinvestigation using the autoradiographic technique. *Journal of Comparative Neurology*, 1974, *173*, 147-164.

Kunzle, H., Akert, K., Wurtz, R. H. Projection of area 8 (frontal eye field) to superior colliculus in the monkey. An autoradiographic study. *Brain Research*, 1976, *117*, 487-492.

Kuypers, H. G. J. M., & Lawrence, D. G. Cortical projections to the red nucleus and the brain stem in the rhesus monkey. *Brain Research*, 1967, *4*, 151-188.

Lund, R. D. Anatomic studies on the superior colliculus. *Investigative Ophthalmology*, 1972, *11*, 434-444. (a)

Lund, R. D. Synaptic patterns in the superficial layers of the superior colliculus of the monkey, *Macaca mulatta*. *Experimental Brain Research*, 1972, *4*, 151-188. (b)

Lynch, J. C., Mountcastle, V. B., Talbot, W. H., & Yin, T. C. T. Parietal lobe mechanisms for directed visual attention. *Journal of Neurophysiology*, 1977, *40*, 362-389.

Matin, L. Eye movements and perceived visual direction. In D. Jameson & L. Hurvich (Eds.), *Handbook of sensory physiology* (Vol. 7). Berlin: Springer, 1972.

Mays, L. E., & Sparks, D. L. Dissociation of visual and saccade-related responses in superior colliculus neurons. *Journal of Neurophysiology*, 1980, *43*, 207-232. (a)

Mays, L. E., & Sparks, D. L. Saccades are spatially, not retinocentrically, coded. *Science*, 1980, *208*, 1163-1165. (b)

Mays, L. E., & Sparks, D. L. The localization of saccade targets using a combination of retinal and eye position information. In A. Fuchs & W. Becker (Eds.), *Progress in oculomotor research*. New York: Elsevier, 1981, pp. 39-47.

McIlwain, J. T. Topographic organization and convergence in corticotectal projections from areas 17, 18, and 19 in the cat. *Journal of Neurophysiology*, 1977, *40*, 189-198.

Mohler, C. W., & Wurtz, R. H. Role of striate cortex and superior colliculus in visual guidance of saccadic eye movements in monkeys. *Journal of Neurophysiology*, 1977, *40*, 74-94.

Optican, L. M., & Robinson, D. A. Cerebellar-dependent adaptive control of primate saccadic system. *Journal of Neurophysiology*, 1980, *44*, 1058-1076.

Pasik, T., Pasik, P., & Bender, M. B. The superior colliculi and eye movements. *Archives of Neurology*, 1966, *15*, 420-436.

Petras, J. M. Connections of the parietal lobe. *Journal of Psychiatric Research*, 1971, *8*, 189-201.

Poirier, L. J., & Bertrand, C. Experimental and anatomical investigation of the lateral spino-thalamic and spinotectal tracts. *Journal of Comparative Neurology*, 1955, *102*, 745-758.

Pollack, J. G. & Hickey, T. L. The distribution of retino-collicular axon terminals in rhesus monkey. *Journal of Comparative Neurology*, 1979, *185*, 587-602.

Ritchie, L. Effects of cerebellar lesions on saccadic eye movements. *Journal of Neurophysiology*, 1976, *39*, 1246-1256.

Robinson, D. A. Eye movements evoked by collicular stimulation in the alert monkey. *Vision Research*, 1972, *12*, 1795-1808.

Robinson, D. A. Models of the saccadic eye movement control system. *Kybernetik*, 1973, *14*, 71-83.

Robinson, D. A. Oculomotor control signals. In G. Lennerstrand, & P. Bach-y-Rita, (Eds.), *Basic mechanisms of ocular motility and their clinical implications.* Oxford: Pergamon Press, 1975, pp. 337-374.

Ron, S., & Robinson, D. A. Eye movements evoked by cerebellar stimulation in the alert monkey. *Journal of Neurophysiology*, 1973, *36*, 1004-1022.

Sakata, H., Shibutani, H., & Kawano, K. Spatial properties of visual fixation neurons in posterior parietal association cortex of the monkey. *Journal of Neurophysiology*, 1980, *43*, 1654-1672.

Schiller, P. H., & Koerner, F. Discharge characteristics of single units in superior colliculus of the alert rhesus monkey. *Journal of Neurophysiology*, 1971, *34*, 920-936.

Schiller, P. H., & Stryker, M. Single-unit recording and stimulation in superior colliculus of the alert rhesus monkey. *Journal of Neurophysiology*, 1972, *35*, 915-924.

Schiller, P. H., Stryker, M., Cynader, M., & Berman, N. Response characteristics of single cells in the monkey colliculus following ablation or cooling of visual cortex. *Journal of Neurophysiology*, 1974, *37*, 181-194.

Schiller, P. H., True, S. D., & Conway, J. L. Effects of frontal eye field and superior colliculus ablations on eye movements. *Science*, 1979, *206*, 590-592. (a)

Schiller, P. H., True, S. D., & Conway, J. L. Paired stimulation of the frontal eye fields and the superior colliculus of the rhesus monkey. *Brain Research*, 1979, *179*, 162-164. (b)

Schiller, P. H., True, S. D., & Conway, J. L. Deficits in eye movements following frontal eye-field and superior colliculus ablations. *Journal of Neurophysiology*, 1980, *44*, 1175-1189.

Schlag, J., Schlag-Rey, M., Peck, C. K., & Joseph, J. -P. Visual responses of thalamic neurons depending on the direction of gaze and the position of targets in space. *Experimental Brain Research*, 1980, *40*, 170-184.

Shebilske, W. L. Visuomotor coordination in visual direction and position constancies. In W. Epsten (Ed.), *Stability and constancy in visual perception: Mechanisms and processes.* New York: Wiley, 1977, pp. 23-63.

Skavenski, A. A. The nature and role of extraretinal eye-position information in visual localization. In R. A. Monty, & J. W. Senders (Eds.), *Eye movements and psychological processes.* New York: Wiley, 1976, pp. 277-287.

Sparks, D. L. Functional properties of neurons in the monkey superior colliculus: Coupling of neuronal activity and saccade onset. *Brain Research*, 1978, *156*, 1-16.

Sparks, D. L., Holland, R., & Guthrie, B. L. Size and distribution of movement fields in the monkey superior colliculus. *Brain Research*, 1976, *113*, 21-34.

Sparks, D. L., & Mays, L. E. Movement fields of saccade-related burst neurons in the monkey superior colliculus. *Brain Research*, 1980, *190*, 39-50.

Sparks, D. L., & Mays, L. E. The role of the monkey superior colliculus in the control of saccadic eye movements: A current perspective. In A. Fuchs & W. Becker (Eds.), *Progress in oculomotor research.* New York: Elsevier, 1981, pp. 137-144.

Sparks, D. L., Mays, L. E., & Pollack, J. G. Saccade-related unit activity in the monkey superior colliculus. In R. Baker & A. Berthoz (Eds.), *Control of gaze by brainstem neurons.* Amsterdam: Elsevier/North-Holland, 1977, pp. 437–444.

Sparks, D. L., & Pollack, J. G. The neural control of eye movements: The role of the superior colliculus. In B. A. Brooks & F. J. Bajandas (Eds.), *Eye movements.* New York: Plenum Press, 1977.

Sprague, J. M. Mammalian tectum: Instrinsic organization, afferent inputs, and integrative mechanisms. Anatomical substrate. *Neurosciences Research Program Bulletin,* 1975, *13,* 204–213.

Stein, B. E., Magalhaes-Castro, B., & Kruger, L. Relationship between visual and tactile representations in cat superior colliculus. *Journal of Neurophysiology,* 1976, *39,* 401–419.

Weber, J. T., & Harting, J. K. Parallel pathways connecting the primate superior colliculus with the posterior vermis. An experimental study using autoradiographic and horseradish peroxidase tracing methods. In C. Roback (Ed.), *Sensory systems of primates.* New York: Plenum Press, 1978, pp. 135–149.

Weber, J. T., Partlow, G. D., & Harting, J. K. The projection of the superior colliculus upon the inferior olivary complex of the cat: An autoradiographic and horseradish peroxidase study. *Brain Research,* 1978, *144,* 369–377.

Whitlock, D. G., & Nauta, W. J. H. Subcortical projections from the temporal neocortex of *Macaca mulatta. Journal of Comparative Neurology,* 1956, *106,* 183–212.

Wurtz, R. H., & Albano, J. E. Visual-motor function of the primate superior colliculus. *Annual Review of Neuroscience,* 1980, *3,* 180–226.

Wurtz, R. H., & Goldberg, M. E. Activity of the superior colliculus in behaving monkey. III. Cells discharging before eye movements. *Journal of Neurophysiology,* 1972, *35,* 575–586. (a)

Wurtz, R. H., & Goldberg, M. E. Activity of the superior colliculus in behaving monkey. IV. Effects of lesions on eye movement. *Journal of Neurophysiology,* 1972, *35,* 587–596. (b)

Young, L. R., & Stark, L. Variable feedback experiments testing a sampled data model for eye tracking movements. *IEEE Transactions on Human Factors in Electronics,* 1963, *HFE-4,* 28–51.

Zahn, J. R., Abel, L. A., & Dell'Osso, L. F. Audio-ocular response characteristics. *Sensory Processes,* 1978, *2,* 32–37.

Zee, D. S., Optican L. M., Cook, J. D., Robinson, D. A., & Engel, W. K. Slow saccades in spinocerebellar degeneration. *Archives of Neurology,* 1976, *33,* 243–251.

Chapter 5

Interface of Visual Input and Oculomotor Command for Directing the Gaze on Target

John Schlag and Madeleine Schlag-Rey

Looking at a target seems to be a trivial task for the nervous system. It is performed almost automatically every second of active life. However, when one tries to explain in terms of pathways and nerve impulses how the gaze can be brought on target, a major difficulty becomes immediately apparent: How is the target selected? It is not a priori obvious in those circumstances in which looking serves to explore, to search, to anticipate something that may not yet be there. The whole scene has to be analyzed, patterns must be isolated and identified, and their significance assessed as a function of various contingencies. This involves a number of processes that are beyond the scope of neurophysiological studies today.

Therefore, wisely, neurophysiologists have focused on the most elementary orientation mechanisms, using paradigms that short-circuit the decision processes. The selection problem is avoided by offering a single stimulus on a blank background. The stimulus is unique, conspicuous, and small, thereby avoiding any ambiguities about its location in space. Often, the experimental subject is reinforced for acquiring this target immediately. The spatial and temporal parameters of the orienting response are thus imposed from outside. The advantage of the paradigm is that the flow chart of nervous operations is collapsed. Only the operations that follow target selection need be considered. They include the transmission of information on target location and the translation of this information into a language that motoneurons can understand.

To such an experimental strategy, we owe, for instance, the foveation hypothesis (Schiller & Koerner, 1971). Superficial neurons of the superior colliculus are retinotopically arranged. They are readily excited by visual events, that is, stimuli turned on or off, or moved. Their responses are usually transient. The intense flurry of activity triggered in a limited set of collicular neurons provides a simple mechanism for coding stimulus location. Deeper in the colliculus lie cells whose discharge accompanies saccades terminating in the same area as the receptive field of the superficial cells. With the assumption that the proper connections exist between superficial and deep layers of the superior colliculus, the essential components of a foveation machine are gathered. The validity of this hypothesis has been examined from different points of view and some of its assumptions have been questioned (Edwards, 1980; Mays & Sparks, 1980; Mohler & Wurtz, 1976; Sparks & Mays, 1980). Nevertheless, the basic principle of the interface is so elegantly simple that one could view targeting movements of the eye as the product of receptive field properties and appropriate wiring.

However, there is a danger inherent in too simple paradigms; they may lead to too simple interpretations of the facts. The properties of neurons may be underestimated, even misconstrued, if their capacities have not been fully explored. A good example is provided by Mays and Sparks' (1980) study of certain "visual" cells of the superior colliculus. These apparently were visual cells since they responded vigorously, reliably, and with constant latency to the appearance of a visual stimulus in the receptive field. A problem arose with the observation that these cells were also activated by a stimulus that was no longer present, when an eye movement brought the site of the extinguished stimulus into the cell's receptive field. Such findings, of course, radically change classical neurophysiological notions. Recent years have seen the emergence of new types of neuronal units with curious properties: "quasivisual cells" (Mays & Sparks, 1980), "visuokinetic units" (Kubota, Iwamoto, & Suzuki, 1974), "visual fixation neurons" (Lynch, Mountcastle, Talbot, & Yin, 1977), and a variety of units that are neither visual nor motor in the usual sense. Under these categorizations lies the reality, still a little vague, of a number of neuronal elements needed at the visuomotor interface. Our task is to determine their actual role. In this presentation, we report on recorded forebrain units that, by conventional standards, would qualify as "visual" units. Observations were made suggesting that rather than indicating target location, they specify the goal to be reached by the eyes.

Internal Medullary Lamina

These observations were made in a central thalamic region of alert cats and, more recently, monkeys (Schlag-Rey & Schlag, 1981). This region is called the internal medullary lamina (IML), previously known as the thalamic nonspecific system, source of cortical recruiting responses. It may be appropriate, first, to recount how the IML came to be known as related to gaze control.

As was the case for the frontal eye field (Ferrier, 1874), the first indication of IML involvement in oculomotor function was provided by electrical stimulation (Schlag & Schlag-Rey, 1971). In the cat, the IML was found to be the part of the thalamus from which saccades could be evoked with short latencies (e.g., 35 msec) at the lowest threshold (e.g., 20 µA). The saccades were contraversive, sometimes goal-directed (Maldonado, Joseph, & Schlag, 1980), with no other movements except, occasionally, head rotation. Conjugate deviation of the eyes has also been obtained in man by electrical stimulation of the intralaminar region (Hassler, 1980). The second indication, again as was the case for the frontal eye field (Kennard & Ectors, 1938), came when lesions were shown to produce a syndrome of contralateral visual neglect characterized by deficits of visual orienting, following, and placing, and by the absence of blinking to visual threat (Orem, Schlag-Rey, & Schlag, 1973). Similar effects have been observed in the monkey (Watson, Miller, & Heilman, 1978), and they are now suspected to occur with thalamic hemorrhages in man (Watson, 1978). Third, recordings made in the IML of alert cats revealed the presence of units exhibiting presaccadic bursts, pauses, and sustained firing related to eye position (Schlag, Lehtinen, & Schlag-Rey, 1974), that is, patterns of activity commonly observed in preoculomotor populations of the brain stem (e.g., in the paramedian pontine reticular formation, Luschei & Fuchs, 1972). The finding of presaccadic bursting neurons in the IML is of particular interest because such neurons are active with all saccades in a specific direction, whether they are visually triggered or self-initiated, in light or dark. The IML differs from the frontal eye field or the parietal lobe in these respects. It still is the only known region of the forebrain where cells discharge before saccades whatever their mechanism of generation. Units firing with head movement or as a function of head position have recently been found in the same region (Maldonado & Schlag, 1980).

The concordance of results obtained by stimulation, lesion, and microelectrode experiments suggests that the intralaminar nuclei, far from being nonspecific, have functions related to gaze control. Therefore, it is significant that many of the cells activated before eye movements were also found to be responsive to visual stimuli and to be so as a function of stimulus location (Schlag-Rey & Schlag, 1977). Stimulus brightness, contrast, shape, or size had little effect, except on response latency. Our experiments were done with extremely dim spots of light, much dimmer than commonly used in studies of visual unit properties, because dim stimuli were more effective than bright ones in attracting the cats' attention. Nevertheless, IML cells always responded vigorously as if they were sensitive to the presence of a stimulus rather than to the photic energy delivered.

In the cat IML there are a variety of cells: some bursting or pausing with saccades (presaccadic units), some tonically firing as a function of eye position (eye position units), some responding to visual stimuli (visual units), and finally those that show a combination of these main properties. We estimate that for each presaccadic-only, we encountered approximately 3 visual-only, 1 eye position, and 7 visual-and-presaccadic cells. Thus, in our sample of more than 400 units the visual-and-presaccadic cells certainly predominated. This type has also been found elsewhere, at sites as different as the superior colliculus (Mohler & Wurtz, 1976; Schiller & Koerner, 1971), nucleus prepositus hypoglossi (Gresty & Baker, 1976), and prefrontal cortex (Kubota et al., 1974). However, they do not seem to be as common at these sites as in the IML. They were found in the monkey's IML also, as shown in Figure 5-1. When the stimulus was presented and some time later the animal made a saccade to it, the unit fired twice: first at stimulus onset and then starting slightly before the saccade.

We have studied visual-and-presaccadic units in great detail and, in all cases, we noticed a strict correspondence in the conditions producing stimulus time-locked and saccade time-locked activities. All cells had a receptive field, generally large and extended preferentially in one direction. The presaccadic burst occurred only with eye movements directed toward and terminating in the field. So consistent was this relation that we can probably refer to it as a law: the *law of directionality*. It is consonant with Schiller and Koerner's (1971) observation of a spatial correspondence of receptive fields and movement fields in the superior colliculus. However, whereas this correspondence concerns cells of different layers in the colliculus, visually triggered and presaccadic patterns were both emitted by the same neurons in the IML. Measurements

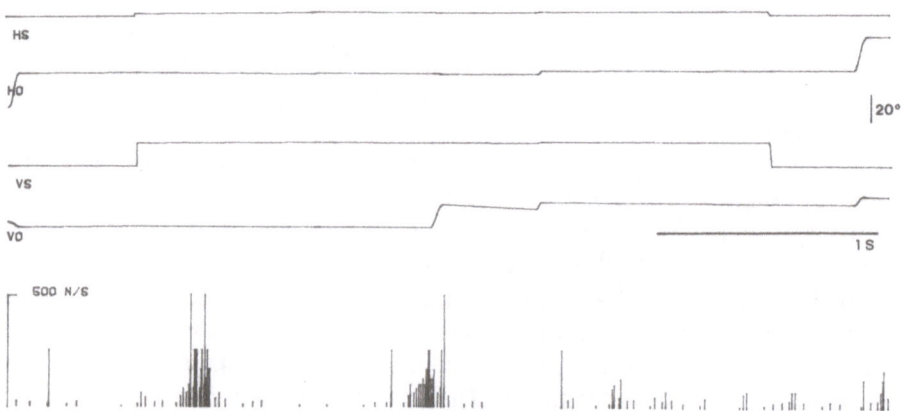

Fig. 5-1. Visual-and-presaccadic unit recorded in IML of monkey. Computer display of horizontal (*HS*) and vertical (*VS*) coordinates of stimulus, horizontal (*HO*) and vertical (*VO*) EOG coordinates, and unit instantaneous frequency. Peaks of firing frequency occurred at stimulus onset and started before large upward saccade of targeting. Calibrations: 20° for stimulus and ocular coordinates, 500 spikes/sec frequency, 1 second.

were made in some units in which receptive fields were so organized that the number of spikes in their on-response increased monotonically with the distance of the stimulus in the preferred direction. In such cases, favorable for quantification, it could be shown that the number of spikes in saccadic bursts also increased with the amplitude of eye displacement in the preferred direction (Schlag & Schlag-Rey, 1977).

The fact that the pattern of firing seen at stimulus onset is repeated at the time of the targeting saccade suggests that the same information is transmitted in both cases. If so, this could not reflect information about the stimulus itself since the signal could also occur before saccades generated in the absence of stimuli, provided that such saccades had the proper direction. It seems, rather, that the information is related to a preparation or intention to move. Indeed, the stimuli that evoked unit responses were precisely those which also could evoke the saccades preceded by firing. This observation has interesting implications for the traditional dichotomy between sensory and motor activities. The usual physiological criteria for this dichotomy are temporal relations (with either stimulus or movement) and information content (about either stimulus parameters or movement parameters). Applied here, however, these criteria lead to different conclusions: whereas the on-responses were stimulus time locked, their occurrences and characteristics appeared to be related to the direction and amplitude of movement that, eventually, the stimulus would induce.

Perhaps it would be useful—and challenging—to introduce in physiology the notion of *adequate response* to parallel the classical notion of adequate stimulus. As the adequate stimulus refers to the natural specific sensory trigger of a neuronal event, the adequate response would refer to the natural specific movement generated. The movement need not be produced, but, were it performed, the adequate response would have the precise characteristics of amplitude and direction encoded by the stimulus time-locked neuronal activity.

Experimental Examination of the Hypothesis of Adequate Response

To return these speculations to the realm of testable hypotheses, we needed an experimental situation in which the target location and the goal of a targeting movement could be dissociated. Predictions of neuronal responses to stimulus onset, based on either target or goal location, would therefore differ. To avoid the issues raised by the plasticity of conditioned neuronal responses, we required a stimulus situation that induced spontaneous rather than conditioned eye movements. For this we exploited the fact that when cats orient their gaze to capture a moving target, they do not necessarily aim at the momentary locus of the target but at some point further along its predicted trajectory.

Such an experiment involved several steps: (a) outlining the receptive field for stationary targets, (b) determining the direction of saccades with which the cells were active, (c) determining the direction of saccades eventually triggered by the moving targets, and (d) analyzing the responses to the moving targets when no saccades occurred. These were long experiments yielding a considerable amount of data. A small sample is presented in Figure 5-2 to illustrate the various steps. This figure shows rasters of the activity of a single IML unit, with the information on stimulus timing under each raster and information on stimulus location on the right of each raster. Stimulus locations are referred to the point of gaze fixation represented by the plus sign. In Figure 5-2h, they are all reproduced on the same diagram for comparison.

This particular unit responded to stationary stimuli (1.5° circles of

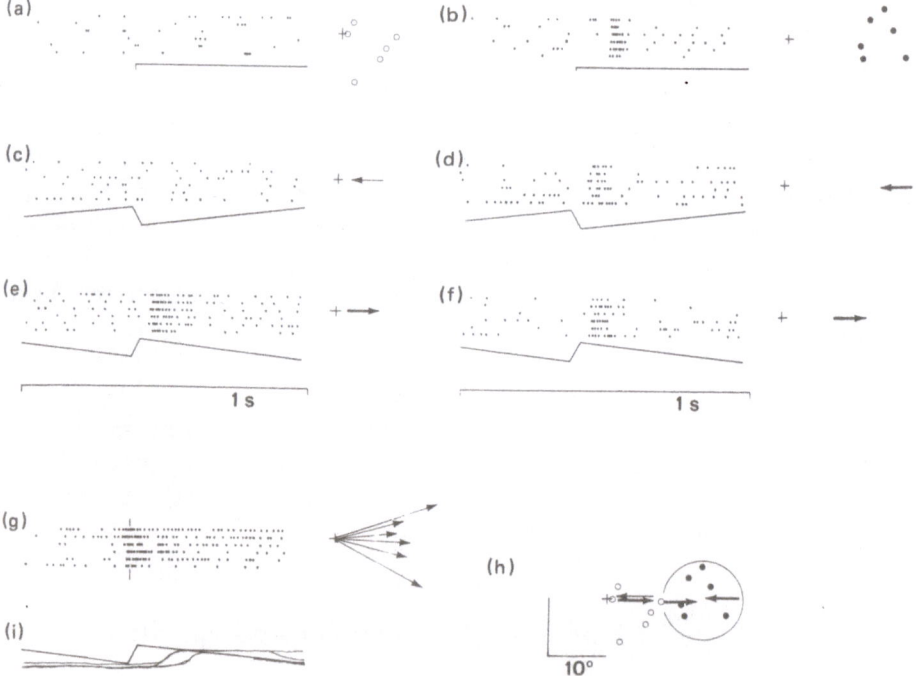

Fig. 5-2. IML unit responses to stationary and moving stimuli in the cat. (a) and (b) Stationary stimuli. (c–f) Moving stimuli. Six samples of unit activity in each raster. Stimulus location with respect to point of fixation (+) is indicated at right of each raster (a–f). *Arrows*: fast component of the stimulus movements. Stimulus time course is indicated under each raster (a–f). No saccades occurred in (a–f). Upward EOG deflection means a displacement of the eyes to the right. (g) Activity of the same unit related to spontaneous rightward saccades represented by vectors; raster synchronized on saccade onsets. (h) All stimulus conditions are gathered together for comparison. *Large circle*: receptive field for stationary stimuli. (i) Samples of saccades triggered in some trials by stimulus moving as in condition (e).

light on a blank background) more than $10°$ degrees on the right side of the point of fixation (Figure 5-2b) but not less than $10°$ on the right side (Figure 5-2a) or anywhere else. After extensive testing, the receptive field for stationary stimuli was outlined as indicated by the circular line in Figure 5-2h.

The next four rasters (Figure 5-2c through f) were obtained while the same $1.5°$ stimulus oscillated across the screen in a sawtooth fashion. Responses were evoked by the fast component ($200°$/sec, direction indicated by arrow) when the stimulus swept across the field in any direction. Only two examples are shown (Figure 5-2d and f). There were no responses when the stimulus trajectories were entirely outside the receptive field (e.g., Figure 5-2c) with one notable exception: when the fast component was directed rightward away from the point of fixation, even though it stopped short of the border of the receptive field (Figure 5-2e).

This was not the first time that receptive fields of visual neurons appeared to be different when mapped with stationary than when mapped with moving stimuli. For instance, the phenomenon has been observed in the superior colliculus (Stein & Dixon, 1979). This may be accepted as a curiosity without further concern. We think, however, that there is a straightforward explanation for the data just described: it resides in the type of saccades that the stimuli may induce. Note that no eye movements occurred in any of the trials in Figure 5-2a through f. Such movements were sometimes made in other trials, however, and, whenever they occurred, they were directed rightward toward the receptive field in the conditions shown in Figure 5-2b, d, e and f, but not in that shown in Figure 5-2c, only in the conditions when bursts occurred. As an example, the horizontal electrooculographic (EOG) recordings in Figure 5-2i show the direction of such saccades when they were produced in a situation corresponding to that in Figure 5-2e.

This unit was active with rightward saccades of $10°$ amplitude, as shown in Figure 5-2g (rasters synchronized on the onset of spontaneous saccades represented by the vectors to the right). Thus, rightward saccades of $10°$ amplitude were the adequate response, according to our definition. In summary, the receptive field for stationary stimuli was located more than $10°$ to the right in the visual field. Nevertheless, responses were also evoked by some moving stimuli even though their course was entirely outside the receptive field. Whereas there seems to be little logic in these properties in terms of stimulus parameters, the occurrence of stimulus time-locked discharges was actually lawful in terms of adequate response: the cells responded to all those stimuli that the cat could acquire with a rightward saccade, even if it did not make such a saccade.

Each time similar tests were made on "visual" IML units using this paradigm, the results were consistent with the prediction based on the notion of adequate response. This implies that the time-locked neuronal activity that we are accustomed to call a sensory response is not simply a passive reaction to external stimulation. There is some similarity with the process of feature extraction; however, whereas feature extraction is usually viewed as a stage in perception, the extraction here seems to be directly geared toward the programing of a movement.

Visual Responses to Target Location in Space

The significance of visually evoked IML unit activities for programing of movement was supported by an additional set of observations. These concerned the processing of spatial coordinates. It is generally recognized that neurons respond to photic stimuli according to their location in retinotopic coordinates. However, it is also a common observation that the visual world does not seem to move when we move in it and that we still can reach a target while displacing our retina with eye, head, or body movement. Therefore, one may pertinently ask whether, somewhere in the brain, visual cells respond to stimuli as a function of their position in space rather than their retinal locus. For any physiological observations relevant to this question it is necessary that the experimental subject be able to fixate in various directions while the same part of the retinal locus is excited.

Our experimental design was favorable for this dissection since our animals were free to look around and occasionally to direct their gaze at the targets. For outlining receptive fields, EOG coordinates were electronically subtracted from target coordinates. Great care was taken to test systematically stimulus positions with respect to the point of fixation (given by the EOG). Since the details of this study have been reported elsewhere (Schlag, Schlag-Rey, Peck, & Joseph, 1980), we shall limit this presentation to the major findings. Three types of IML units were distinguished on the basis of the location and size of their receptive fields. Indeed, these parameters imposed different requirements regarding the need for target fixation.

The first type of units responded only when the cat looked directly at the target or within a few degrees of it. These cells had what we call a *central receptive field*, that is a receptive field located at the center of the retina. Nevertheless, retinal receptive field characteristics did not suffice to account for the unit responsiveness. In addition, it was necessary for the target to be located in some particular region of the screen, for instance, on the cat's left, as depicted in Figure 5-3a. The sustained firing frequency (top trace) was lower when the target was in front of the cat (Figure 5-3b) and minimal when it was on the cat's right (Figure 5-3c). There was no response when the cat did not look at the target (Figure 5-3d). Most units had a preference for the contralateral side of

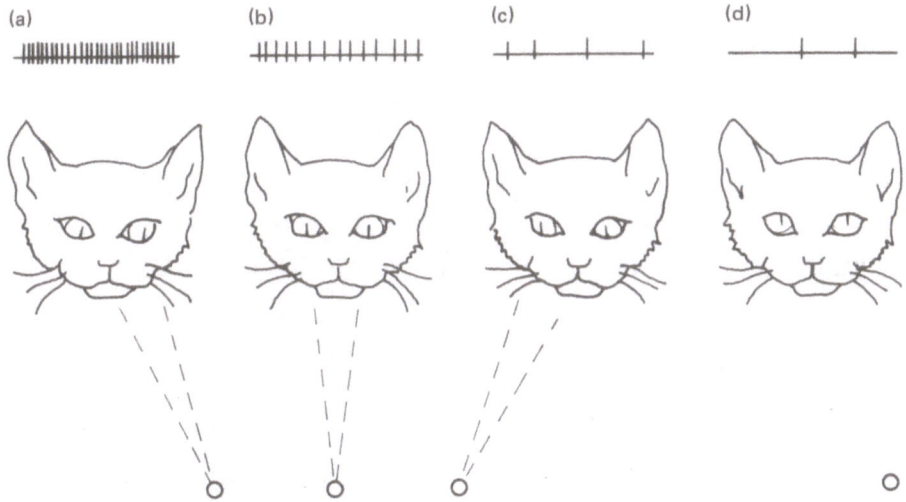

Fig. 5-3. Differential stimulus conditions affecting activity of an IML unit, absolute position Type I (small central receptive field). Tonic firing is shown (*top*) during stimulus presentation when the cat fixated a stimulus on its left (a), on center (b), and on its right (c), and when it did not fixate the stimulus (d).

the screen, but some had up, down, or ipsilateral preferences. The discharge rate displayed either a continuous, smooth function of stimulus position along one axis, or an abrupt, almost stepwise transition close to the midline. Units with small, central receptive fields behaved as if they emitted a frequency signal related to eye position when they had the target caught in their viewfinder (i.e., their receptive field). In fact, most of these units emitted a signal related to eye position even in the dark, although signal frequency was much less (one-fourth at the most) than when there was a target to fixate. Quantitative analysis of 47 units of this type was made. Similar results were also obtained by Sakata, Shibutani, and Kawano (1980) for visual fixation neurons of the posterior parietal cortex in the monkey.

The second type of units (16 cells) differed from the first by the larger size of their receptive field, which also included the point of fixation. This difference had dramatic consequences because target fixation was no longer required for unit activation. The gaze could be directed as much as $10°$ from the target without affecting the activity of the unit. The frequency of firing was almost solely a function of target position on the screen, with coefficients of correlation as high as .88. For instance, the cell schematized in Figure 5-4 responded maximally to a target on the cat's right (Figure 5-4a), minimally to a target on the cat's left (Figure 5-4d), and at an intermediate level to a target at the center of the screen (Figure 5-4c). The responses were the same whether the animal fixated (Figure 5-4a) or did not fixate the target (Figure 5-4b).

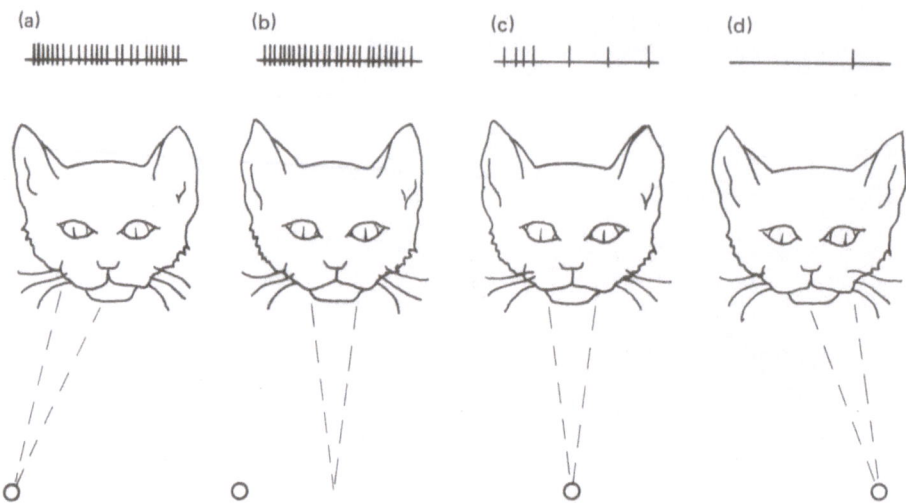

Fig. 5-4. Differential stimulus conditions affecting activity of an IML unit, absolute position Type II (large receptive field). Same presentation as in Figure 5-3.

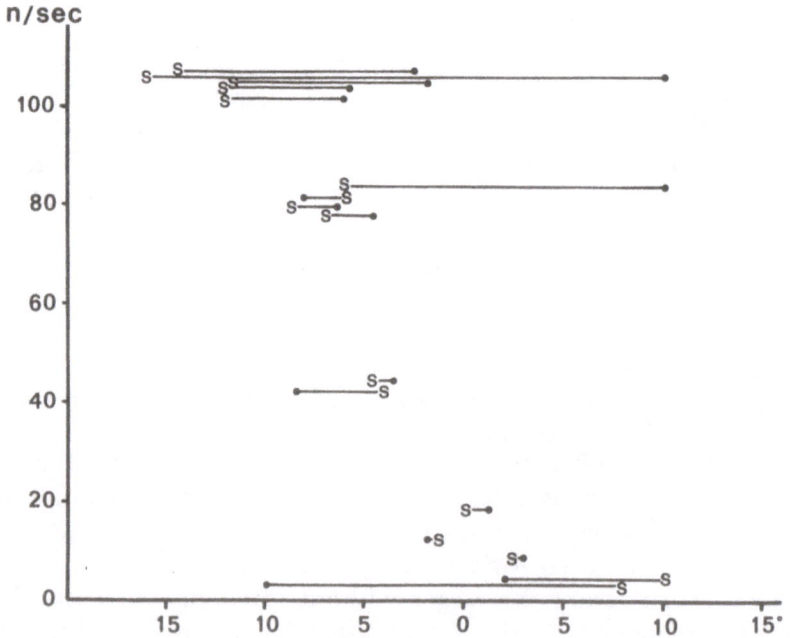

Fig. 5-5. Frequency of firing of an IML unit, absolute position Type II, as a function of stimulus (*S*) and eye (*dot*) positions in the unit preferred direction, which in this case was horizontal and contralateral to the recorded side. On the abscissa, $0°$ corresponds to the center of the screen. The length of the line between *S* and *dot* represents the magnitude of retinal error in each individual measurement. Firing frequency was measured on samples of at least .5 seconds.

The type of analysis routinely performed on this type of unit is presented in Figure 5-5. Firing frequency is plotted against the coordinates of stimulus (*S*) and point of fixation (*dot*) measured along the axis of the preferred direction of the cell. For each trial, the length of the line between S and the dot represents the amplitude of the retinal error along the same axis. Only a few trials are included to avoid obscuring the diagram. This example shows that the firing rate was more related to absolute position of the stimulus in space (*S*) than either to eye position (*dot*) or retinal error (*line*). In our opinion, units with large receptive fields, similar to those described by Wiersma (1966) in crustacea but yet unknown in vertebrates, corresponded most closely to the type of units postulated to account for the phenomenon of space constancy. Indeed, space constancy implies, first, some way of coding position and, second, an invariance of this signal when the direction of gaze changes.

The third type of unit had an eccentric receptive field. A typical example is depicted in Figure 5-6. The field never included the point of fixation, and therefore no responses or only weak ones (Figure 5-6c) occurred when the animal fixated the target (which it did spontaneously in most trials). These units were difficult to study since many trials were needed to provide a sufficient number of tests of various target–eye positions with no saccades. Thus, only 3 of 14 cells of this type were extensively tested, but the results were revealing. Only stimuli

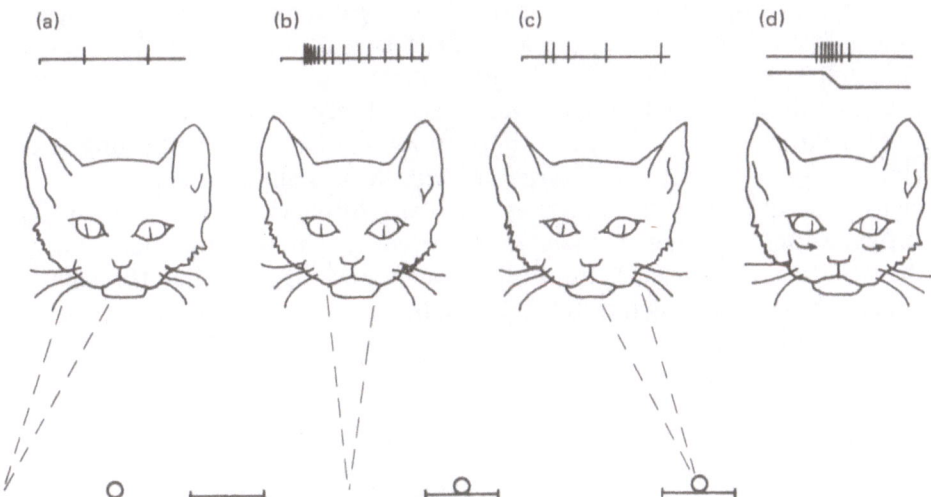

Fig. 5-6. Differential stimulus conditions affecting activity of an IML unit, absolute position Type III (eccentric receptive field). Same presentation as in Figure 5-3. *Bar*: position on the screen where stimulus must appear to trigger a response (absolute field). There was a presaccadic discharge of the same unit during a spontaneous leftward saccade (d).

presented within a limited area of the screen, generally circular and at least $10°$ degrees from the center (marked by bar), elicited responses. In addition, the animal's gaze had to be pointed close to the center of the screen. These two conditions are met in Figure 5-6b. In Figure 5-6a, the stimulus was in the same retinotopic position as in Figure 5-6b, but not in the critical area of the screen. Conversely, in Figure 5-6c, the stimulus was in the critical area of the screen but not in the proper retinotopic position. There was no strong unit response in either Figure 5-6a or Figure 5-6c. Here, again, it is appropriate to bring in the notion of adequate response to explain the results, for the same cell was also active in the absence of a target when spontaneous saccades drove the gaze into the same area of the screen (Figure 5-6d). In other words, the activity time locked to the stimulus was evoked by the targets so located as to produce eye movements in that direction for which the cell was bursting. The stimulus evoked the time-locked activity even in the absence of saccades.

The behavior of these three types of units can only be accounted for by assuming that the cells received information on eye position. Actually, some of them (Type I) were weakly sensitive to eye position in the dark and other cells in the IML were pure eye position units (Schlag et al., 1974; Schlag et al., 1980). In their own way, all three types of units were sensitive to stimulus absolute position. We have speculated about the possible functions that they could subserve (Schlag et al., 1980). Many hypotheses are open, with a number of options to consider. For instance, do the units studied play a role in perception of the location of objects in space, or do they serve mainly in guiding movements toward targets, assuming that the two functions are distinguishable (e.g., see Schott & Jeannerod, 1966)? In the latter case, are the movements those of the eyes, head, gaze, body, and/or limbs? Assuming a role in gaze control, are these units involved in maintaining fixation or in acquiring new targets? Are these cells sending command signals or are they transmitting corollary discharge or proprioceptive information? Further studies are needed to narrow the spectrum of possibilities. At present, the bulk of evidence suggests that the significance of signals emitted by IML cells has to be sought in terms of guidance of the gaze.

Discussion

In this presentation, we have shown several examples of "visual" responses that did not passively reflect stimulus characteristics as impressed directly on the retinal receptors. A last point should be made in this respect. To determine the extent of a receptive field, stimuli were successively presented at random sites around the point of fixation.

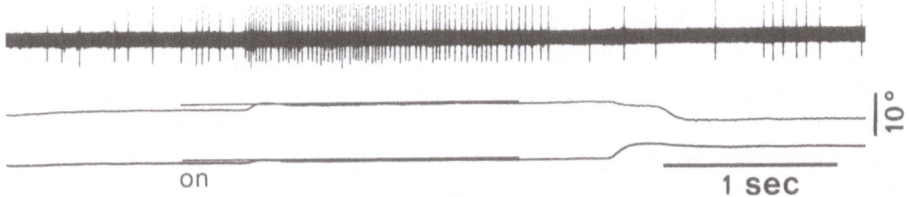

Fig. 5-7. Tonic firing of an IML unit during stimulus fixation in the cat: vertical (a) and horizontal (b) coordinates of stimulus and eye. Note that the unit started discharging at a high rate before the targeting saccade.

Sometimes, the animal made an orienting response. In these cases, units with small central receptive fields began to discharge tonically. In a majority of cells, however, the tonic discharge started before the onset of the saccade, that is, before the stimulus was brought into the field, as shown in Figure 5-7. If the whole train of spikes were a visual response, a latency of at least 50 msec would be expected from the time when the stimulus entered the field (i.e., when the saccade was completed). Conceivably, the initial part of the train might have a different mechanism from the later part. It could be a saccadic discharge progressively mingling with a visual response. This seems very unlikely, however, because, first, there never was any sign of an hiatus between the period of the saccade itself (whatever the size of the saccade) and the subsequent period, which lasted as long as the stimulus was present; second, even cells that had no change of activity with spontaneous saccades showed this very early activation or inhibition when targeting was initiated. We would rather think that the unit change of firing was related to an intention or preparation to foveate. In units with an absolute field, such as those that we have just described, the frequency of discharge depended on the location of the stimulus on the screen. Such a signal could represent an intended position of the gaze. In current models of goal-directed movements (Bizzi, Polit, & Morasso, 1976; Robinson, 1975), the specification of the goal is assumed to be the primary operation around which all the others are organized. Thus, Robinson (1975) specifically proposed that the spatiotemporal course of saccades is obtained by subtracting a signal of actual position of the eyes from a signal of intended position, thereby driving the eyes until the difference becomes zero. The signal of intended position should appear before the movement is performed, and, very likely, it should be frequency coded since it has to be compared to actual position, which is frequency coded. This is the hypothesis that we are presently entertaining in planning future experiments. This hypothesis is a specific interpretation of the general idea of adequate response.

Although the place where IML neurons fit in the circuitry controlling the gaze is not yet clear, recent findings of Schiller, True, and Conway

(1980) offer suggestions. These investigators showed that eye movements, still possible after either colliculectomy or ablation of the frontal eye field, disappeared when both regions were destroyed in monkey.
Thus, both structures seem needed. Their roles are probably not redundant, however. Since collicular neurons respond best to events and are
topographically arranged to register their locations, the colliculus seems
superbly equipped to organize responses of orientation. On the other
hand, the operations that require pattern analysis, recognition of objects, evaluation of their significance, and, finally, target selection are
more likely to depend on forebrain mechanisms. This is still pure speculation (but see Chapter 3, this volume). However, if there are two poles
in the apparatus controlling gaze, each one with its independent access
to the brain stem oculomotor centers (Schiller et al., 1980), it appears
logical to assume that they are specialized in the type of information
that they process: stimuli that impose themselves by their physical
saliency in the case of the superior colliculus and image features of subjective significance in the case of the forebrain. Often, there are con-
flicts between these two types of solicitation.

Anatomically, the IML is located at the confluence between the two
poles. It receives afferents from the deep layers of the superior colliculus (Harting, Huerta, Frankfurter, Strominger, & Royce, 1980). It projects to the basal ganglia and to several cortical areas including the
frontal eye field (Kievit & Kuypers, 1977) and parietal lobe (Kasdon &
Jacobson, 1978). In return, it receives projections from the frontal eye
field, apparently as collaterals of fibers reaching down to the superior
colliculus (Leichnetz, 1980). These paths place the IML at the intersection of two-way communications between forebrain and tectum. In
addition, the IML receives inputs from a number of other structures,
for instance, the nucleus prepositus hypoglossi (Kotchabhakdi, Rinvik,
Yingchareon, & Walberg, 1980), vestibular nuclei (Kotchabhakdi,
Rinvik, Walberg, & Yingchareon, 1980), ventral lateral geniculate
nucleus (Magnin & Kennedy, 1979), pretectum (Benevento, Rezak, &
Santos-Anderson, 1977), and area 19 (Galletti, Squatrito, & Battaglini,
1979). Most of this anatomical information has been acquired in recent
years with tracing techniques in the cat and monkey and its relevance
to our neurophysiological observations needs to be verified. At the
present time, the overall organization of forebrain gaze mechanisms is
still very much a mystery. Unit activity that meets the requirements for
command signals driving the gaze in any type of saccade, whether visually triggered, visually guided, or anticipatory, has only been seen in the
middle of the thalamus. There is no clear evidence that these signals are
transmitted directly to the brain stem. It is likely that they reach the
telencephalon, probably at the level of the cerebral cortex rather than
the basal ganglia whose involvement in gaze control has not yet been
supported experimentally. We are inclined to think that it is at the level

of the cerebral cortex that the trace of the missing signals now has to be sought.

Acknowledgments

This work was supported by USPHS Grants NS04955 and EY02305. Different parts of the work discussed here have been performed in collaboration with Drs. C. K. Peck, H. Maldonado, J. -P. Joseph, and I. Lehtinen. We are also grateful to Drs. H. Noda and B. Merker for discussion of some of the material, and J. Mason, M.D. Schlag, V. Pagano, D. Blanco, and L. Carver for their assistance.

References

Benevento, L. A., Rezak, M., & Santos-Anderson, R. An autoradiographic study of the projections of the pretectum in the rhesus monkey (*Macaca mulatta*): Evidence for sensorimotor links to the thalamus and oculomotor nuclei. *Brain Research*, 1977, *127*, 197-218.

Bizzi, E., Polit, A., & Morasso, P. Mechanisms underlying achievement of final head position. *Journal of Neurophysiology*, 1976, *39*, 435-444.

Edwards, S. B. The deep cell layers of the superior colliculus: Their reticular characteristics and structural organization. In A. Hobson & M. Brazier (Eds.), *The reticular formation revisited: Specifying functions for a nonspecific system.* New York: Raven Press, 1980, pp. 193-209.

Ferrier, D. The localization of function in the brain. *Proceedings of the Royal Society of London; B.*, 1874, *22*, 229-232.

Galletti, C., Squatrito, S., & Battaglini, P. P. Visual cortex projections to thalamic intralaminar nuclei in the cat. An autoradiographic study. *Archives Italiennes de Biologie*, 1979, *117*, 280-285.

Gresty, M., & Baker, R. Neurons with visual receptive field, eye movement and neck displacement sensitivity within and around the nucleus prepositus hypoglossi in the alert cat. *Experimental Brain Research*, 1976, *24*, 429-433.

Harting, J. K., Huerta, M. F., Frankfurter, A. J., Strominger, N. L., & Royce, G. J. Ascending pathways from the monkey superior colliculus: An autoradiographic analysis. *Journal of Comparative Neurology*, 1980, *192*, 853-882.

Hassler, R. Brain mechanisms of intention and attention with introductory remarks on other volitional processes. In *Motivation, motor and sensory processes of the brain. Progress in Brain Research*, 1980, *54*, 585-614.

Kasdon, D. L., & Jacobson, S. The thalamic afferents to the inferior parietal lobule of the rhesus monkey. *Journal of Comparative Neurology*, 1978, *177*, 685-706.

Kennard, M. A., & Ectors, L. Forced circling in monkeys following lesions of the frontal lobes. *Journal of Neurophysiology*, 1938, *1*, 45-54.

Kievit, J., & Kuypers, H. G. J. M. Organization of the thalamo-cortical connexions of the frontal lobe in the rhesus monkey. *Experimental Brain Research*, 1977, *29*, 299-322.

Kotchabhakdi, N., Rinvik, E., Walberg, F., & Yingchareon, K. The vestibulothalamic projections in the cat studied by retrograde axonal transport of horseradish peroxidase. *Experimental Brain Research*, 1980, *40*, 405-418.

Kotchabhakdi, N., Rinvik, E., Yingchareon, K., & Walberg, F. Afferent projections to the thalamus from the perihypoglossal nuclei. *Brain Research*, 1980, *187*, 457–461.

Kubota, K., Iwamoto, T., & Suzuki, H. Visuokinetic activities of primate prefrontal neurons during delayed-response performance. *Journal of Neurophysiology*, 1974, *37*, 1197–1212.

Leichnetz, G. R. An anterogradely-labeled prefrontal cortico-oculomotor pathway in the monkey demonstrated with HRP gel and TMB neurohistochemistry. *Brain Research*, 1980, *198*, 440–445.

Luschei, E. S., & Fuchs, A. F. Activity of brain stem neurons during eye movements of alert monkeys. *Journal of Neurophysiology*, 1972, *35*, 445–461.

Lynch, J. C., Mountcastle, V. B., Talbot, W. H., & Yin, T. C. Parietal lobe mechanisms for direct attention. *Journal of Neurophysiology*, 1977, *40*, 362–389.

Magnin, M., & Kennedy, H. Anatomical evidence of a third ascending vestibular pathway involving the ventral lateral geniculate nucleus and the intralaminar nuclei of the cat. *Brain Research*, 1979, *171*, 523–529.

Maldonado, H., Joseph, J. -P., & Schlag, J. Types of eye movements evoked by thalamic microstimulation in the alert cat. *Experimental Neurology*, 1980, *70*, 613–625.

Maldonado, H., & Schlag, J. Thalamic neurons related to head and eye movements. *Neuroscience Abstracts*, 1980, *6*, 476.

Mays, L. E., & Sparks, D. L. Dissociation of visual and saccade-related responses in superior colliculus neurons. *Journal of Neurophysiology*, 1980, *43*, 207–232.

Mays, L. E., & Sparks, D. L. Saccades are spatially, not retinocentrically, coded. *Science*, 1980, *208*, 1163–1165. (b)

Mohler, C. W., & Wurtz, R. H. Organization of monkey superior colliculus: Intermediate layer cells discharging before eye movements. *Journal of Neurophysiology*, 1976, *39*, 722–744.

Orem, J., Schlag-Rey, M., & Schlag, J. Unilateral visual neglect and thalamic intralaminar lesions in the cat. *Experimental Neurology*, 1973, *40*, 784–797.

Robinson, D. A. Oculomotor control signals. In G. Lennerstrand & P. Bach-y-Rita (Eds.), *Basic mechanisms of ocular motility and their clinical implications*. Oxford: Pergamon Press, 1975, pp. 337–374.

Sakata, H., Shibutani, H., & Kawano, K. Spatial properties of visual fixation neurons in posterior parietal association cortex of the monkey. *Journal of Neurophysiology*, 1980, *43*, 1654–1672.

Schiller, P. H., & Koerner, F. Discharge characteristics of single units in superior colliculus of the alert rhesus monkey. *Journal of Neurophysiology*, 1971, *34*, 920–936.

Schiller, P. H., True, S. D., & Conway, J. L. Deficits in eye movements following frontal eye-field and superior colliculus ablations. *Journal of Neurophysiology*, 1980, *44*, 1175–1189.

Schlag, J., Lehtinen, I., & Schlag-Rey, M. Neuronal activity before and during eye movements in the thalamic internal medullary lamina of the cat. *Journal of Neurophysiology*, 1974, *37*, 982–995.

Schlag, J., & Schlag-Rey, M. Induction of oculomotor responses from thalamic internal medullary lamina in the cat. *Experimental Neurology*, 1971, *33*, 498–508.

Schlag, J., & Schlag-Rey, M. Visuomotor properties of cells in the cat thalamic internal medullary lamina. In R. Baker & A. Berthoz (Eds.), *Control of gaze by brain stem neurons*. Amsterdam: Elsevier/North-Holland, 1977, pp. 453–462.

Schlag, J., Schlag-Rey, M., Peck, C. K., & Joseph, J. -P. Visual responses of thalamic neurons depending on the direction of gaze and the position of targets in space. *Experimental Brain Research*, 1980, *40*, 170–184.

Schlag-Rey, M., & Schlag, J. Visual and presaccadic neuronal activity in the thalamic internal medullary lamina of cat: A study of targeting. *Journal of Neurophysiology*, 1977, *40*, 156–173.

Schlag-Rey, M., & Schlag, J. Eye movement-related neuronal activity in the central thalamus of monkeys. In A. Fuchs & W. Becker (Eds.), *Progress in oculomotor research*. New York: Elsevier/North-Holland, 1981.

Schott, B., & Jeannerod, M. L'agnosie spatiale unilatérale: Pertubation en secteur des mécanismes d'exploration et de fixation du regard. *Journal de Médecine de Lyon*, 1966, *47*, 169–195.

Sparks, D. L., & Mays, L. E. Movement field of saccade-related burst neurons in the monkey superior colliculus. *Brain Research*, 1980, *190*, 39–50.

Stein, B. E., & Dixon, J. P. Properties of superior colliculus neurons in the golden hamster. *Journal of Comparative Neurology*, 1979, *183*, 269–284.

Watson, R. T. Thalamic neglect. *Neurology*, 1978, *28*, 396.

Watson, R. T., Miller, B. D., & Heilman, K. M. Nonsensory neglect. *Annals of Neurology*, 1978, *3*, 505–508.

Wiersma, C. A. G. Integration in the visual pathway of crustacea. In C. A. G. Wiersma (Ed.), *Nervous and hormonal mechanisms of integration*. Chicago: University of Chicago Press, 1966, pp. 151–177.

Chapter 6

Coordination of Eye-Head Movements in Alert Monkeys: Behavior of Eye-Related Neurons in the Brain Stem

Francis Lestienne, Doug Whittington, and Emilio Bizzi

Usually, fixation of a target appearing in the visual field is accomplished by coordinated eye and head movements. A series of studies has shown that this spatially coordinated behavior is a reflex that utilizes the sensory feedback generated by head movement (Bizzi, 1974; Bizzi, Kalil, & Tagliasco, 1971). This reflex interacts with centrally initiated motor programs for moving the head and eyes. To maintain target fixation during head turns, activation of vestibular receptors initiates signals that lead to diminished saccade size and generates compensatory eye movements (Dichgans, Bizzi, Morasso, & Tagliasco, 1973; Morasso, Bizzi, & Dichgans, 1973). During these combined eye–head movements, the sum of the eye position within the orbit and of the head position is equal to the required total gaze shift. When the head is held fixed, total gaze shift is accomplished by the eye saccade. This orderly sequence of movements is presented in Figure 6-1.

Considering that modification of saccade characteristics is a clear manifestation of the interaction between eye and head movement, we investigated the potential role of the pontine reticular formation (PRF) for this interaction. Study of the PRF was suggested by several considerations. This region of the brain stem is critically involved in the generation of eye movements (Bender & Shanzer, 1964; Büttner-Ennever & Henn, 1976; Cohen & Henn, 1972; Eckmiller, Blair, & Westheimer, 1980; Graybiel, 1977; Keller, 1974, 1977; Luschei & Fuchs, 1972; Scheibel & Scheibel, 1958). The PRF is also known as an area with projections to the neck muscles (Peterson, Maunz, Pitts, & Mackel, 1975; Peterson, Pitts, Fukushima, & Mackel, 1978). Finally, this region receives input from receptors in neck muscles and in the vestibular system (Brodal, 1974; Ladpli & Brodal, 1968; Peterson, Filion, Felpel, & Abzug, 1975; Pompeiano & Swett, 1963). The studies reported in this

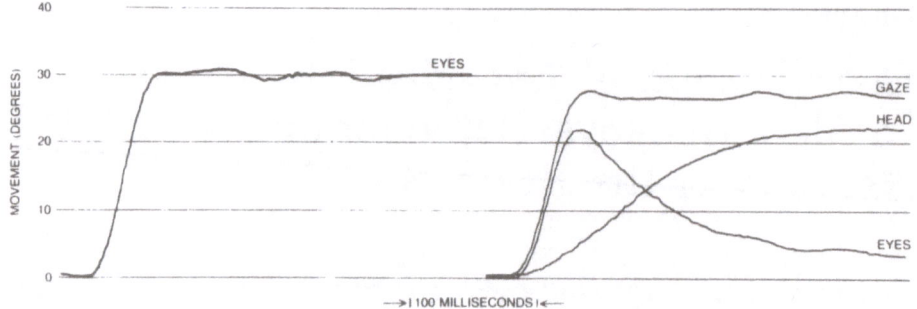

Fig. 6-1. Head-fixed and head-free responses to an identical 30° gaze shift. (a) The size of the gaze shift equals the size of the saccade. (b) The saccade is smaller than the gaze shift because the turning of the head accomplishes part of the shift in the gaze. Notice the compensatory eye movement after the saccade. (From Bizzi, E., The coordination of eye–head movement. *Scientific American*, 1974, *231*, 100–106.)

chapter were designed to clarify the functional characteristics of the PRF that underlie gaze shifting behavior in alert monkeys. Specifically, we studied the contribution of head movements to the activity of neurons that are responsible for generating and controlling eye movements.

Before presenting our results, we summarize the behavior of the PRF neurons that are involved in the control of eye movements when the head is fixed. On the basis of the firing patterns of the PRF units re-

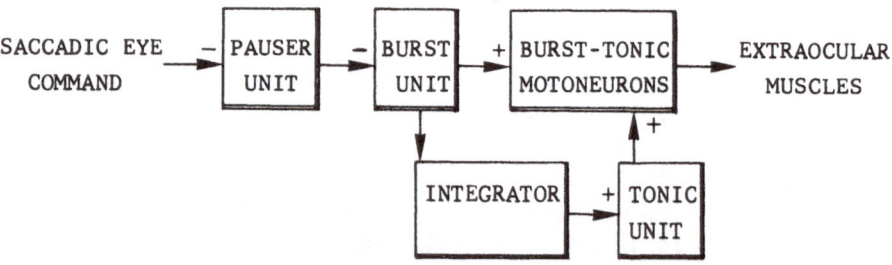

Fig. 6-2. Simplified scheme of saccadic eye control. The size of a desired saccade is determined at a higher level (*saccadic eye command*). The *pauser unit* is inhibited for a time proportional to the planned size of the saccade. During this interval the *burst unit* is disinhibited. The resulting burst signal is integrated to indicate the position of the eye in the orbit. The cell that carries this signal is called a *tonic unit*. The burst and tonic signals feed the *burst–tonic motoneurons*, thereby generating the appropriate pulse-step control signal to the extraocular musculature. (Adapted from Robinson, D. A. Oculomotor control signals. In G. Lennerstrand & P. Bach-y-Rita (Eds.), *Basic mechanisms of ocular motility and their clinical implications.* Oxford: Pergamon Press, 1974, pp. 337–374.)

lated to eye movement and/or position (Keller, 1974; Luschei & Fuchs, 1972), four main categories have been identified: (a) burst-tonic cells that exhibit a burst of firing before and during the saccade with a tonic component corresponding to steady eye position; (b) tonic cells whose firing frequency corresponds with eye position in the orbit; (c) burst cells that produce high-frequency bursts related to saccade duration; and (d) pausing cells that fire during fixation and cease firing during saccades. From the basic models of saccadic eye control proposed by Robinson (1974) these four categories of neurons can be combined into the simple scheme shown in Figure 6-2.

Experimental Method

In order to explore the PRF structures and to record the activity of units related to eye movement during eye-head coordination, we trained monkeys to foveate a visual target. The size and direction of the gaze shift were controlled by the experimenter's choice of target position. On some trials the monkey's head was restrained, forcing the animal to make gaze shifts utilizing saccades alone. This procedure also allowed accurate calibration of eye position.

The monkeys were trained until they became proficient at the visual fixation task. They were then anesthetized with sodium pentobarbital (Nembutal) and electrodes were implanted for recording horizontal and vertical electro-oculograms. Screws were anchored to the skull to serve as connectors to the head movement apparatus. Electromyographic electrodes were placed into the left and right splenii capitis muscles. A stainless steel recording well was attached to the skull over an opening that had been located stereotaxically to permit an obliquely driven microelectrode to reach the brain stem in the area of the sixth nucleus. Appropriately designed tungsten microelectrodes allowed recording of the activity of single brain stem neurons when the monkey was free to move his head. The locations of recording tracks and of small lesions placed at the tip of each track were histologically determined when recording was completed.

During the experimental sessions we monitored horizontal eye-head movements in addition to recording brain stem neurons. We also stimulated the vestibular system by rotating the animal, and the muscle joint receptors of the neck by rotating the torso while the head was fixed. We examined each cell encountered to determine whether its firing rate was (a) related to eye movements or to head movements, and (b) modulated during stimulation of the vestibular system and/or the muscle joint receptors of the neck.

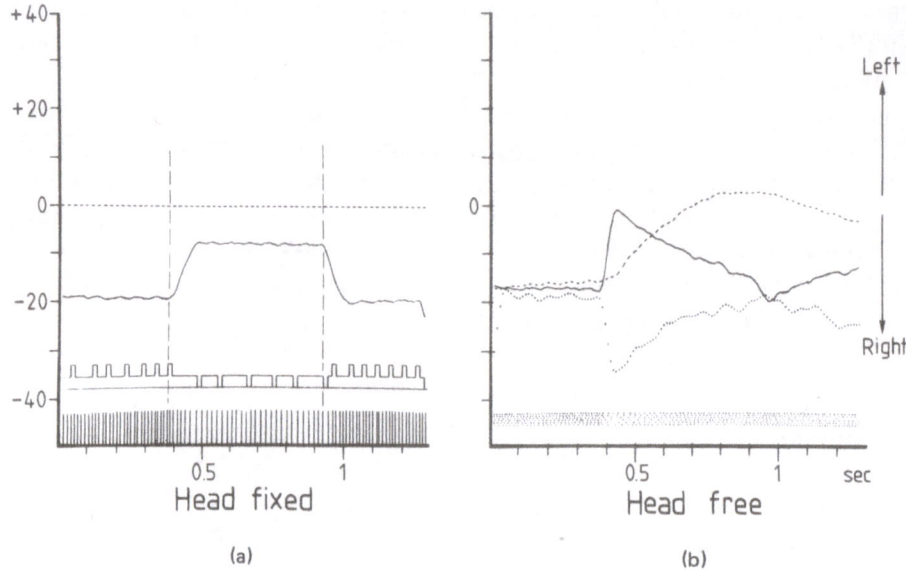

Fig. 6-3. Behavior of a T cell. (a) Response of a T cell for different steady horizontal eye positions with head fixed. *H*, position of the head; *HE*, horizontal eye movement. Spike histograms (cumulated discharges in bins of 10 msec) and standardized single-unit discharges are shown at the bottom. *Vertical broken lines*, onset of the saccade. (b) Response of the same T cell during six eye and head movements. *H*, averaged data of head movements; *HE*, averaged data of horizontal eye movements. *S*, tracings of the average value of the instantaneous spike frequency of the T cell. The successive lines of the raster are shown at the bottom.

Behavior of PRF Preoculomotor Neurons During Eye-Head Coordination

In this study, the behavior of more than 140 eye-related neurons were examined during eye–head coordination. They have been categorized as tonic, pausing, or burst cells.

Tonic Cells

All the tonic (T) cells discharged with regular interspike intervals during fixation periods. With the head fixed the discharge rate of the majority of the T cells was unaltered until a given position of the eye in the orbit was achieved. Beyond this point the instantaneous frequency of T cells was linearly related to eye position (Figure 6-3a). The firing characteristics of these neurons were unaffected by stimulation of the neck muscle and joint receptors or by the head position. When the head was free, there was a significant correlation between T cell instantaneous firing frequency and instantaneous position of the eye in the orbit, whether the eye movement was saccadic or compensatory for head movement. This correlation was maintained for every change in head position. It is demonstrated most clearly in Figure 6-3b, which displays the mean instantaneous frequency of a representative T cell as a function of the mean instantaneous eye position within saccades. These data were recorded by averaging six eye–head movements that had different initial head positions and neck muscle activities. These results fit well with Robinson's (1970, 1971) model of saccadic eye control and make these cells an ideal candidate for carrying the information produced by a putative neural integrator, which presumably receives input from burst neurons (Figure 6-2).

Pausing Cells

Among pausing cells, *omnipauser* (OP) *cells*, whose activity ceases during all saccades, were chosen for consideration. The cessation of OP cell discharge leads the saccadic eye movement by about 20 msec. We segregated the OP cells into two groups, depending upon whether or not their activity was modified in response to vestibular stimulation.

With the OP cells that were not modulated by vestibular stimulation, there were no surprises. With the head free these cells behaved as expected, pausing for all saccades. The length of the pause was always equal to the saccade duration (Figure 6-4). For the OP cells whose intersaccadic discharge was modulated by vestibular stimulation, cessation in tonic activity did not necessarily correlate with the duration of the saccade. The behavior of one of these vestibularly modulated OP cells is shown in Figure 6-5. In the absence of head movement these units discharged at a relatively constant rate, which was interrupted by a cessation of activity locked to saccade duration (Figure 6-5a). When the head was free to move, and for an appropriate direction of head movement (rightward rotation in this case), the pause was considerably longer than the accompanying saccade (Figure 6-5b). Vestibular stimulation tests provided an explanation for this pause: with adequate horizontal vestibular stimulation, these OP units exhibited an increase in activity during head turns in one direction, and a decrease during turns in the other direction. The record shows that the pause was related to high, rightward head velocity. This particular cell's firing dropped to zero as head movement velocity approached 50°/sec. Furthermore, we found no evidence that the activity of any OP cell was affected by rotating the torso with the head position fixed. In agreement with Robinson's scheme of saccadic eye control, and under the head-free condition, it is tempting to put these cells in the command chain. This requires an assumption that they are just upstream of the burst cells

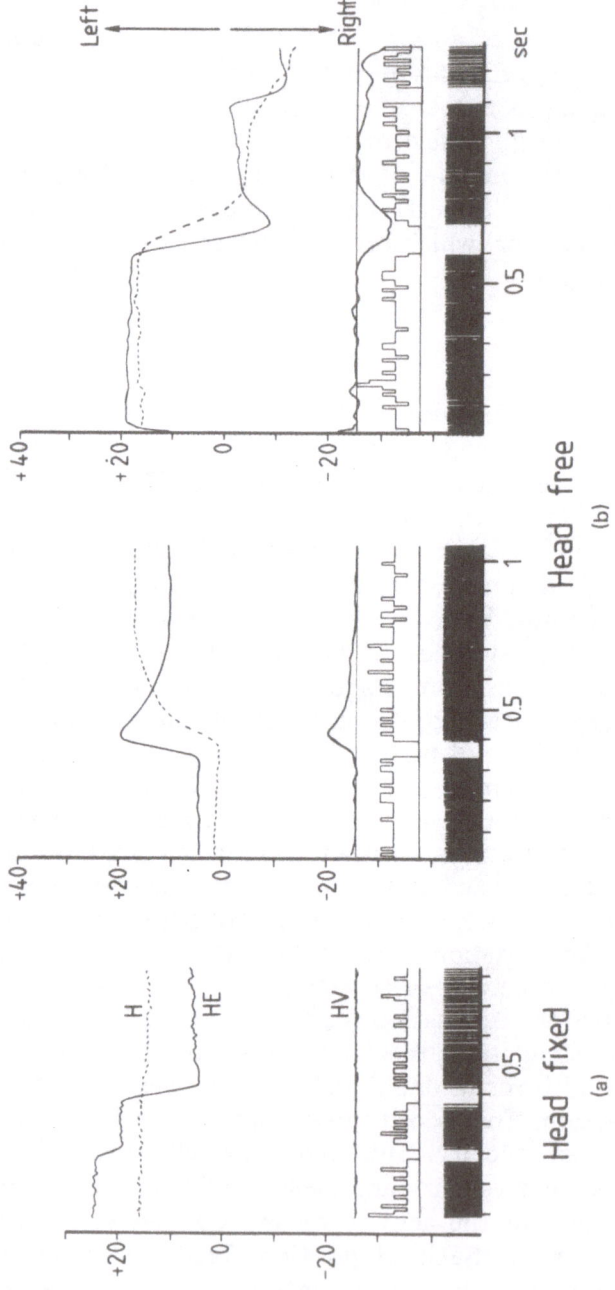

Fig. 6-4. Behavior of an OP cell that is not modulated by vestibular stimulation. *H*, head movement; *HE*, horizontal eye movement; *HV*, head velocity. Standardized single-unit discharges and spike histograms are shown at the bottom. The two experimental conditions were head fixed (a) and head free (b).

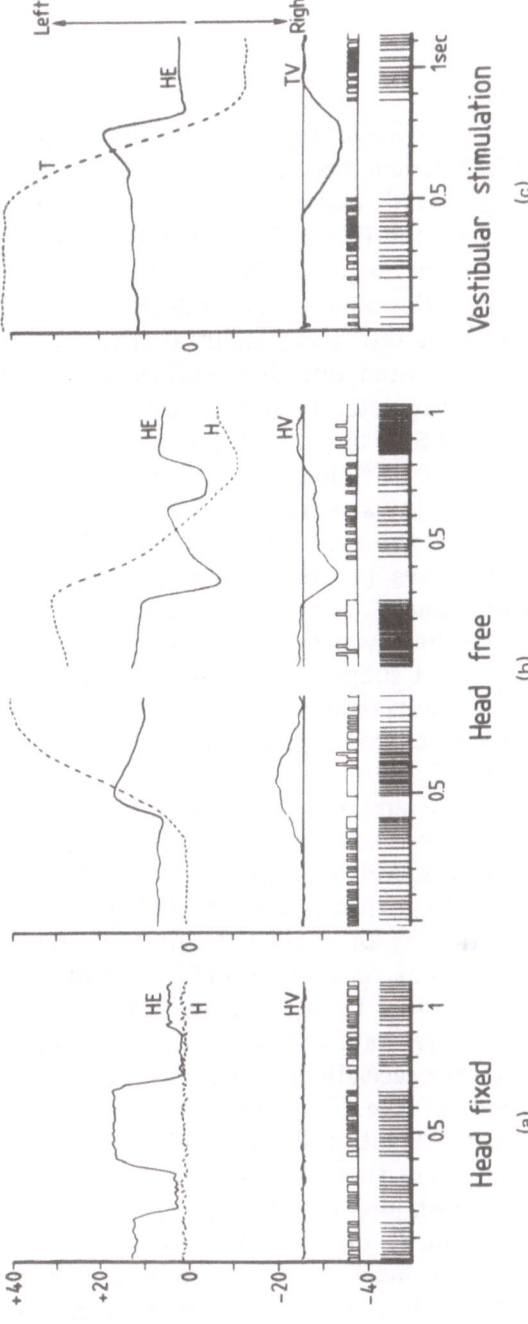

Fig. 6-5. Behavior of a vestibular-modulated OP cell. The three experimental conditions were head fixed (a), head free (b), and vestibular stimulation (c). *HE*, horizontal eye movement; *H*, head movement; *HV*, head velocity; *T*, turnchair movement; *TV*, turnchair velocity.

and that their silence causes a disinhibition of the burst cells (Figure 6-2). In contrast, the vestibularly modulated OP cells cannot be part of the chain response for producing saccades, because their pause is not always locked to the duration of the saccade.

Burst Cells

We focus now on one particular type of PRF burst neurons, called short-lead bursters (SLB), which discharge at a high frequency (approximately 500 Hz) during saccadic eye movement in a particular direction. The time interval between the first spike of the burst and the onset of the saccadic eye movement ranged from 5 to 20 msec. Most of these units produced a burst of the same duration as the fast eye movement, and the intraburst discharge rate was roughly constant until the end of the burst. It should be pointed out that SLB cells do not modulate when neck receptors are stimulated by torso rotation.

Quantitative analysis of SLB units, with the animal's head restrained, confirmed that the number of spikes per burst was proportional to the duration, and therefore to the size, of the saccade and did not depend on the starting position of the eye. However, the tight correlation between the number of spikes per burst and the size of the saccade, demonstrated when the head was fixed, was not found for all the units of this category when the head was free to move. In fact, with head free we found two quite distinct subgroups of units that had been indistinguishable when the head was restrained.

When the head is free to move, saccades during coordinated eye–head movements contribute only part of the total gaze shift, the rest being provided by the head. One subgroup of SLB units (labeled "S" units) included those units whose discharges were closely related to the size of saccadic eye movements and were independent of head movement (Figure 6-6a). The second subgroup (labeled "G" units) included those SLB units whose firing rates were correlated with the total shift in gaze, including the head movement contribution (Figure 6-6b). In Figure 6-7 these differences are illustrated by the behavior of two typical S and G units and the regression lines that fit the relation between the number of spikes per burst and either saccadic gaze shifts with the head fixed or saccade sizes with the head free. The S unit had similar regression lines for both saccadic gaze shifts with the head fixed and saccade sizes with the head free; however, these two regressions lines were quite different from the regression line obtained for the total gaze shifts with the head free (Figure 6-7b). For the G unit, the converse was true; regression lines for saccades with the head fixed and total gaze shifts with head free were identical, and both of these differed from the regression line of saccades with the head free (Figure 6-7a). The results of this test indicate that the S units encode information only about the saccadic

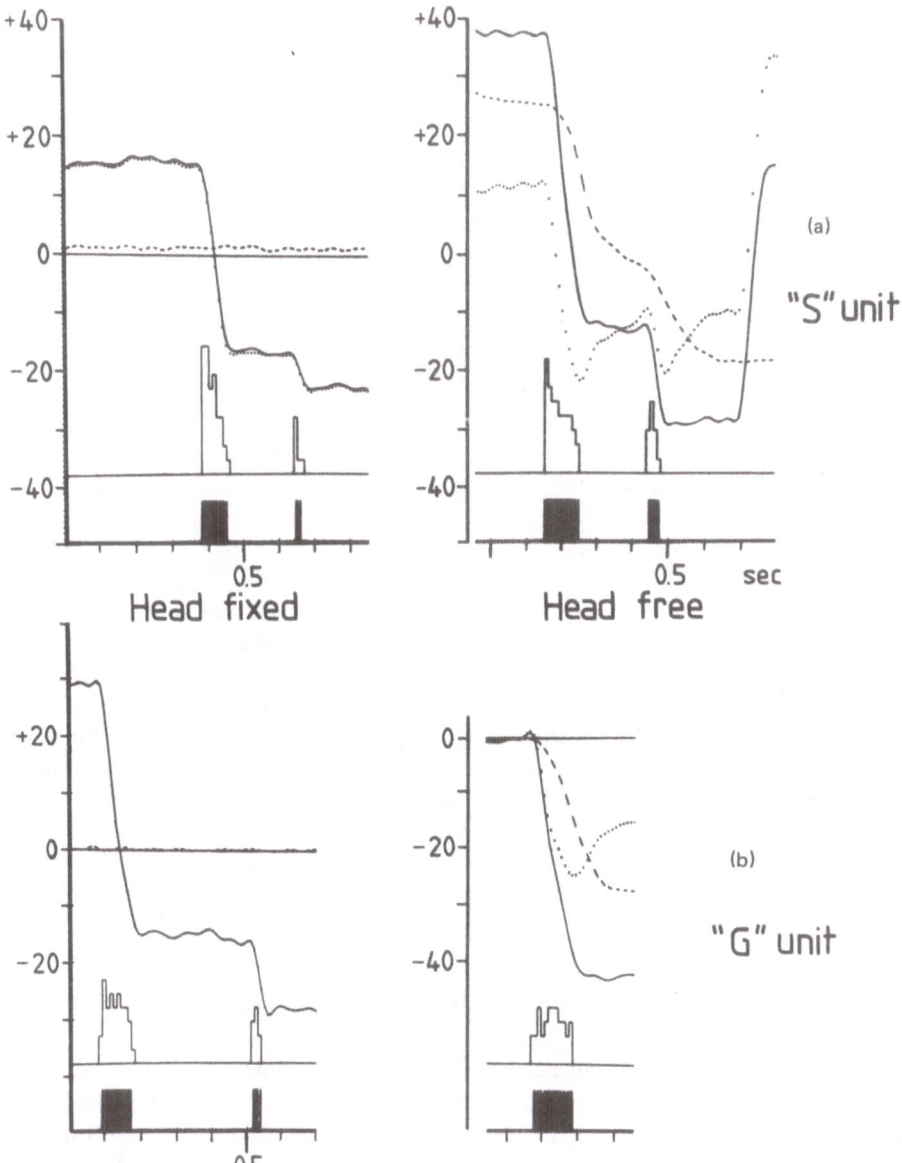

Fig. 6-6. Behavior of two different types of SLB units. The two conditions were head fixed and head free. The three tracings are head movement (*dashed line*), horizontal eye movement (*dotted line*), and total gaze shift (*solid line*). Spike histograms (cumulated discharges in bins of 10 msec) and standardized single-unit discharges are shown at the bottom. (a) Response of an S unit, whose discharges are related to the size of the saccadic eye movement. *Head fixed*: size of the first saccade was $32°$; number of spikes was 42. *Head free*: size of the first saccade was $36°$; size of the total gaze shift was $50°$; number of spikes was 41. (b) Response of a G unit, whose discharges are related to the size of the total gaze shift. *Head fixed*: size of the first saccade was $44°$; number of spikes was 46. *Head free*: Size of the first saccade was $24°$; size of the total gaze shift was $41°$; number of spikes was 47.

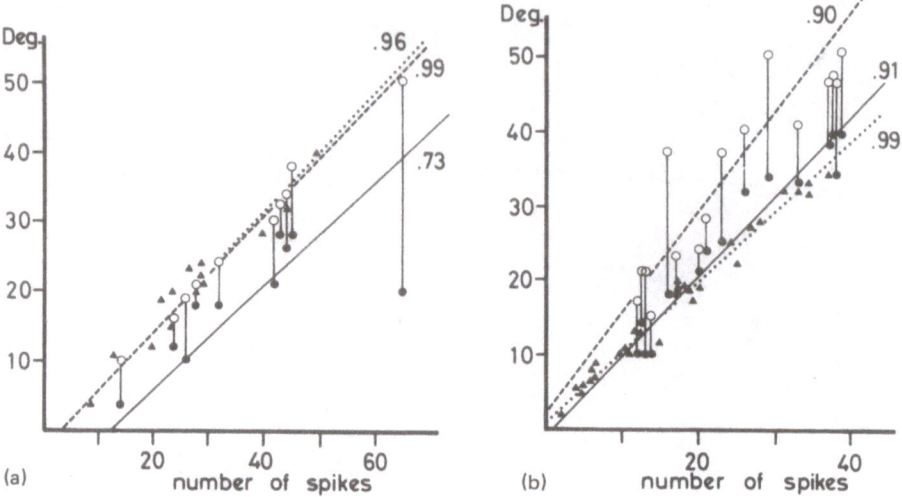

Fig. 6-7. Relationship between three types of movement size and number of spikes per burst for a typical G unit (a) and S unit (b) among the SLB units. The three different types of movements are saccade gaze shift with the head fixed (*filled triangles and dotted regression lines*), total gaze shift including the head movement with the head free (*open circles and dashed regression lines*), and saccade size with the head free (*solid circles and solid regression lines*). *Vertical lines*: amplitude of the head contribution during eye and head movement.

behavior of the eye in the orbit, whereas G units specify the total shift in gaze, including the head movement contribution.

Coordination of Head and Eyes in the Saccadic Changes of Gaze

The results reported above revealed prominent features of the activity of PRF units related to horizontal eye movements when the monkey is free to move its head. Some types of preoculomotor cells behave in ways that might have been predicted from previous single-cell recordings with the head fixed, from behavioral experiments, and from models of eye movements. The best example is provided by the behavior of T cells.

Similarly, in the head-free condition, the role of the OP cells that were not modulated by vestibular stimulation was not different from the one deduced from previous studies in which the head was restrained and from saccadic eye control models. For example, stimulation of unmodulated pausers isolated in clusters along the midline of the brain stem at the pontine level had been found to abolish all saccades (Keller, 1974). Our results lend further support to the suggestion that these neurons could provide a gating function for the occurrence of all fast eye movements (Cohen & Henn, 1972; Keller, 1974). In agreement with Robinson's model, the pause of the tonic activity of these cells would be the signal that specifies saccade duration. In fact, these cells are inhibited for a time proportional to the intended size of the saccade. Our results demonstrated that the duration of this pause is also tightly coupled to the duration of the saccade when the saccadic eye movements are coordinated with voluntary head movements. On the other hand, in the case of the vestibularly modulated OP cells, such predictions were shown to be erroneous. Indeed, it is impossible for these cells to code saccade size by the duration of their pause.

The other important result of this study concerns the SLB units. Previous neurophysiological (Cohen & Henn, 1972; Sparks & Travis, 1971) and anatomical (Büttner-Ennever & Henn, 1976; Graybiel, 1977) studies have led to the acceptance of the idea that SLB units provide excitatory input to the oculomotor neurons, specifying saccade parameters. However, our experiments showed that this class of cells can actually be partitioned into S bursters and G bursters according to the cell's behavior during coordinated eye–head movements. This differential behavior of S bursters and G bursters raises the question of the location of these two classes of bursters in the control scheme of saccadic eye control during eye–head coordination.

At this point, it may be worthwhile to attempt to organize all our results into a general model of coordinated eye–head movements. This scheme is presented in Figure 6-8. The main characteristic of this model is that it assumes an independent head control system that impinges upon the preoculomotor control system only by vestibular feedback (Bizzi et al., 1971; Dichgans et al., 1973; Morasso et al., 1973). The general structure of the preoculomotor control system is consistent with the simplified scheme of saccadic eye control adapted from Robinson (Figure 6-2). Within the oculomotor control system, the first block assumes a neural substrate responsible for generating saccadic eye com-

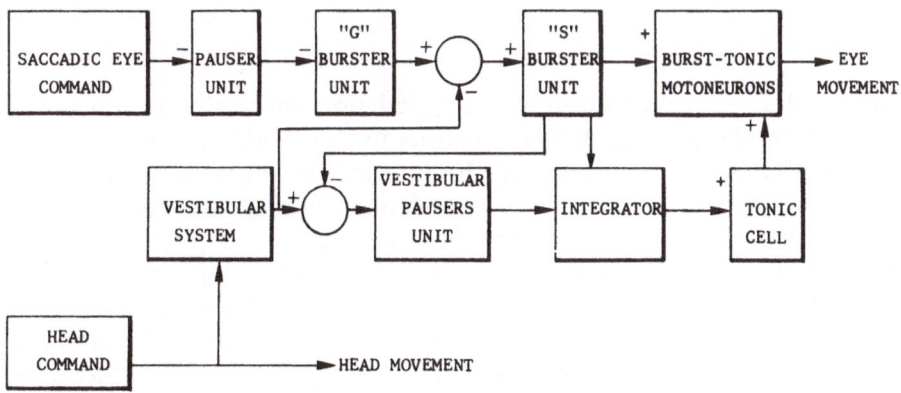

Fig. 6-8. Model of preoculomotor function in coordinated eye–head movement.

mand. According to the foveation hypothesis (Schiller & Stryker, 1972) and the assumption of independent eye and head motor systems, this substrate would code the size of the planned saccade or the total shift in gaze as a function of the size of the retinal error independent of the intended head movement. In the next block, pauser units that are not modulated by vestibular stimulation provide the signal that specifies saccade duration, that is, the size of the planned saccade. Assuming that the vestibular system ensures adequate eye–head coordination by reducing the size of the saccade and generating compensatory eye movements, we have tentatively placed the G burster unit at a level where the total shift of gaze is specified. In other words, the G burster units are upstream in preoculomotor processing, at a point where the input of vestibular feedback does not yet impinge on preoculomotor structures. In contrast, the S burster units are further downstream, at a point where the contribution of head movement has been subtracted from the total shift in gaze.

Given the common assumption that the contribution of the vestibular system occurs at the level of the oculomotor neurons (Büttner, Hepp, & Henn, 1977; Robinson, 1971, 1974), S units from which the contribution of head movement has already been substracted are logically inappropriate for projection onto oculomotor neurons. By assuming, as in this model, that the vestibular input to the motoneurons would be silent during a saccade, the problem of the double subtraction would be solved. This is precisely the pause signal carried by the vestibularly modulated pausers. With respect to the T cells, our results confirmed that they carry the tonic information to produce the appropriate step control signal of the position of the eye in the orbit. If the tonic activity of this group of cells derives from an integration of SLB units, then the candidate cells should be the group that we have identified as G burster units. Clearly, integration of S burster units would lead to an incorrect eye position signal.

Conclusions

The results reported here have provided new information concerning the role of the preoculomotor neurons in the modification of saccade characteristics as one aspect of the interaction between eye and head movement during spatially oriented behavior. Perhaps the most interesting result concerns the existence in the PRF of a population of SLB units whose activity encodes the total shift in gaze independently of the intended head movement. Furthermore, it has been demonstrated that vestibular input strongly influences the second class of SLB units that specify only the shift in eye position. Finally, we have found no evidence that afferent and/or efferent activity of neck muscles affects the behavior of preoculomotor cells.

Taken together, these findings provide neurophysiological support for the idea of independent eye and head motor programs coordinated by sensory feedback generated by the vestibular system. The next area requiring investigation is the relationship of brain stem activity to head motor behavior. As indicated previously, these brain stem structures are known to have projections to neck motoneurons. The important questions are whether the activity of the oculomotor system influences the head-related areas of the brain stem, and how activity in these regions represents various parameters of motor activity such as force, velocity, direction, and postural maintenance of the head.

Acknowledgments

This work was supported by National Institute of Neurological Disease and Stroke Research Grant NS09343, National Aeronautics and Space Administration Grant NGR 22-009-798, National Eye Institute Grant EY02621, and a grant from the DGRST (génie biologique) 80.7.0248.

References

Bender, M. R., & Shanzer, S. Oculomotor pathways defined by electrical stimulation and lesions in the brainstem of monkey. In: M. B. Bender (Ed.), *The oculomotor system.* New York: Harper & Row, 1964, pp. 81–140.

Bizzi, E. The coordination of eye-head movement. *Scientific American*, 1974, *231*, 100–106.

Bizzi, E., Kalil, R. E., & Tagliasco, V. Eye-head coordination in monkeys: Evidence for centrally patterned organization. *Science*, 1971, *173*, 452–454.

Brodal, A. Anatomy of vestibular nuclei and their connections. In H. H. Kornhuber (Ed.), *Handbook of sensory physiology* (Vol. VIII),: *Vestibular system* (Part 1), *Basic mechanisms.* New York: Springer-Verlag, 1974, pp. 239–352.

Büttner, U., Hepp, K., & Henn, V. Neurons in the rostral mesencephalic and paramedian pontine reticular formation generating fast eye movements. In R. Baker & A. Berthoz, (Eds.), *Control of gaze by brainstem neurons.* Amsterdam: Elsevier/North-Holland, 1977, pp. 309–318.

Büttner-Ennever, J. A., & Henn, V. An autoradiographic study of the pathways from the pontine reticular formation involved in horizontal eye movements. *Brain Research*, 1976, *108*, 155–164.

Cohen, B., & Henn, V. Unit activity in the pontine reticular formation associated with eye movements. *Brain Research*, 1972, *46*, 403–410.

Dichgans, J., Bizzi, E., Morasso, P., & Tagliasco, V. Mechanisms underlying recovery of eye-head coordination following bilateral labyrinthectomy in monkeys. *Experimental Brain Research*, 1973, *18*, 548–562.

Eckmiller, R., Blair, S. M., & Westheimer, G. Fine structure of saccade bursts in macaque pontine neurons. *Brain Research*, 1980, *181*, 460–464.

Graybiel, A. M. Direct and indirect preoculomotor pathways of the brainstem: An autoradiographic study of the pontine reticular formation in cat. *Journal of Comparative Neurology*, 1977, *175*, 37–78.

Keller, E. L. Participation of medial pontine reticular formation in eye movement generation in monkey. *Journal of Neurophysiology*, 1974, *37*, 316–332.

Ladpli, R., & Brodal, A. Experimental studies of the commissural and reticular formation projections from the vestibular nuclei in cat. *Brain Research*, 1968, *8*, 65–96.

Luschei, E. S., & Fuchs, A. F. Activity of brainstem neurons during eye movements of alert monkeys. *Journal of Neurophysiology*, 1972, *35*, 445–461.

Morasso, P., Bizzi, E., & Dichgans, J. Adjustments of saccade characteristics during head movements. *Experimental Brain Research*, 1973, *16*, 497–500.

Peterson, B. W., Filion, M., Felpel, L. P., & Abzug, C. Responses of medial reticular neurons to stimulation of the vestibular nerve. *Experimental Brain Research*, 1975, *22*, 335–350.

Peterson, B. W., Maunz, R. A., Pitts, N. G., & Mackel, R. G. Patterns of projection and branching of reticulospinal neurons. *Experimental Brain Research*, 1975, *23*, 333–351.

Peterson, B. W., Pitts, N. G., Fukushima, K., & Mackel, R. Reticulospinal excitation and inhibition of neck motoneurons. *Experimental Brain Research*, 1978, *32*, 471–489.

Pompeiano, O., & Swett, J. E. Actions of graded cutaneous and muscular afferent volleys on brainstem units in the decerebrate cerebellectomized cat. *Archives Italaliennes de Biologie*, 1963, *101*, 584–613.

Robinson, D. A. Oculomotor unit behavior in the monkey. *Journal of Neurophysiology*, 1970, *33*, 393–404.

Robinson, D. A. Models of oculomotor neural organization. In P. Bach-y-Rita & C. C. Collins (Eds.), *The control of eye movements*. New York: Academic Press, 1971, pp. 519–538.

Robinson, D. A. Oculomotor control signals. In G. Lennerstrand & P. Bach-y-Rita (Eds.), *Basic mechanisms of ocular motility and their clinical implications*. Oxford: Pergamon Press, 1974, pp. 337–374.

Schiller, P. H., & Stryker, M. Single unit recording and stimulation in superior colliculus of the alert rhesus monkey. *Journal of Neurophysiology*, 1972, *35*, 401–419.

Scheibel, M. E. & Scheibel, A. B. Structural substrates for integrative patterns in the brainstem reticular core. In H. H. Jaspers, L. D. Proctor, R. S. Knighton, W. C. Noshay, & R. T. Costello (Eds.), *Reticular formation of the brain*. Boston: Little, Brown, 1958, pp. 31–35.

Sparks, D. L., & Travis, R. P. Firing patterns of reticular formation neurons during horizontal eye movements. *Brain Research*, 1971, *33*, 477–481.

Chapter 7

Contribution of Eye Movement to the Representation of Space

Alan Hein and Rhea Diamond

In a series of studies with kittens we have explored the way in which movements of the body come to be guided by visual information, a process that has been described as requiring the formation of a representation of visual space. These studies have indicated that the information incorporated in this representation is derived from visual feedback from self-produced movements. Visually coordinated behavior has been segregated into several components, each of which may be acquired separately, according to the opportunities for motor-visual feedback that are provided. However, the sequence of these acquisitions is constrained, in that information must be extracted from certain feedback loops before that available in others can be utilized.

The initial investigation in this series demonstrated that dark-reared kittens required exposure in light, under conditions that provided correlated visual feedback from self-produced locomotion, in order to acquire the capacity to move appropriately with respect to visual depth. Kittens exposed in light under conditions in which their own body movements were decorrelated with visual input failed to develop visually guided locomotion (Held & Hein, 1963). Later studies confirmed the importance of visual feedback from movement by exposing freely moving animals only under strobe illumination and demonstrating that this interruption of the motor-visual feedback loop precluded development (Hein, Gower, & Diamond, 1970). Another study demonstrated that visual feedback from movements of the forelimb was essential to acquisition of the ability to guide the limb toward small visual targets (Hein & Held, 1967). Subsequently, the capacity for visually guided reaching was shown to be acquired separately with respect to movements of each of the forelimbs and with respect to control of each forelimb by each eye; only the eye exposed with visual feedback from movements of a particular limb became able to guide that limb (Hein & Diamond, 1971b; Hein, Gower, & Diamond, 1970).

Visually guided behaviors were also shown to be separable into discrete components with regard to the level of illumination at which feedback from movement was provided. When visual feedback from the forelimb was provided in light so dim as to stimulate the rods exclusively, the animal subsequently displayed guided reaching at levels of illumination in the scotopic but not in the photopic range (Hein & Diamond, 1971a). In one experiment in that study, although forelimb movements were visible only at levels stimulating the rods, the source of illumination itself provided a target that stimulated the cones and thus provided a source of photopic feedback from locomotion. Unlike visually guided reaching, visually guided locomotion was subsequently displayed in bright as well as dim light. When the source of illumination was itself in the scotopic range, both behaviors were restricted to light levels in that range.

A final study in this series established a constraint on the sequence in which visually guided locomotion and visually guided reaching may be acquired (Hein & Diamond, 1972). Dark-reared kittens that were provided visual feedback from a forelimb in the absence of visual feedback from locomotion failed to develop visually guided reaching. Later they were given the opportunity for visual feedback from locomotion, without view of the forelimb, whereupon they acquired visually guided locomotion. Then the original exposure condition was reconstituted, this time with visual feedback from the forelimb contralateral to that in the first condition. At this point the kittens acquired visually guided reaching, with the capacity restricted to the forelimb from which they had received visual feedback following their acquisition of visually guided locomotion (Hein & Diamond, 1972).

How Is Visual Direction Given?

Recent work focused on the initial steps in the extraction of spatial information from visual input (Hein, Vital-Durand, Salinger, & Diamond, 1979). Although dark-reared kittens turned their eyes toward eccentric high-contrast targets when first brought into light, they could not then move their bodies toward such targets. For this truly spatial behavior to be displayed, it appeared to be necessary for the animal to know the position of the eye in its orbit at the moment the target was seen. It seemed that fixating movements of the eyes and the compensatory head movements that accompany them might provide the opportunity for visual feedback that would establish the three-way correspondence among points in visual space, retinal loci of images, and the set of positions the eye can assume.

Kittens Exposed Only with an Immobilized Eye

These results suggested that, in the absence of eye movements, the opportunity to locomote freely in light would not enable kittens to develop visually guided locomotion. In order to examine this possibility, we immobilized one eye of a kitten prior to its first exposure in light. Cranial nerves III, IV, and VI, which innervate the ocular muscles, were sectioned on one side. We used the surgical procedure developed by Berlucci, Munson, and Rizzolatti (1966): An approach was made through the roof of the mouth, the soft palate was sectioned, and the overlying bone was removed to reveal the optic chiasm and portions of one side of the optic nerve and optic tract. Lateral to these structures, cranial nerves III, IV, and VI appeared just caudal to the orbit. Careful dissection permitted nerve V to be isolated and nerves III, IV, and VI to be sectioned.

Four dark-reared kittens were operated on at 6–8 weeks of age. The contralateral eye was sutured shut at the time of cranial nerve section and the kittens were kept continuously in light thereafter. None of the kittens was able to move the operated eye toward high-contrast targets. When the animals were placed on a platform that was spun and then abruptly stopped, they showed no postrotatory nystagmus. Repeated tests with the head in various orientations confirmed the absence of postrotatory nystagmus. Both observations are consistent with paralysis.

After the kittens had been provided 2 weeks of exposure in light, formal testing of visually guided behaviors was begun. Intact animals of this age display visually guided responses after a much shorter perior in light (e.g., Riesen, 1961). Two tests were used to assess visually guided behaviors. The first, a test of visually guided locomotion, consists of observing the animals in an obstacle course. Wooden blocks inserted into a perforated base in an irregular arrangement provide narrow passageways. The kitten, with vibrissae trimmed short, is placed at various points in the course and observed as it walks about. Normal animals traverse the course rapidly, avoiding collisions by appropriately sinuous movements of the body. Deficiencies are revealed by slow movement, frequent collisions, and startle responses. The second test evaluated visually guided reaching. The apparatus is referred to as a bridge box (Hein & Diamond, 1971b) and is similar to a devise used by Dews and Wiesel (1970) as an alternative to the simple interrupted surface used in previous work (Hein & Held, 1967). The kitten is required to step from a small start box across a gap onto a narrow bridge leading to a platform baited with food. The behavior is first shaped without a gap, with the bridge positioned directly ahead. The animal is required to take a discrete, committed step, without swiping. The size of the gap is gradually increased to the maximum the animal will reach without jumping. In eight test trials the start box is positioned randomly to left and right of the bridge. Interspersed trials with the bridge straight ahead (which are not scored) help to maintain the behavior. The score is the number of hits, with seven of eight considered criterial for guided reaching. Animals that fail to show guided reaching typically grope straight ahead when the bridge is moved to the side.

The animals with an immobilized eye and the contralateral intact eye sutured shut failed to show visually guided locomotion or visually guided reaching. Their scores on the bridge box test are shown in Table

Table 7-1 Visually Guided Reaching Tested in the Bridge Box Following Exposure in Light.

Exposure condition	Animal no.	Reaches (hits/misses)	
Immobilized eye alone	1	0/8	
	2	0/8	
	3	0/8	
	4	1/7	
With atropine	5	7/1	
	6	8/0	
	7	8/0	
Following vestibulectomy	8	8/0	
	9	7/1	
	10	7/1	
Alternating intact and immobilized eye	11	2/6	8/0[a]
	12	0/8	8/0[a]
	13	2/6	7/1[a]
	14	2/6	8/0[a]
	15	0/8	8/0[a]
	16	1/7	7/1[a]
Binocular preoperatively	17	7/1[b]	8/0[c]
	18	8/0[b]	7/1[c]
	19	7/1[b]	8/0[c]
Binocular postoperatively	20	8/0	
	21	7/1	
	22	7/1	
Alternating monocular preoperatively	23	7/1	
	24	8/0	
	25	8/0	

Note. Eight trials were scored with the bridge positioned randomly to the right and left of the animal.
[a] Intact eye.
[b] Testing prior to surgery.
[c] Testing following surgery.

7-1 (animals 1-4). Both behaviors were reexamined periodically. When the experiment was terminated 1 year later, neither capacity was present.

This result suggests that, although the eye remains in a single position (because of the denervation of the extraocular muscles), the visual feedback available from head and body movements does not support formation of a representation of visual space. The reason might be that visually elicited movements of the head, for example, are not related to target location in the precise way in which fixating movements of the eyes come to be very early in life. To draw this conclusion about the role of visual feedback from eye movements, however, it is necessary to eliminate certain possible confounding effects of the eye immobilization surgery.

Some Necessary Controls

First, sectioning of cranial nerve III paralyzes the intraocular muscles that subserve accommodation and changes in pupillary diameter. These losses would degrade the retinal image for near objects, perhaps to a degree that would interfere with the utilization of visual feedback from head movement. To evaluate this possibility we examined a new group of animals in which accommodative and pupillary reflexes were eliminated by a topical administration of atropine, leaving eye movements unimpaired. The procedure was as follows: three dark-reared kittens 6-8 weeks of age had one eye sutured shut. Subsequent to the surgery they remained continuously in the dark except for a period of 3 hours each day in light; 20 minutes before being brought into light and again immediately before the exposure period began, ophthalmic atropine was applied to the eye. The pupil remained fully dilated during the period in light. After 7 days, at the end of the daily exposure period, testing of visually guided behaviors was begun. At this time the animals displayed visually guided locomotion in the obstacle course and within 1 day had been shaped and tested in the bridge box (all exposure and testing with the pupil dilated with atropine). Because the immobilized eye animals had been first tested 2 weeks after initial exposure, the atropine kittens were continued in their daily exposure with atropine for another week of exposure; scores in the bridge box are given in Table 7-1 (animals 5-7). We concluded that whatever degradation of the optics of the eye is consequent on the absence of pupillary and accommodative responses does not interfere with acquisition of the capacities tested. Therefore, absence of visually guided locomotion and visually guided reaching in the animals with an immobilized eye cannot be attributed to this factor.

Second, paralysis of the extraocular muscles eliminates not only visually elicited eye movements, but vestibularly elicited eye movements as well. It appeared possible that blurring of the retinal image when the animal moved, consequent on loss of the vestibulo-ocular reflex, might have interfered with utilization of motor visual feedback from head movements. To examine this possibility, three dark-reared kittens 6 weeks of age were deprived of vestibularly initiated eye movements by bilateral vestibulectomy. The surgical procedure was as follows: the vestibular apparatus was approached ventrally through the acoustic bulla, and the petrous portion of the petromastoid bone was excised. Following surgery the animals were kept continuously in light and were repeatedly observed when placed in various positions on a platform that was rotated and then abruptly stopped. Postrotatory nystagmus remained absent, indicating that the vestibular apparatus was no longer functioning.

The vestibulectomized kittens were difficult to test, initially appearing both hyperactive and ataxic. However, these behaviors gradually subsided so that 3 months after surgery it was possible to conduct formal tests. At this time they were able to traverse the obstacle course without colliding with the wooden blocks. They were shaped and then tested in the bridge box, where guided reaching was displayed. Their scores on the latter test are given in Table 7-1 (animals 8–10). The ability of the vestibulectomized kittens to pass these tests indicated that the absence of the vestibulo-ocular reflex is not inimical to acquisition of visually guided behaviors. Therefore, the absence of these behaviors in the animals with an immobilized eye cannot be attributed to this factor.

These controls lent support to the view that denervation of the eye muscles precludes the acquisition of visually guided behavior because it eliminates visual feedback from eye movements. However, it remained possible that the surgery accomplished this by interfering with the animal's ability to respond to visual input to the immobilized eye. In order to examine this issue a new group of kittens was subjected to the eye immobilization surgery but the intact contralateral eye was exposed in light rather than being sutured shut. The procedure was as follows: six dark-reared kittens 6–8 weeks of age were subjected to unilateral cranial nerve section to denervate the muscles of one eye. The animals recovered from the surgery in the dark (3–4 days). Then they were exposed in light while locomoting freely for 3 hours daily with one eye covered with a scleral occluder. On alternate days the immobilized and intact eye were exposed monocularly. Visually guided behaviors were tested monocularly beginning after 21 hours of exposure. At this time the kittens displayed visually guided locomotion using the intact eye and were shaped and subsequently tested in the bridge box. They displayed visually guided reaching when using the intact eye (Table 7-1, animals 11–16). When tested using the immobilized eye, they were indistinguishable from the first group of immobilized eye animals (those that had received exposure in light with the paralyzed eye only). The contrast between the animals' capacities when tested using the two eyes indicates that their failure to display visually guided behaviors with the immobilized eye was not due to some generalized inhibition of response to visual input.

These animals permitted another observation to be made concerning the processing of visual input to the immobilized eye. When the intact eye was shielded and the immobilized eye stimulated with a small bright light, the pupil of the intact eye was seen to constrict. This indicated that the retina of the immobilized eye remained responsive to light; the presence of the consensual pupillary response of the intact eye indicated that this input was processed. Another observation, made with both groups of immobilized eye animals, was also consistent with this interpretation. When the kittens were using the immobilized eye and they were lowered toward a broad horizontal surface, the forelimb was extended at a time appropriate to contact the surface. Visually triggered extension of the forelimb is a component of visually coordinated behavior that is acquired in the absence of motor-visual feedback and that, unlike guided behaviors, transfers intraocularly (Hein, Held, & Gower, 1970). Its mediation by an eye that has not received visual feedback from eye movements is consistent with its status as a component distinct from visually guided reaching and serves in the present case to confirm that the animal remains responsive to visual input to the immobilized eye.

Immobilization of the Eye in Animals
That Have Acquired Guided Behaviors

The question of whether some unknown effect of eye muscle denervation might preclude mediation of visually guided behaviors by an immobilized eye was addressed in the following way: Three light-reared kittens 6-8 weeks of age were tested monocularly for visually guided locomotion and visually guided reaching. The presence of both behaviors was confirmed. Then the kittens were subjected to denervation of the muscles of one eye. The kittens recovered from surgery (3-4 days in the dark) and were then retested monocularly. Guided behaviors were mediated by the immobilized eye as well as by the intact eye. Preoperative and postoperative scores for tests using the immobilized eye in the bridge box are given in Table 7-1 (animals 17-19). Thus, in animals that have already acquired visually guided behaviors, immobilization of the eye does not interfere with mediation of guided behaviors by that eye. This confirms that it is the acquisition phase of visuo-motor coordination to which eye movements are crucial, and this result is consistent with the readily made observation that eye movements need not occur during the execution of visually guided behaviors.

Role of Binocular Exposure

The results for the light-reared kittens raised the question of whether movements of the intact contralateral eye in any way contribute to the mediation of guided behaviors with the immobilized eye. In these kittens, preoperative exposure in light had been binocular. It was possible that this factor was crucial to the apparent postoperative sparing of behaviors mediated by the immobilized eye. Perhaps dark-reared kittens with the muscles of one eye denervated might become able to mediate guided behaviors with that eye if they were provided postoperative binocular exposure with the intact and immobilized eyes. This procedure was followed with a new group of dark-reared kittens. Three animals 6–8 weeks of age were subjected to unilateral denervation of the eye muscles and remained thereafter in light. Tests of guided behavior 2 weeks after the surgery revealed that guided behaviors were mediated by either eye. These data are given in Table 7-1 (animals 20–22). In contrast to the kittens exposed monocularly with the immobilized and intact eyes on alternate days, these animals displayed guided behaviors under control of an eye immobilized prior to first exposure in light. These results do not, however, point unambiguously to the role of binocular exposure, whether preoperative or postoperative, in enabling the mediation of guided behaviors by an immobilized eye. It remains possible that while binocular exposure in visually naive animals was indeed essential, postoperative sparing was determined by whether or not the eye to be immobilized had previously mediated guided behaviors.

Alternating Monocular Exposure Before Eye Immobilization

We approached this question with a new group of dark-reared kittens. Three animals 6–8 weeks of age were provided alternating monocular exposure for 2 weeks and were then tested for mediation of guided behaviors by each eye. When the presence of these capacities had been confirmed, the muscles of one eye were surgically denervated. The animals recovered from the surgery in the dark (3–4 days) and were then retested monocularly. No deficits in the mediation of guided behaviors by the denervated eye were observed. Scores in the bridge box using the denervated eye are given in Table 7-1 (animals 23–25). We concluded that preoperative acquisition of guided behaviors mediated by the to-be-immobilized eye was critical to the sparing, and that binocular preoperative exposure was not a factor. This was also confirmed with another preparation—kittens in which one eye was enucleated prior to the animal's first exposure in light. These animals were permitted to acquire visually guided behaviors monocularly and the eye was then surgically immobilized. Following recovery from surgery, the kittens traversed the obstacle course and displayed guided reaching in the bridge box with no impairment.

Contribution of Binocular Exposure

Given that binocular preoperative exposure is not necessary for post-operative sparing of guided behaviors mediated by an immobilized eye, what might be the contribution of binocular exposure following the denervation surgery in visually naive animals? The results for kittens exposed monocularly with the intact and immobilized eye on alternate days illustrated the independence of the two eyes with respect to the mediation of visually guided behaviors. This is consistent with our previous studies of intact kittens, in which the two eyes were exposed under conditions differing in opportunities for visual feedback from movement. As previously noted, the capacities mediated by each eye subsequently were dependent on the feedback opportunities that had been provided to that eye (Hein & Diamond, 1972). For example, kittens exposed binocularly while locomoting without view of the forelimbs but provided an opportunity to view the limbs mediated guided locomotion with either eye; however, they displayed guided reaching only when using the eye that had viewed the limbs. The kittens with one eye immobilized prior to exposure in light that were then exposed binocularly, however, acquired the capacity to mediate guided behaviors with an eye that appeared not to have received visual feedback from eye movements.

How might we understand this apparent exception? Specifically, why should monocular and binocular exposure conditions differ in their results for animals with one eye immobilized and the other eye intact? We suggest that the difference may lie in the reliability of visual feedback from head movement in the two conditions. During binocular exposure, the intact eye moves to fixate eccentric visual targets, and these eye movements come to be precisely related to target location. All eye movements are accompanied by compensatory head movements. Such head movements produce changes on the retina of the immobilized eye that are as systematically related to the signal to move the eyes as are those occurring on the retina of the intact eye. Thus, the conditions necessary for correlation of the output signal with movement-produced visual feedback are met for each eye. During monocular exposure with the immobilized eye, on the other hand, head movements are not precisely related to target location. Signals to move the eyes initiated in response to images falling eccentrically on the retina of the immobilized eye, while they may be expected to produce movements of the (occluded) intact eye and accompanying compensatory head movements, fail to exhibit the precise relation to target location that holds during normal fixation. During monocular exposure with the intact eye, of course, the eye receives visual feedback from eye and head movements that is precisely related to target location and that eye becomes able to mediate guided behaviors. The occluded immobilized eye, however, receives no visual feedback, and, in agreement with previous studies of intact cats, does not gain the capacity to mediate visual guidance.

What Is the Source of Knowledge of Eye Position?

All of the foregoing studies addressed the role of eye movements in the initial acquisition of visually guided behaviors. These behaviors imply formation of a representation of visual space that incorporates information derived from various visuo-motor feedback loops. It has been suggested that this representation must incorporate the correspondence between the direction of visual targets and the retinal loci of their images, given the set of positions that the eye can assume. How, in fact, is the necessary information about the posture of the viewing eye provided? In particular, is proprioceptive input from the extraocular muscles essential during the period when the representation is being formed?

This question in its most general form has a long history. Two suggestions have been made about the source of information about eye posture. The first, the position taken by Helmholtz (1866), is that a corollary of the efferent signal to move the eye is the source. The second, the position taken by Sherrington (1918), is that afference from the eye muscles provides this information. In order to determine whether afference from the eye muscles might be critical to the acquisition of visually guided behaviors, we conducted a series of studies of kittens deprived of this input.

Kittens Exposed with Deafferented Extraocular Muscles

An initial experiment assessed whether kittens deprived of proprioceptive input from the extraocular muscles could acquire visually guided behaviors when there was no input from the contralateral eye. Six animals were reared in the dark until they were 2–4 weeks of age, at which time one eye was enucleated and the kittens were returned to the dark. When they were 10 weeks old, the extraocular muscles of the remaining eye were deafferented. The surgical procedure was as follows: an approach was made through the soft palate and sphenoid bone, and the ophthalmic branch of cranial nerve V was sectioned as it emerges from the semilunar ganglion. In the cat, the majority of afferent fibers from the extraocular muscles run in this branch after leaving the orbit (Batini

& Buisseret, 1974). Following the deafferentation surgery, the kittens remained continuously in light. The eye remained mobile and the cornea remained clear. However, when tested in the obstacle course and bridge box after 2 weeks, the kittens lacked visual guidance. Their scores in the bridge box are given in Table 7-2 (animals 1–6). Two of the animals were retested after an additional month in light, but no improvement was shown (Table 7-2, animals 5 and 6, retest). These results suggest that the exclusion of proprioceptive input from the eye muscles precludes the acquisition of visually guided behaviors.

Table 7.2. Visually Guided Reaching Tested in the Bridge Box Following Various Treatments.

Treatment	Animal no.	Reaches (hits/misses)
One eye enucleated,	1	0/8
extraocular muscles of	2	2/6
fellow eye deafferented,	3	1/7
2 weeks of monocular	4	0/8
exposure in light	5	0/8 0/8[a]
	6	1/7 0/8[a]
Test 1: one eye sutured	7	8/0
shut, extraocular	8	7/1
muscles of fellow eye	9	7/1
deafferented, 2 weeks	10	7/1
of monocular exposure		
in light		
Test 2: enucleation of	7	7/1
previously sutured	8	8/0
eye	9	7/1
	10	8/0
Test 3: remained in light	7	8/0
Returned to dark	8	0/8
2–3 months	9	0/8
	10	0/8
Test 4: returned to light	8	7/1[b]
	9	8/0[c]
	10	7/1[d]

Note. Eight trials were scored with the bridge positioned randomly to the right and left of the box.
[a] Retest after an additional month in light.
[b] After 1 week in light.
[c] After 3 weeks in light.
[d] After 4 weeks in light.

Input from the Fellow Eye

This outcome paralleled what had been observed in the study of the effect of eye immobilization on acquisition of guided behaviors; that is, when visual input was provided to the operated eye only, guided behaviors were not acquired. In the case of the kittens with an immobilized eye, postoperative binocular exposure with the immobilized and intact eyes had enabled the mediation of guided behaviors by the denervated eye. This suggests that in the present case such as result might also obtain with respect to proprioceptive input provided by a fellow eye. We hypothesized that a kitten exposed with a deafferented eye, with the contralateral eye intact but occluded, might gain the capacity to mediate guided behaviors with the deafferented eye. In order to examine this, a new group of four dark-reared kittens 10 weeks of age was subjected to unilateral deafferentation of the extraocular muscles with the contralateral eye left intact but sutured shut. After the surgery the animals were kept continuously in light. Visually guided behaviors were displayed when the animals were tested 2 weeks later. Their scores in the bridge box are given in Table 7-2 (Test 1, animals 7-10).

We concluded that proprioceptive input from the consensual movements of the (sutured) contralateral eye provided the necessary information about eye posture. Thus, visual feedback from movements of the deafferented eye could be informative about the direction of objects in space in a way that it could not be when the deafferented eye was exposed in the absence of proprioceptive input. The question arises as to whether input from the (sutured) contralateral eye is essential for the maintenance of this capacity once it has been acquired. This was addressed by enucleating the sutured eye after this first series of tests. The kittens recovered from the enucleation surgery in the dark (3 days) and were then retested. Guided locomotion and guided reaching remained intact. Scores in the bridge box are given in Table 7-2 (Test 2, animals 7-10). This outcome led to the conclusion that, whereas proprioceptive information about eye posture appears to be essential to the initial acquisition of visually guided behaviors, once acquired, those behaviors can be maintained in the absence of such input. This suggests that an animal with one eye enucleated prior to exposure in light that had acquired visually guided behaviors monocularly would retain those behaviors if the eye were subsequently deafferented.

Contribution of Visual Input

It appeared possible to us that the maintenance of visually guided behaviors mediated by a deafferented eye, in the absence of any proprioceptive input, might require continued exposure in light. The supposition was that visual feedback from eye movement might serve continually to update the program for control of those movements. In order to assess this possibility, three of the kittens that had acquired visually guided behaviors with a deafferented eye were returned to the dark while the fourth remained in light. The animal remaining in light was retested periodically and showed no loss of visually guided behaviors over a period of 3 months. Scores on the bridge box are given in Table 7-2, (Test 3, animal 7). Animal 8 was removed from the dark after 2 months and animals 9 and 10 after 3 months. All three showed deficiencies in guided locomotion and guided reaching at this time. Their scores at the end of the prolonged period in the dark are given in Table 7-2 (Test 3, animals 8-10). Intact kittens of this age show no impairment of guidance after comparable periods in the dark. Thus, the degredation of behavior in the case of deafferented eye kittens is consistent with the view that proprioception and visual feedback can be complementary in maintaining these capacities. In contrast, however, the initial formation of a representation of visual space seems to require information from both sources.

Recovery Without Proprioceptive Input

The kittens that had lost visually guided behaviors when they were kept in the dark were afterward kept in light and retested weekly. In all cases they regained visually guided behaviors. The exposure durations required for recovery are shown in Table 7-2 (Test 4, animals 8-10). It appears that the kittens kept longer in the dark recovered more slowly. The recovery of the behaviors that had been lost in the dark presents a puzzle. When they emerged from the dark, these kittens with a deafferented eye appeared to be in the same situation as the kittens with only one eye that had been deafferented prior to exposure in light. Unlike the visually naive animals, however, kittens that had acquired visually guided behaviors and then lost them were able to reacquire those behaviors, without proprioceptive input from eye muscles, when visual input was again made available.

This difference between kittens that had once acquired a representation of visual space and those that had not is suggestive of just what the critical aspect of proprioceptive input from the extraocular muscles might be. During the period when eye movements are relatively imprecise, inflow from the eye muscles indicating the starting position of the eye might be essential to modulate properly the signal to execute a movement. Once precise control of eye movements has been attained, monitoring of the efferent signal alone could provide the information about eye position utilized in localizing objects. At this point, loss of afference from eye muscles might have a negligible effect. If control of eye movements were to deteriorate, however, monitoring of the efferent signal would no longer give sufficiently accurate information about eye posture. There is some direct evidence that visual feedback and proprioceptive input from the eye muscles have a complementary relationship in maintaining normal eye movements. Adult cats with eye muscles deafferented showed normal saccades and nystagmic responses while in light but repetitive pendular movements of the eyes and abnormal vestibularly induced eye movements while in the dark (Fiorentini & Maffei, 1977).

Why a prolonged period in the dark should so degrade the program for eye movements (in the absence of eye muscle afference) that a considerable period in light should be required for recovery is not at all clear. Of course, intact kittens first acquire a representation of visual space during a period of exposure in which proprioceptive input from the eye muscles and visual feedback from eye movements are both available and give redundant information. Extraction of spatial information under these conditions takes time, and it is perhaps not remarkable that its reconstruction should take time as well. The relatively long time required when proprioceptive input is not available might simply emphasize the importance of inflow information about eye posture during normal neonatal development.

Conclusions

The two sets of studies discussed here extend our understanding of the formation of the representation of visual space that underlies visuomotor coordination; backward, as it were, to the way in which spatial referents are assigned to retinal input. We have presented evidence that eye movements and proprioceptive feedback from the eye muscles are fundamental to this process. The two factors are intimately related: without inflow from the eye muscles a mobile eye is not localizable in its orbit; without eye movement any proprioceptive input that remains available from the paralyzed eye seems insufficiently informative about eye posture. Thus, as in the accounts we had previously offered for the acquisition of all components of visually guided behavior, self-produced movement, in this case eye movements, is the basis for the organism's knowledge of its environment.

Acknowledgments

This work was supported by the Office of Naval Research Contract N00014-80-K-0243, National Eye Institute Grant 1 P30-EY02621 and the Undergraduate Research Opportunities Program of the Massachusetts Institute of Technology.

References

Batini, C., & Buisseret, P. Sensory peripheral pathway from extrinsic eye muscles. *Archives Italiennes de Biologie*, 1974, *112*, 18–38.
Berlucchi, G., Munson, J. B., & Rizzolatti, G. Surgical immobilization of the eye

and pupil, permitting stable photic stimulation of freely moving cats. *Electroencephalography and Clinical Neurophysiology*, 1966, *21*, 504-505.

Dews, P., & Wiesel, T. Consequences of monocular deprivation on visual behavior in kittens. *Journal of Physiology* (London), 1970, *206*, 437-455.

Fiorentini, A., & Maffei, L. Instability of the eye in the dark and proprioception. *Nature*, 1977, *269*, 330-331.

Hein, A., & Diamond, R. M. Contrasting development in kittens of visually triggered and guided movements with respect to interocular and interlimb equivalence. *Journal of Comparative and Physiological Psychology*, 1971, 76, 219-224. (a)

Hein, A., & Diamond, R. M. Independence of scotopic and photopic systems in acquiring control of visually guided behavior. *Journal of Comparative and Physiological Psychology*, 1971, 76, 31-38. (b)

Hein, A., & Diamond, R. M. Locomotory space as a prerequisite for acquiring visually guided reaching in kittens. *Journal of Comparative and Physiological Psychology*, 1972, *81*, 394-398.

Hein, A., Gower, E. C., & Diamond, R. M. Exposure requirements for developing the triggered component of the visual-placing response. *Journal of Comparative and Physiological Psychology*, 1970, *73*, 188-192.

Hein, A., & Held, R. Dissociation of the visual placing response into elicited and guided components. *Science*, 1967, *158*, 390-392.

Hein, A., Held, R., & Gower, E. C. Development and segmentation of visually-controlled movement by selective exposure during rearing. *Journal of Comparative and Physiological Psychology*, 1970, *73*, 181-187.

Hein, A., Vital-Durand, F., Salinger, W., & Diamond, R. Eye movements initiate visual-motor development in the cat. *Science*, 1979, *204*, 1321-1322.

Held, R., & Hein, A. Movement-produced stimulation in the development of visually guided behavior. *Journal of Comparative and Physiological Psychology*, 1963, *56*, 872-876.

Helmholtz, H. von. [*Treatise on physiological optics* (Vol. 3)] (J. P. C. Southall, Ed. and trans.). New York: Dover, 1962 (Originally published, 1866.)

Riesen, A. H. Stimulation as a requirement for growth and function in behavioral development. In D. W. Fiske & S. R. Maddi (Eds.), *Functions of varied experience*. Homewood, Ill.: Dorsey Press, 1961.

Sherrington, C. S. Observations on the sensual role of the proprioceptive nerve-supply of extrinsic ocular muscles. *Brain*, 1918, *41*, 332-343.

Chapter 8

Control of the Optokinetic Reflex by the Nucleus of the Optic Tract in the Cat

Klaus P. Hoffman

The nucleus of the optic tract in the pretectum has been identified as the sensorimotor link between the retina and the premotor structures in the pathway mediating the optokinetic reflex in mammals (Collewijn, 1975a, 1975b; Hoffmann, Behrend, & Schoppmann, 1976; Hoffmann & Schoppmann, 1975; Precht & Strata, 1980). In the cat the nucleus of the optic tract (NOT) is a diffuse cell aggregation in the pretectum with most of the cells embedded in fibers from the retina and the cortex in the brachium of the superior colliculus. NOT cells are optimally located in the midbrain to receive both strong direct retinal and extensive diffuse cortical projections (Berman, 1977; Updyke, 1977). The cortical projection provides a binocular input, whereas the retinal input is mainly from the contralateral eye. One target area of the NOT output fibers is the dorsal cap of the inferior olive (Hoffmann et al., 1976; Maekawa & Simpson, 1973; Mizuno, Nakamura, & Iwahori, 1974; Takeda & Maekawa, 1976; Walberg, Nordby, Hoffmann, & Holländer, 1981). We recently presented quantitative data on the response properties of direction selective cells in the NOT of normal adult cats (Hoffmann & Schoppmann, 1981) and discussed the possible role of the output of these cells for the control of optokinetic nystagmus (OKN).

Another way to demonstrate the relationship between NOT and OKN is to change OKN by lesions or developmental manipulations and to see whether or not these changes can be explained by alterations in the response properties of the direction-selective cells in the NOT. Early monocular deprivation has been shown to have a drastic effect on visuomotor behavior as well as on performance in pattern recognition tasks when the cat is forced to use the deprived eye. Neuronal changes in the lateral geniculate nucleus, in area 17 and 18 of the visual cortex, and in the superior colliculus have been well documented and attempts have been made to relate behavioral deficits to specific neuronal changes in these structures (Hoffmann & Cynader, 1977; Wiesel & Hubel, 1963a,

1963b). Early monocular deprivation also has a clear effect on OKN van Hof-van Duin, 1978; Hoffmann, 1981) and on the NOT (Hoffmann, 1979, 1981). In the normal cat, OKN to horizontal stimulus movement is slightly asymmetrical. This asymmetry of OKN is such that for each eye the temporal-to-nasal direction of the stimulus elicits a better response. The monocular response can be rendered more and more asymmetrical by lesion of the visual cortex or by early monocular deprivation due to lid closure. Temporonasal stimulus movement, that is, from left to right through the left eye or from right to left through the right eye, is then much more effective in eliciting the slow phase of OKN. This clear behavioral effect is precisely correlated with changes in response properties of NOT cells.

Most of the changes in response properties of cells in the superior colliculus that follow visual deprivation can be related to changes in the connectivity of the visual cortex (Hoffmann & Sherman, 1974, 1975). We may also ask how visual deprivation alters connectivity in central visual pathways, particularly NOT subserving OKN.

Functional Anatomy of NOT

A number of methods have been applied (a) to identify anatomically the connectivity of retinal ganglion cells with NOT; (b) to describe the response properties of the NOT cells to visual stimulation; and (c) to assess changes in OKN due to lesions and developmental manipulations.

Connections Between Retinal Ganglion Cells and NOT

In experiments with normal adult cats, horseradish peroxidase (HRP) was injected iontophoretically into the inferior olive or by microsyringe into the eyeball. The pretectum was cut into 80-μm transversal sections on a freezing microtome and processed with Hanker-Yates reagent or tetramethylbenzidine (Hanker, Yates, Metz, & Rustioni, 1977; Mesulam, 1978). By this procedure, NOT cells could be labeled retrogradely by the HRP injected into the inferior olive, and their relationship to retinal terminals could be visualized by the anterograde transport of HRP from the injected eye. The goal of this experiment was to determine whether NOT cells could be reached by retinal fibers from the ipsilateral eye. A definite answer would have simplified the interpretation of the effects of cortical lesions and of visual deprivation.

As shown in Figure 8-1, the transport of HRP to retinal terminals was about equally strong in the pretectum ipsilateral and contralateral to the injected eye (see Berman, 1977). The surprising result of the double injection of HRP (Figure 8-1) was that NOT cells do not lie in the conspicuous terminal zones of the retinofugal fibers in the pretectum, but are distributed along the fibers of the optic tract, where they may be reached by short axon collaterals or by axodendritic synapses of the optic tract fibers directly. This puts the NOT cells in an optimal location to summate the input from a great number of retinal fibers over large retinal areas.

Figure 8-2 shows the distribution of microlesions at the recording sites of direction-specific cells that could be antidromically activated from the inferior olive to the location of the cells labeled retrogradely by HRP injection into the inferior olive. The superimposition of microlesions and labeled cells indicates that we were recording from this anatomically demonstrated population of cells, but it seems impossible to decide whether these NOT cells have direct retinal input from the

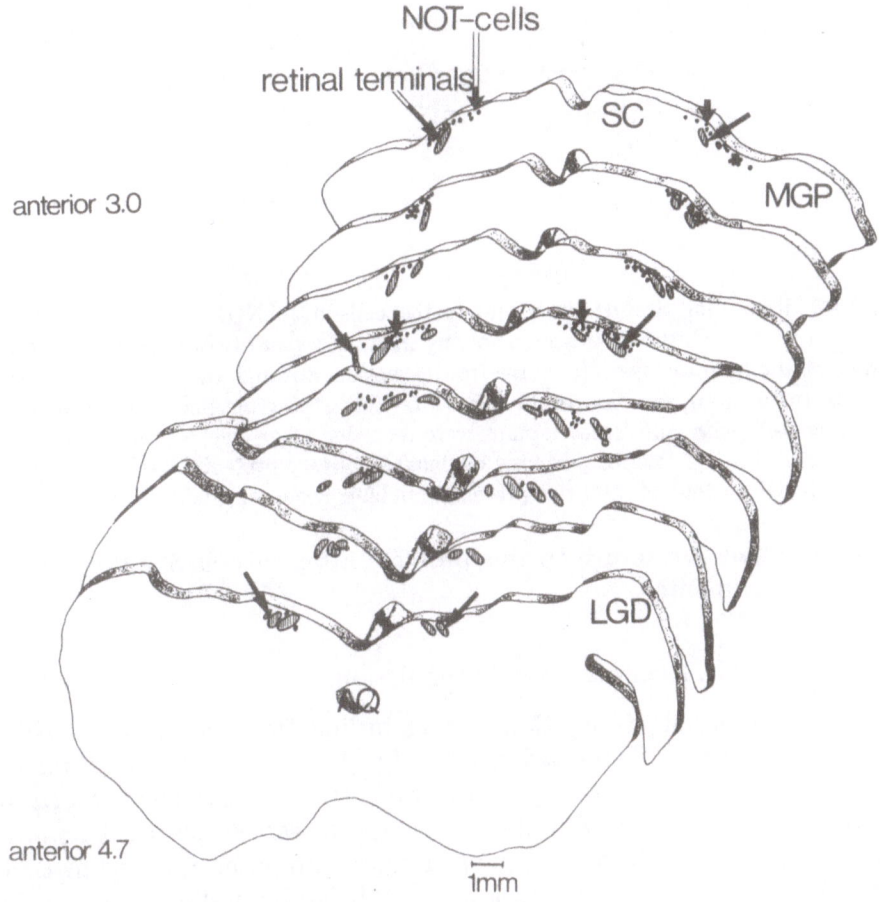

Fig. 8-1. Frontal sections showing the relationship between NOT cells labeled retrogradely by HRP injections bilaterally into the inferior olives and retinal terminals labeled by anterograde transport of HRP from the left retina. Each labeled cell in the NOT is marked by a dot (*short arrows*). Retinal terminals were found within the shaded area (*long arrows*). Sections are presented at intervals of 240 μm from anterior 4.7 to anterior 3.0. *SC*, superior colliculus; *MGB* and *LGB*, medial and lateral geniculate body; *AQ*, aqueduct.

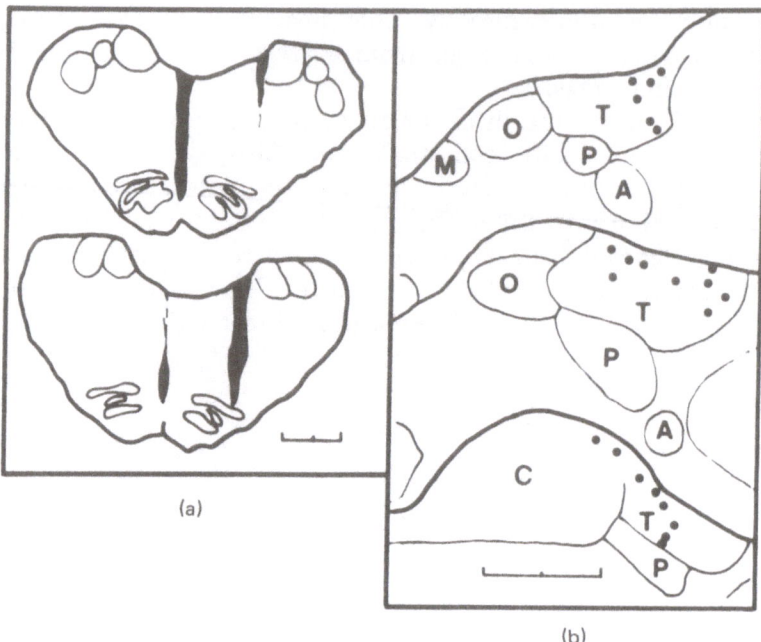

(a)

(b)

Fig. 8-2. Recording sites of direction-selective cells in the NOT of the cat. (a) Frontal sections through the brain stem showing tracts of a pair of stimulating electrodes aimed at the inferior olive. (b) Three frontal sections through the pretectum: posterior on the bottom, anterior on the top. *Dots*, sites of microlesions where direction-selective cells with antidromic spikes were recorded after electrical stimulation of the inferior olive; *M*, medial pretectal nucleus; *O*, olivary pretectal nucleus; *P*, posterior pretectal nucleus; *A*, anterior pretectal nucleus. Scale represents 2 mm.

ipsilateral eye. We return to this problem after we describe the physiological experiments.

Response of NOT cells to Visual Stimulation

Methods of visual stimulation and recording from units in the NOT were the same as described previously (Hoffmann & Schoppmann, 1981). Large random noise patterns projected onto a tangent screen by a slide projector via a double-mirror system served as visual stimuli. These patterns were moved along a circular path without changing their orientation. The cells' responses were analyzed by on-line computation of average response histograms, which were then displayed in polar coordinates. The responses of the cells were pooled to represent the response characteristics of the total population of one nucleus. One pair of stimulating electrodes was stereotaxically inserted into the inferior olive ipsilateral to the NOT in which recordings were to be made. A second pair of stimulating electrodes was placed in the optic chiasm to measure the conduction velocity of retinal fibers projecting to the NOT.

Patterns rich in contour and covering a large area of the visual field were significantly more effective stimuli for NOT cells than single spots or bars. We assume this criterion is unaffected by experimental manipulation such as cortical lesions or visual deprivation. As expected from the anatomical experiment, these cells can be antidromically activated by electrical stimulation of the inferior olive, and they receive direct retinal input from the contralateral eye. The conduction velocity of the retinal fibers to NOT cells, measured between the optic chiasm and the optic tract near the lateral geniculate nucleus, indicates that cells with slowly conducting axons (W cells) from the retina innervate the NOT cells.

The outstanding property of NOT cells is their directional specificity. All cells studied in the right nucleus preferred movements from left to right and those in the left nucleus preferred movement from right to left. In the visual field contralateral to the recording site the preferred movements were from the periphery toward the vertical meridian. In binocular cells the preferred direction was identical whichever eye was stimulated.

As shown in Figure 8-3 all units can be influenced by the contra-

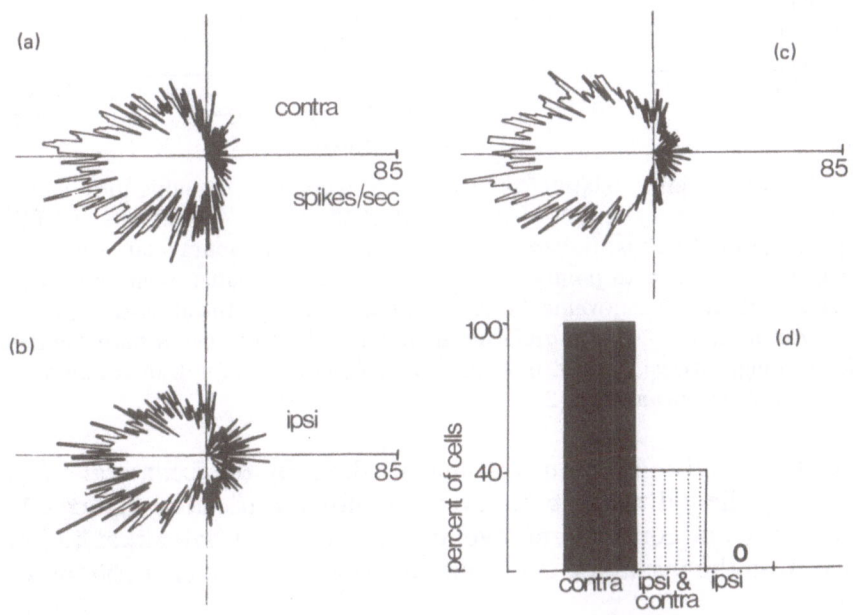

Fig. 8-3. Polar histograms showing the response strength and directional tuning of NOT cell after stimulation of the contralateral eye (a), ipsilateral eye (b), or both eyes (c). Response strengths in relation to an extensive random dot pattern moving on a circular path are given for the various directions by the vector from the origin to the curve. Ocular dominance distribution of 120 NOT cells (d) shows that all cells recorded were influenced by the contralateral eye (*Contra*): all cells also influenced by the ipsilateral eye (40%) were therefore binocular (*Ipsi & Contra*). No cell was driven exclusively by the ipsilateral eye (*Ipsi*).

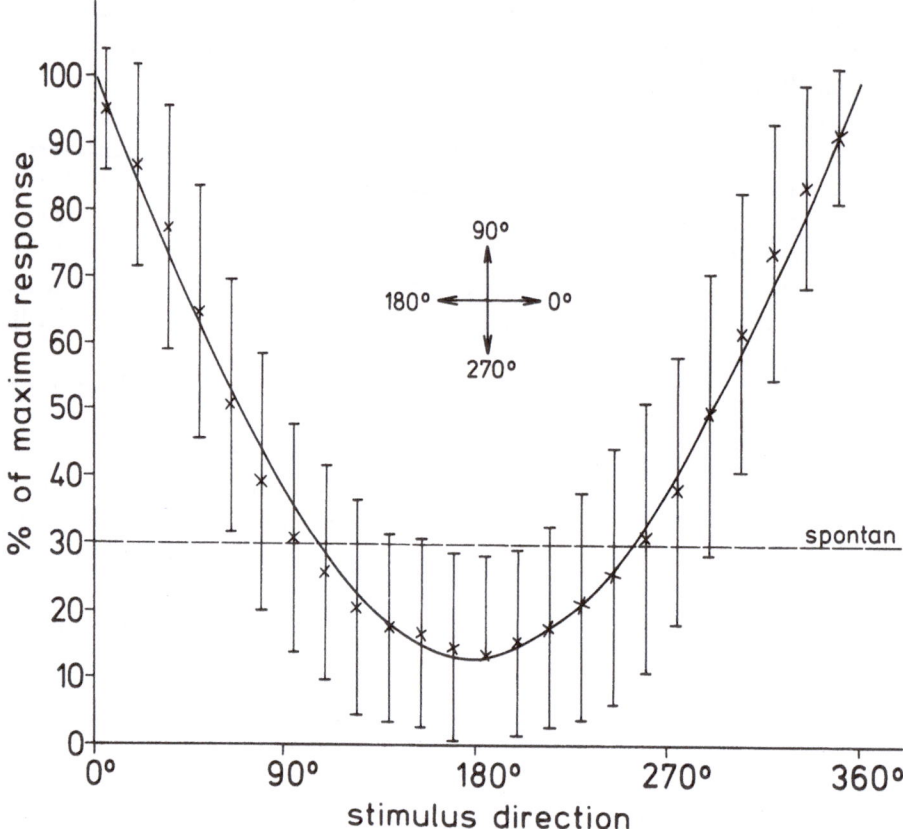

Fig. 8-4. Quantitative relationship between direction of stimulus movement and normalized response strength of the output of the NOT. *Vertical bars*, ± 1 SD in a population of 49 cells; *horizontal line*, mean level of spontaneous activity. The curves through the data points were calculated using the equation given in the text. Direction of stimulus movement α is presented linearly on the abscissa starting with a horizontal movement in a preferred direction at 0° and then deviating anticlockwise through vertical upward movement at 90°, horizontal null direction at 180°, and downward movement at 270°.

lateral eye and only about one-half of them by the ipsilateral one. The response elicited from both eyes was not significantly stronger than that from the contralateral eye alone. It is shown later that the influence from the ipsilateral eye comes by way of binocular corticopretectal fibers.

Direction specificity of the units' responses was tested by continuously moving a large pattern on a circular path across a receptive field. The average activity of 49 cells in relation to different directions is shown in Figure 8-4. The curve fitted to the points is given by the equation

$$f\alpha = a - m \cdot \left(\sin \frac{\alpha}{2} \right)$$

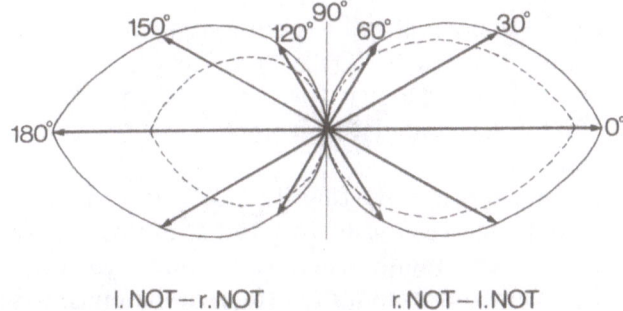

I. NOT − r. NOT   r. NOT − I. NOT

Fig. 8-5. NOT output signal, defined as the difference between the activity of the left and right NOT plotted in polar coordinates. *Solid line*, values for binocular stimulation; *broken line*, values for stimulation of the left eye alone. Movements from left to right (0°) increase the activity maximally in the right NOT and decrease it in the left NOT. The difference in activity between both NOTs is maximal. This difference is zero with vertical movements and maximal again with movements from right to left, but now the activity in the left nucleus is increased and that in the right is decreased. The curves are constructed from the direction-specificity tuning curves in Figure 8-4. Maximal activity in the horizontal direction is set to 1.

where $f\alpha$ is the discharge rate as a function of α, the angle of deviation from the horizon, a is the response in the preferred direction, and m is the difference between the response in the preferred and null directions. Binocular stimulation leads to a direction-specific difference signal between the two nuclei as represented in Figure 8-5. We assume that stimulus movement in a particular direction in visual space is coded by the activity in one NOT minus the activity in the other NOT. For vertical directions the NOT output is zero because vertical movements do not influence the NOT cells; for horizontal movement the NOT output is maximal. This signal is called the *NOT output*, and it is this signal that we want to relate to OKN.

Relationship Between NOT Output and OKN

OKN was measured by implanting a magnetic search coil in one eye according to the technique of Robinson (1963). In addition, head restraining bolts were attached to the skull. Some days after surgery eye movements elicited by the movement of a random dot pattern across a 90° by 90° screen 40 cm in front of the animal were recorded. This stimulus is different from the more commonly used optokinetic drum, providing full-field stimulation. We chose this condition for several reasons. First, and most importantly, we wanted to test OKN with stimuli more or less identical to those used to test the neuronal responses in the NOT. Second, it would be easy to test OKN in directions other than horizontal. Third, more subtle changes in the central pathways mediating OKN could be revealed because the contrast of the pattern could be easily varied. Of course, the same stimulus parameters were used to compare normal and experimental animals. For a comparison of OKN with responses of NOT cells we investigated the relation of eye velocity to stimulus velocity for the two eyes separately, and together. Eye velocity was calculated from the eye position signal by on-line computer analysis.

The most striking result in testing the OKN of normal adult cats was the asymmetry between responses to nasal stimulus movement and to temporal stimulus movement when only one eye was stimulated. Temporonasal movements were more effective than nasotemporal. With binocular stimulation, of course, no such asymmetry was seen. The varying degree of this superiority in the gain of response to nasal stimulus movement in 10 cats is shown in Figure 8-6. We have no explanation for this variation between cats, but as is shown later, decortication as well as visual deprivation or diverging strabismus significantly increases this asymmetry.

From the neurophysiological analysis a straightforward explanation for this asymmetry can be provided. As stated previously, each eye alone strongly activates all units in the contralateral NOT when the visual stimulus moves temporonasally, but only 40% of the units are activated in the ipsilateral nucleus when the stimulus moves nasotemporally. This differential activation in the number of units and response strength in the contralateral NOT and ipsilateral NOT could lead to the

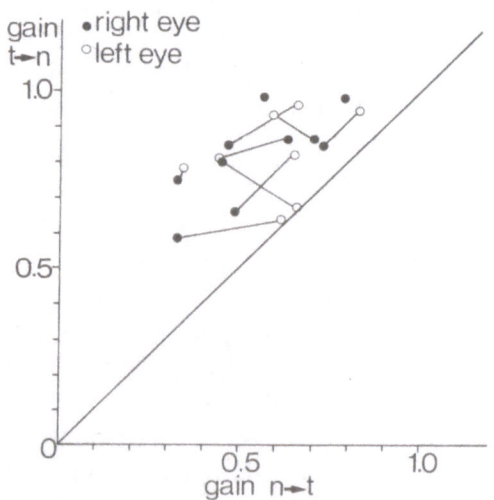

Fig. 8-6. Superiority of temporonasal (*T-N*) over nasotemporal (*N-T*) stimulus movement to elicit OKN. Gain is defined as the ratio of eye velocity to stimulus velocity. When both eyes were measured in the same cat the values are joined by lines. It was our observation in 10 cats that the two eyes could have quite different gains.

different quality of OKN (frequency and gain) elicited monocularly in the nasotemporal and in the temporonasal directions, respectively.

Our quantitative neurophysiological results, combined with what is known about the anatomy of the input to NOT, form the basis for the model proposed in Figure 8-7. Two components normally contribute to the NOT activity: a strongly contralaterally dominated retinal input and binocular input from the visual cortex. In this model each eye projects to the ipsilateral NOT only via the binocular cells in the cortex, whereas the contralateral eye has two ways to influence NOT cells, one by direct retinal fibers and the other through binocular cells in the visual cortex. We assume that in monocular units the retinal input is dominant and in binocular units the cortical input is dominant. The ratio of monocular to binocular units is .6:.4. This is taken as the relative strength of retinal versus cortical input. With horizontal movement 80% of the NOT output is supplied by the activation of one nucleus and 20% by the inhibition of the other nucleus (Figure 8-5). Thus, the

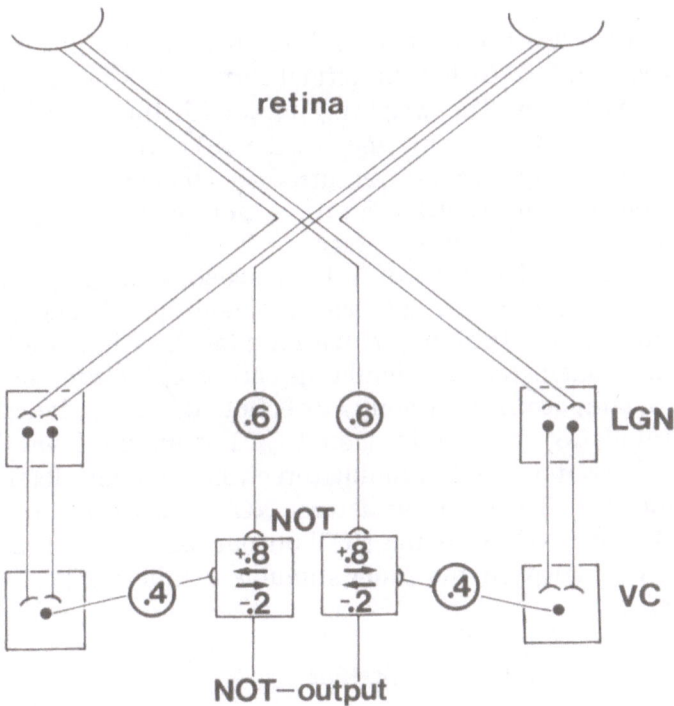

Fig. 8-7. Input–output model for NOT. Each NOT is reached by direct retinal fibers from the contralateral eye (.6) and by binocularly driven axons from the ipsilateral visual cortex (.4). The activity of NOT can increase (+.8) with temporonasal movement or decrease (−.2) with nasotemporal movement presented to the contralateral eye. Tables 8-1–8-3 are calculated from this model. *LGN*, lateral geniculate nucleus; *VC*, visual cortex.

Table 8-1. Predicted NOT Input and Output and Measured OKN Gain

Stimulation	Left NOT		Right NOT		L − R NOT Output	OKN gain
	Cort.	Ret.	Cort.	Ret.		
Binocular						
←	.3	.5	−.1	−.1	1.0	1.0
→	−.1	−.1	.3	.5	−1.0	1.0
Right eye						
←	.3	.5	−.1	−	.9	.9
→	−.1	−.1	.3	−	− .5	.6
Left eye						
←	.3	−	−.1	−.1	.5	.6
→	−.1	−	.3	.5	− .9	.9

Abbreviations: Cort., cortical contribution; Ret., retinal contribution; L − R NOT, discharge rate in the left NOT minus discharge rate in the right NOT as absolute values; arrows, direction of stimulus movement; −, decrease in NOT activity.

weighting factor for the contribution to the NOT output made up by excitation is .5 (.8 × .6) for the retinal and .3 (.8 × .4) for the cortical input from each eye. The weighting factor for the contribution from inhibition is .1 (.2 × .6) retinal and .1 (.2 × .8) cortical for each eye. To keep the model simple we do not introduce changing weighting factors as a function of stimulus velocity; for a more complete description of the system, however, this is necessary.

The maximal NOT output of 1.0 is reached with binocular stimulation only. Then one NOT is fully activated and the other fully inhibited by the retinal and cortical inputs. With monocular stimulation the NOT output is dependent on the stimulus direction and becomes asymmetrical. With a temporonasal stimulus direction the contralateral NOT is fully activated (.8) and the ipsilateral NOT is inhibited via the cortex (.1). With a nasotemporal stimulus direction the contralateral NOT is fully inhibited (.2) and the ipsilateral NOT is activated only via the cortex (.3). These values of the NOT output are matched more or less by the gain of OKN in the same stimulus conditions (Table 8-1 and Figure 8-5).

OKN in Animals with Cortical Lesions

This model can now be used to predict the effect of adult cortical lesions or developmental manipulation on the OKN of the cat. Lesions of the visual cortex were performed in six adult cats. All of areas 17, 18, 19, and the suprasylvian gyrus were removed. Gain of the Clare Bishop area OKN as a function of stimulus velocity is presented for one animal in Figure 8-8. In this animal OKN was tested before and after

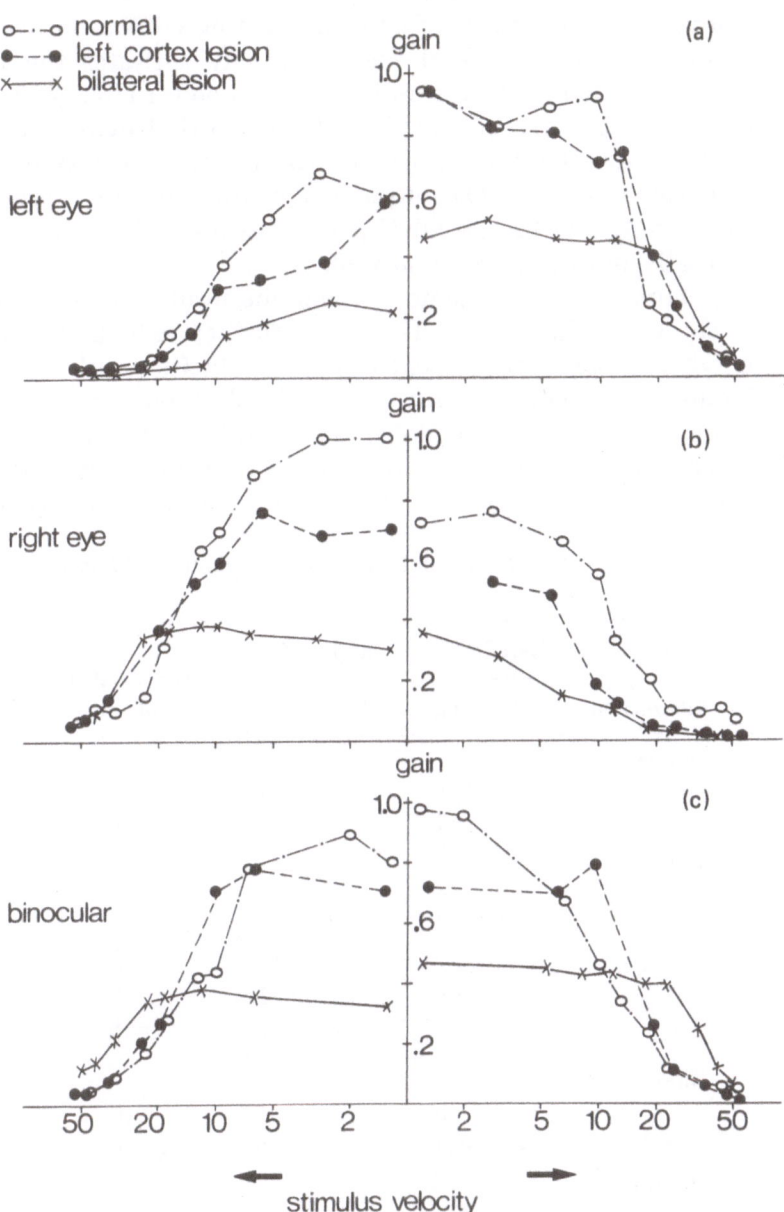

Fig. 8-8. Gain of OKN in a normal, first unilaterally and then bilaterally decorticated cat. Movement of the pattern from right to left is plotted to the left, and pattern movement from left to right is presented on the right. (a) Stimulus presented to the left eye. (b) Stimulus presented to the right eye. (c) Binocular stimulation.

the lesions, and later the animal was used for neurophysiological analysis of the NOT responses. With the lesion of the left cortex, leftward stimulus movement was not as effective in eliciting OKN as in the normal cat. This direction in visual space normally activates the NOT ipsilateral to the lesion. The clearest effect was seen for leftward movement presented to the eye ipsilateral to the lesion (left eye). The right eye, which has a direct retinal projection to the left NOT, was also less effective. In Table 8-2 the NOT output values from our model are summarized and compared to the OKN gains measured after left cortex lesion. Qualitatively they correspond well.

The right cortex was also removed, and the results of the bilateral lesion are shown in Figure 8-8. There is a further reduction in OKN gain, and OKN is practically absent at higher velocities with nasotemporal stimulus movement. As with the unilateral lesion, the additional lesion of the right cortex has a global effect on both directions tested through each eye. The NOT values predicted from our model for this preparation are given in Table 8-2. In summary, decortication strongly

Table 8-2. Effect of Cortical Lesions on Predicted NOT Values and Measured OKN Gains

Cortical lesion	Stimulation	Left NOT		Right NOT		L − R NOT output	OKN gain
		Cort.	Ret.	Cort.	Ret.		
Left	Binocular						
	←	—	.5	−.1	−.1	.7	.7
	→	—	−.1	.3	.5	−.9	.8
	Right eye						
	←	—	.5	−.1	—	.6	.7
	→	—	−.1	.3	—	−.4	.5
	Left eye						
	←	—	—	−.1	−.1	.2	.3
	→	—	—	.3	.5	−.8	.8
Bilateral	Binocular						
	←	—	.5	—	−.1	.6	.4
	→	—	−.1	—	.5	−.6	.5
	Right eye						
	←	—	.5	—	—	.5	.4
	→	—	−.1	—	—	−.1	.2
	Left eye						
	←	—	—	—	−.1	.1	.2
	→	—	—	—	.5	−.5	.5

Abbreviations: See Table 8-1.

reduces the nasotemporal direction of monocularly elicited OKN, which coincides with the preferred direction of cells in the ipsilateral NOT. The pathway activating the ipsilateral NOT from each eye seems entirely dependent on a relay through the cortex. In agreement with this, in six cats with large lesions of the visual cortex we found that only 1 of 40 units in the NOT could be activated from both eyes. Normal direction-specific responses could be elicited only at low velocities less than $10°$/sec) through the contralateral eye. The main differences between normal and decorticated animals was therefore the loss of the ipsilateral input and of the sensitivity to visual stimuli moving faster than $10°$/sec. These properties seem to require an intact corticopretectal projection.

OKN in Visually Deprived Cats

To determine how visual deprivation influences the development of properties of NOT cells, experiments were carried out on six cats that had one eye closed by lid suture during the first year after birth. The properties of NOT neurons remained largely unaltered, except that neurons in each NOT were now driven only through the contralateral eye regardless of whether it was the deprived or the nondeprived eye. In the normally reared animal, in contrast, one-half of the units can be driven from either eye. Direction specificity was examined in the responses to either the deprived or nondeprived eye. Responses were pooled from all NOT cells recorded in one cat. When the NOT cells were stimulated through the contralateral eye there was a clear direction-specific response which was absent when the ipsilateral eye was stimulated (Figure 8-9).

Why does the nondeprived eye not have access to the ipsilateral NOT? In a monocularly deprived cat the nondeprived eye controls most of the units in area 17 and 18 in the ipsilateral cortex and, by means of the corticotectal fibers, most the the units in the ipsilateral superior colliculus as well (Hoffmann & Cynader, 1977; Hoffmann & Sherman, 1974, 1975; Wiesel & Hubel, 1963b). Furthermore, we have shown that in normal cats the visual cortex provides a strong excitatory input to NOT cells, probably via axon collaterals of the so-called corticotectal axons (Palmer & Rosenquist, 1974; Schoppmann, 1981). In the present study, electrical stimulation of the visual cortex ipsilateral to the nondeprived eye also yielded a clear, strong spike discharge in all NOT units, even though they could be driven visually only by the contralateral, deprived eye. Double stimulation with intervals of 4–100 msec showed no sign of inhibition following the short-latency excitation. Therefore, an excitatory corticopretectal pathway must have been functioning as in the normal animal, and we have to postulate that the dis-

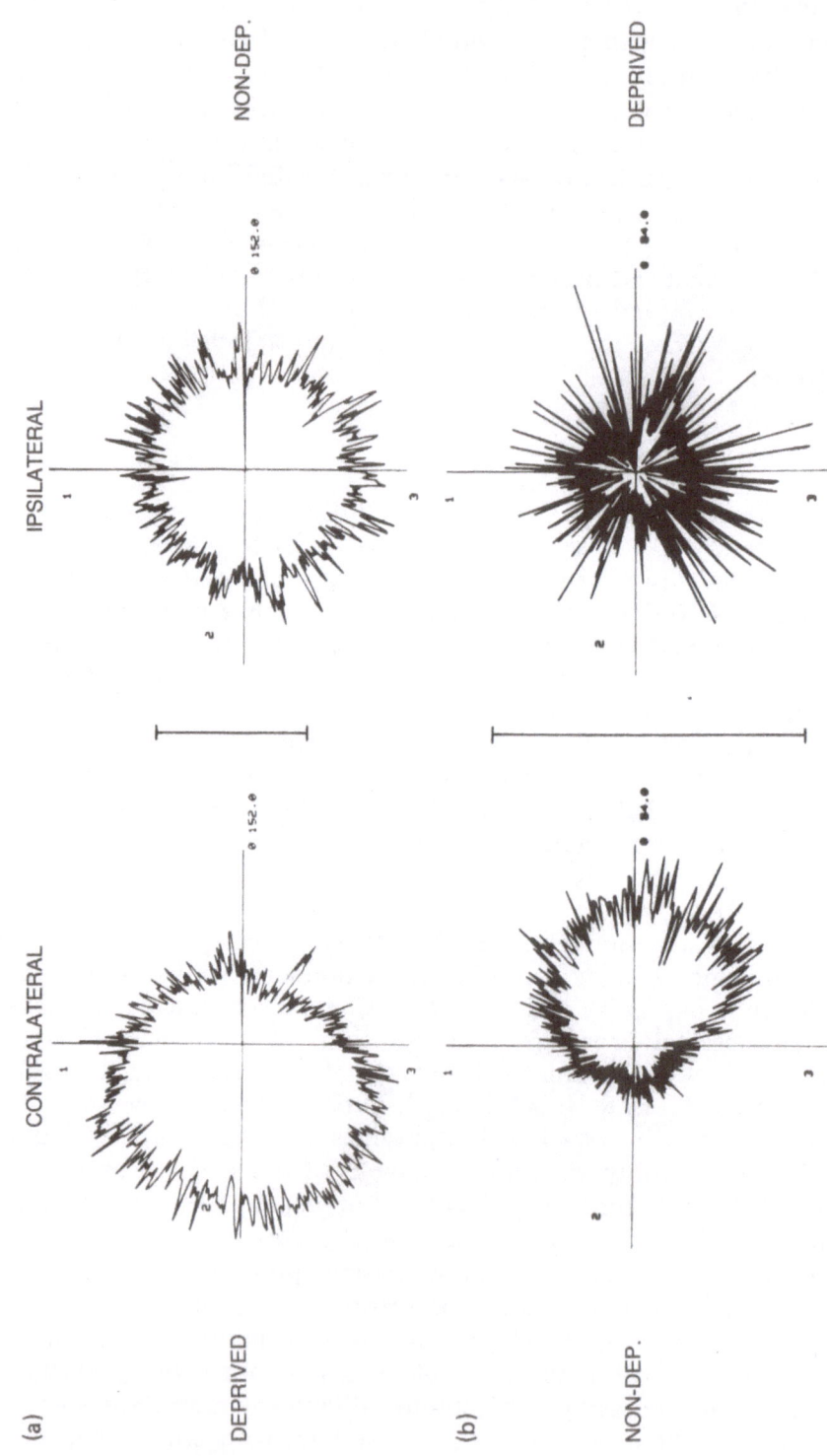

ruption of the pathway from the nondeprived eye to the ipsilateral NOT was in the visual cortex or that the corticopretectal synapses were not driving the NOT cells above threshold during visual stimulation.

No qualitative difference in direction specificity between the deprived eye and the nondeprived eye was observed. NOT cells driven by the deprived eye failed to respond to stimulus velocities higher than $10°/sec$. NOT cells driven by the nondeprived eye showed direction-specific responses to stimuli moving with velocities of up to $100°/sec$, as in normal cats.

From these neurophysiological results one might predict that monocular visual deprivation has the same effect on OKN as a bilateral cortical lesion, because in both NOTs no cells could be driven by the ipsilateral eye, indicating a lack of visual cortex input. The data presented in Figure 8-10 show that the results are somewhat different than expected. The nondeprived eye is more effective in eliciting OKN than the deprived eye, and the asymmetry between the nasotemporal direction and the temporonasal direction is more pronounced in the deprived than in the nondeprived eye. This can be interpreted in terms of our model in the following way (Table 8-3): the nondeprived eye has connections to the contralateral NOT via the direct retinal fibers and through the visual cortex, but it cannot inhibit the ipsilateral NOT—we did not find any NOT cells influenced by the ipsilateral eye. This permits an almost normal OKN to temporonasal stimulus movement presented to the nondeprived eye. In contrast, nasotemporal stimulus movement presented to the nondeprived eye cannot activate the ipsilateral NOT because there is no cortical input, however, it may strongly inhibit the contralateral NOT, thus creating a NOT output signal of .2. The deprived eye only has its retinal connections to the contralateral NOT and thus can elicit an OKN to temporonasal stimulus directions. The gain of this OKN is low because the cortical contribution to the output of NOT is completely missing. The response of the deprived eye to nasotemporal stimulus movement is very low, as expected, because the only effect this stimulus movement can have on the NOT output is the inhibition of the contralateral NOT via the retinal fibers.

Monocular visual deprivation can thus be seen as equivalent to an ipsilateral decortication for the nondeprived eye and to a bilateral decor-

---

Fig. 8-9. Average responses displayed in polar coordinates for neurons recorded in the two NOTs in a monocularly deprived animal. Response strength in relation to a large-area, random-dot pattern moving on a circular path is given for the various directions by the vector from the origin to the curve. Vertical calibration bars: 100 spikes/sec. (a) Pooled responses of 17 units in the left NOT. (1) Stimulation of the contralateral (deprived) eye. (2) Stimulation of the ipsilateral (nondeprived) eye (not different from the spontaneous activity). (b) Pooled responses of 13 units in the right NOT. (1) Stimulation of the contralateral (nondeprived) eye. (2) Stimulation of the ipsilateral (deprived) eye (not different from the spontaneous activity).

tication for the deprived eye. Note, however, that the OKN elicited through the deprived eye is much stronger than the OKN in response to a stimulus in the temporonasal direction elicited through either eye after bilateral decortication. This may indicate that some corticopretectal cells in the cortex contralateral to the deprived eye can still be driven from the deprived eye.

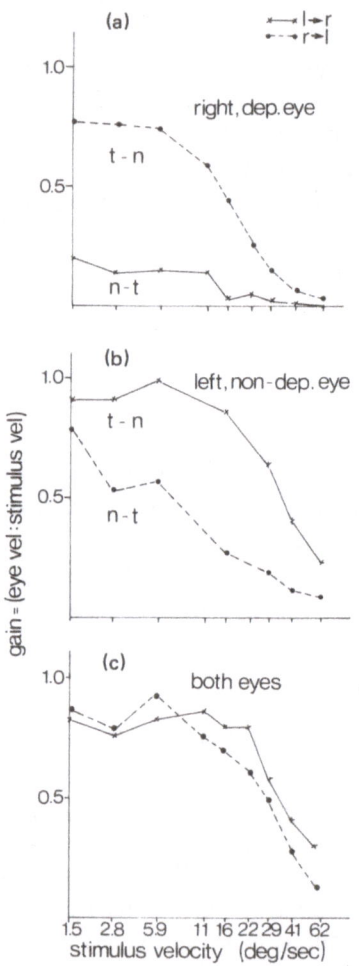

Fig. 8-10. Gain of OKN in monocularly deprived cats. (a) Stimulation of right (deprived) eye. (b) Stimulation of left (nondeprived) eye. (c) Stimulation of both eyes *T–N* and *N–T*: temporonasal and nasotemporal movement of the pattern. Rightward movement (*solid line*) is optimal for the left eye and leftward movement (*broken line*) is optimal for the right eye. The inferiority of the nasotemporal pattern movement in eliciting OKN can be seen regardless of which eye views the moving stimulus. The deprived eye, however, is weaker in driving OKN in both directions.

Table 8-3. Effect of Monocular (Right Eye) Visual Deprivation on Predicted NOT Values and Measured OKN Gain

Stimulation	Left NOT Cort.	Ret.	Right NOT Cort.	Ret.	L − R NOT Output	OKN gain
Binocular						
←	—	.5	−.1	−.1	.7	.7
→	—	−.1	.3	.5	−.9	.8
Right eye						
←	—	.5	—	—	.5	.6
→	—	.1	—	—	−.1	.1
Left eye						
←	—	—	.1	.1	.2	.3
→	—	—	.3	.5	−.8	.8

Abbreviations: See Table 8-1.

Conclusions

We have presented a model in which the activity difference in the left and the right NOT is related to the gain of OKN. The activity difference between the two nuclei always reflects an increase in activity in one nucleus and a decrease in the opposite one. A decrease in activity in one nucleus compared to the spontaneous activity in the other seems to be sufficient to elicit OKN at very low gains. We have tested this model with binocular and monocular stimulation in the normal animal, in animals with unilateral and bilateral cortical lesions, and in animals with monocular deprivation throughout the first year of life. The resultant characteristic modifications of OKN have their counterparts in changes of activity differences between left and right NOT. Retinal and cortical contributions play a different role in temporonasal and nasotemporal directions of stimulus movement when presented to one eye. Cortical lesions or visual deprivation reveal a retinal dominance during temporonasal stimulus movement and a cortical dominance for the slow phase of OKN during nasotemporal stimulus movement.

Acknowledgments

I would like to thank Ms. C. Markner and Dr. A. Meyer-Koll, who recorded and analyzed most of the data in the eye movement experiments. Mr. H. P. Huber wrote the computer programs. I am thankful to Ms. L. Harrison for secretarial assistance.

This work was supported by Grants Ho 450/7+12 of the Deutsche Forschungsgemeinschaft.

References

Berman, N. Connections of the pretectum in the cat. *Journal of Comparative Neurology*, 1977, *174*, 227–254.

Collewijn, H. Direction-selective units in the rabbit's nucleus of the optic tract. *Brain Research*, 1975, *100*, 489–508. (a)

Collewijn, H. Oculomotor area in the rabbit's midbrain and pretectum. *Journal of Neurobiology*, 1975, *6*, 3–22. (b)

Hanker, J. S., Yates, P. E., Metz, C. B., & Rustioni, A. A new specific, sensitive and non-carcenogenic reagent for the demonstration of horseradish peroxidase. *Histochemistry*, 1977, *9*, 789–792.

Hoffmann, K. -P. Optokinetic nystagmus and single cell responses in the nucleus tractus opticus after early monocular deprivation in the cat. In R. D. Freeman (Ed.), *Developmental neurobiology of vision*. New York: Plenum, 1979, pp. 63–72.

Hoffmann, K. -P., Neuronal responses related to optokinetic nystagmus in the cat's nucleus of the optic tract. In Fuchs & Becker (Eds.), *Progress in oculomotor research*. Amsterdam: Elsevier North-Holland, 1981.

Hoffmann, K. -P., Behrend, K., & Schoppmann, A. A direct afferent visual pathway from the nucleus of the optic tract to the inferior olive in the cat. *Brain Research*, 1976, *115*, 150–153.

Hoffmann, K. -P., & Cynader, M. Functional aspects of plasticity in the visual system of adult cats after early monocular deprivation. *Philosophical Transactions of the Royal Society of London; B.*, 1977, *278*, 411–424.

Hoffmann, K. -P., & Schoppmann, A. Retinal input to direction selective cells in the nucleus tractus opticus of the cat. *Brain Research*, 1975, *99*, 359–366.

Hoffmann, K. -P., & Schoppmann, A. A quantitative analysis of the direction-specific response of neurons in the cat's nucleus of the optic tract. *Experimental Brain Research*, 1981, *42*, 1–12.

Hoffmann, K. -P., & Sherman, S. M. Effects of early monocular deprivation on visual input to cat superior colliculus. *Journal of Neurophysiology*, 1974, *37*, 1276–1286.

Hoffmann, K. -P., & Sherman, S. M. Effects of early binocular deprivation on visual input to cat superior colliculus. *Journal of Neurophysiology*, 1975, *38*, 1049–1059.

Maekawa, K., & Simpson, J. I. Climbing fiber responses evoked in vestibulocerebellum of rabbit from visual system. *Journal of Neurophysiology*, 1973, *36*, 649–666.

Mesulam, M. -M. Tetramethylbenzadine for horse-radish peroxidase neurohistochemistry: Incubation parameters and visibility. *Journal of Histochemistry and Cytochemistry*, 1978, *26*, 106–107.

Mizuno, N., Nakamura, Y., & Iwahori, N. An electron microscope study of the dorsal cap of the inferior olive in the rabbit, with special reference to the pretecto-olivary fibers. *Brain Research*, 1974, *77*, 385–395.

Palmer, L. A., & Rosenquist, A. C. Visual receptive fields of single striate cortical units projecting to the superior colliculus in the cat. *Brain Research*, 1974, *67*, 27–42.

Precht, W., & Strata, P. On the pathway mediating optokinetic responses in vestibular nuclear neurons. *Neuroscience*, 1980, *5*, 777–787.

Robinson, D. A. A method of measuring eye movement using a scleral search coil in a magnetic field. *IEEE Transactions on Biomedical Electronics*, 1963, *BME-10*, 137-145.

Schoppmann, A. Projections from areas 17 and 18 of the visual cortex to the nucleus of the optic tract. *Brain Research*, 1981, *223*, 1-18.

Takeda, T., & Maekawa, K. The origin of the pretecto-olivary tract. A study using the horseradish peroxidase method. *Brain Research*, 1976, *117*, 319-325.

Updyke, B. V. Topographic organization of the projections from the cortical areas 17, 18 and 19 onto the thalamus, pretectum and superior colliculus in the cat. *Journal of Comparative Neurology*, 1977, *173*, 81-122.

van Hof-van Duin, J. Direction-preference of optokinetic responses in monocularly tested normal kittens and light deprived cats. *Archives Italiennes de Biologie*, 1978, *116*, 471-477.

Walberg, F., Nordby, T., Hoffmann, K. -P., & Holländer, H. Olivary afferents from the pretectal nuclei in the cat. *Anatomy and Embryology*, 1981, *161*, 291-304.

Wiesel, T. N., & Hubel, D. H. Effects of visual deprivation on morphology and physiology of cells in the cat's lateral geniculate body. *Journal of Neurophysiology*, 1963, *26*, 978-993. (a)

Wiesel, T. N., & Hubel, D. H. Single-cell responses in striate cortex of kittens deprived of vision in one eye. *Journal of Neurophysiology*, 1963, *26*, 1003-1017. (b)

Chapter 9

Development of Optokinetic Nystagmus and Effects of Abnormal Visual Experience During Infancy

Janice R. Naegele and Richard Held

Shortly after birth, a human infant exhibits binocular optokinetic nystagmus (OKN) in response to patterns moving horizontally in both directions (Fantz, Ordy, & Udelf, 1962; Gorman, Cogan, & Gellis, 1957). When one eye of the infant is covered, however, OKN can only be elicited by a stimulus moving in the temporonasal direction (Atkinson, 1979). This early OKN asymmetry diminishes over the first 3–5 months until OKN following is elicited equally well by stimuli moving either away from or toward the midline (Atkinson, 1979; Naegele & Held, 1980, 1982). A similar progression toward symmetrical monocular OKN occurs in normal kittens (van Hof-van Duin, 1976, 1978). As in the case of normal human infants, the early deficit consists of an impairment in the ability of the organism to exhibit optokinetic following when patterns move in the nasotemporal direction. In contrast to species that show this pattern of optomotor development, other species, such as the rabbit, guinea pig, and some birds and lizards, exhibit a permanent optokinetic bias for movement in the temporonasal direction (Fukuda & Tokita, 1957; Huizinga & Meulen, 1951; Tauber & Atkin, 1968).

The existence of temporonasal OKN in all species at birth, in combination with the postnatal development of nasotemporal OKN in some species, suggests an ontogenetic recapitulation of evolutionary events that led to separate visual pathways subserving the visual component of visuovestibular reflexes such as OKN. Several lines of research have contributed to our understanding of the phylogeny and ontogeny of OKN asymmetry. Current knowledge of developmental anatomy and visual and vestibulo-ocular behaviors in the human supports but cannot yet distinguish among the various proposals to account for OKN asymmetry.

Distinctions between the cortical and subcortical pathways subserving the optomotor reflex date from the observations of Ter Braak, who found residual binocular OKN after decortication in rabbits, dogs, and monkeys (Rademaker & Ter Braak, 1948). However, direction-specific optokinetic deficits occur after removal of the visual cortex in the cat when monocular stimulation is used (Wood, Spear, & Braun, 1973). With bilateral destruction of the visual cortex and transection of one optic tract, cats exhibit monocular OKN only in the temporonasal direction at low stimulus velocities (Montarolo, Precht, & Strata, 1981). The presence of the visual cortex improves the match between eye velocity and stimulus velocity (i.e., the gain) with higher velocity stimuli moving in either direction.

Manipulations of the visual environment early in life have led to the proposal that the development of OKN symmetry is in some way dependent upon the development of cortical binocularity or stereopsis (van Hof-van Duin, 1976, 1978). The OKN reflex has been studied in visually deprived animals and humans with a history of interocular misalignment from childhood, with the finding that these early insults to the visual system may impair only nasotemporal OKN (Crone, 1977; Schor & Levi, 1980; van Hof-van Duin, 1978). Early disruption of visual input can reduce or eliminate cortical binocularity in the cat (Hubel & Wiesel, 1965) and monkey (Von Noorden & Crawford, 1978), and binocular convergence onto pretectal neurons known to mediate aspects of OKN (Harris, Lepore, Guillemot, & Cynader, 1980; Hoffman, Behrend, & Schoppmann, 1977). (A more specific discussion of physiological changes in the cat's pretectum following monocular deprivation and the relation of these changes to modifications of the optokinetic reflex appears in Chapter 8, this volume).

Human OKN Development

In order to define some aspects of early visual experience that contribute to the achievement of OKN symmetry in the human, we recorded the eye movements of normal infants and of infants and young children with impaired interocular alignment (strabismic esotropia and strabismic exotropia) or blurred retinal images (unilateral cataract). A cross-sectional comparison of the developmental time course for monocular OKN in the normal and clinical populations allowed us to ask the following: What aspect of the OKN pattern of eye movements accounts for increasing OKN symmetry? Does abnormal visual experience alter this aspect of the eye movement pattern? Does any disruption of visual experience suffice to alter the development of OKN symmetry? What are the critical periods during which the OKN reflex may be influenced by interactions with the environment?

Our study was designed to measure the relative gain of the slow phases of the OKN reflex. Velocity changes in OKN eye movements were not evaluated in Atkinson's (1979) study of infant OKN because the data were collected by an observer who watched the infants' eyes during optokinetic stimulation.

In order to assess optokinetic responses, we projected vertical gratings moving at a velocity of 25°/sec onto a semicylindrical screen placed around the infant (Figure 9-1). Each infant was tested once, and trials in which the infant was fussy or drowsy were discarded since it is known that changes in state of alertness markedly affect the slow-phase gain of OKN (Collins, Schroeder & Elam, 1975). Of 48 infants tested, 24 were eliminated from the final analysis for this reason.

EOG recordings of the eye movements were made using silver–silver chloride electrodes placed bitemporally with the reference electrode placed on the forehead. The direct current EOG signal was monitored and simultaneously recorded on frequency-modulated tape. Testing was performed in a semidarkened room with the infant facing the screen,

Fig. 9-1. OKN apparatus with infant seated in reclining seat. The infant sat with the eyes centered horizontally and vertically with respect to the screen. At the infant's viewing distance, the stimulus subtended 170° horizontally and 60° vertically. Moving, high-contrast vertical gratings were rear projected onto the semicircular Polacoat screen.

seated on the parent's lap or on an upright infant seat, with a pacifier, bottle, or the parent's finger serving to stabilize head movements and eliminate vestibulo-ocular interactions. Details of this procedure were described by Naegele and Held (1982). The taped sessions were later scored by a naive experimenter who counted the number of nystagmic beats during 10-second intervals and calculated their frequency. We found that absolute calibration of the eye movements during EOG recording was unreliable because of poor fixation by the infant. Therefore, we limited our comparisons to within each infant's record. This approach assumes that equal voltage changes are produced by equal displacements of the eye within the range over which eye movements were recorded, irrespective of direction. Cumulative slow-phase displacement (velocity) records were constructed by removing fast-phase eye movements (saccades) from the EOG records. This allowed a comparison of the relative velocities (slopes of the curves) of the slow phases of monocular OKN under conditions of temporonasal and nasotemporal stripe motion. To this end, the slow-phase slopes from trials with nasotemporal and temporonasal directions of motion were used to calculate symmetry ratios (slow-phase nasotemporal slope/slow-phase temporonasal slope).

Normative Data

In Figure 9-2 the data for four representative infants illustrate the appearance of the EOG traces in infants of different ages. In Figure 9-2a, each eye of an 8-week-old was tested under monocular conditions. The EOG traces show a marked asymmetry between the temporonasal and nasotemporal directions of stripe movement. With temporonasal stripe movement, the OKN was vigorous (72 beats/min), in contrast to trials in which the stripes moved nasotemporally (17 beats/min). The slopes of the slow phase also varied during each 30-second trial, suggesting a poor match between the eye velocity and the constant stimulus velocity.

Two binocular trials and two monocular trials for each infant appear in Figures 9-2b through d. In Figure 9-2b the EOG traces of a 14-week-old are shown; the binocular frequency and slow-phase slopes do not differ with stimulus direction. When this infant was tested monocularly, OKN was less frequent than during binocular stimulation and was asymmetrical. The temporonasal frequency was 70 beats/min and the nasotemporal frequency was 60 beats/min, on the average. The small saccades that occur in the direction of the slow phases have previously been noted in very young infants during tests of smooth pursuit (Aslin, 1981). Aslin found that a low gain in infant smooth pursuit eye movements was often associated with the presence of "catch-up" saccades. Our observations, along with those of Aslin, suggest that the saccades in

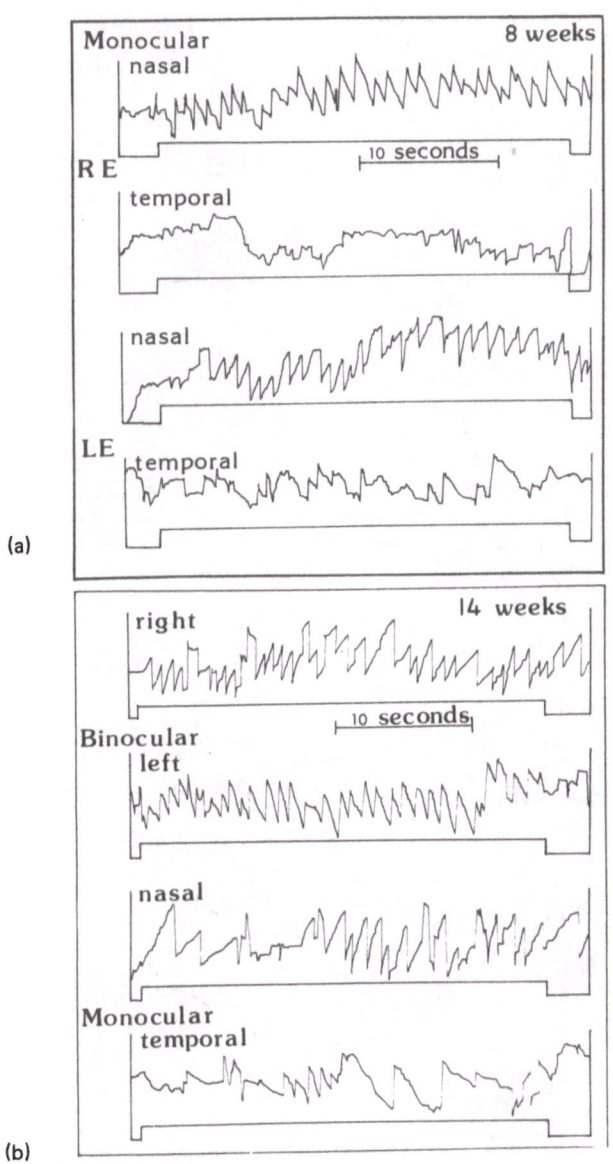

Fig. 9-2. Sample EOG traces from four infants of different ages. (a) Monocular EOG records from both the left and right eyes of an 8-week-old infant. In the two upper traces, the right eye alone was tested with temporonasal (*T-N*) and naso-temporal (*N-T*) motion. In the two lower traces the left eye alone was stimulated with each direction of stimulus motion. (b) Traces from a 14-week-old infant. The two upper traces show OKN with both eyes viewing either rightward or leftward moving gratings. The two lower traces show monocular trials with the left eye view-ing each direction of stimulus motion. (c) Traces from a 17-week-old infant with right eye viewing stimulus (next page). (d) Traces from a 19-week-old infant with left eye viewing stimulus.

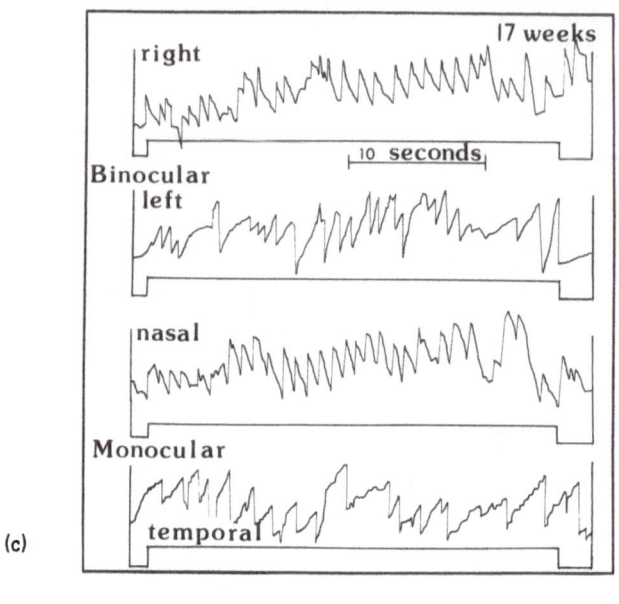

(c)

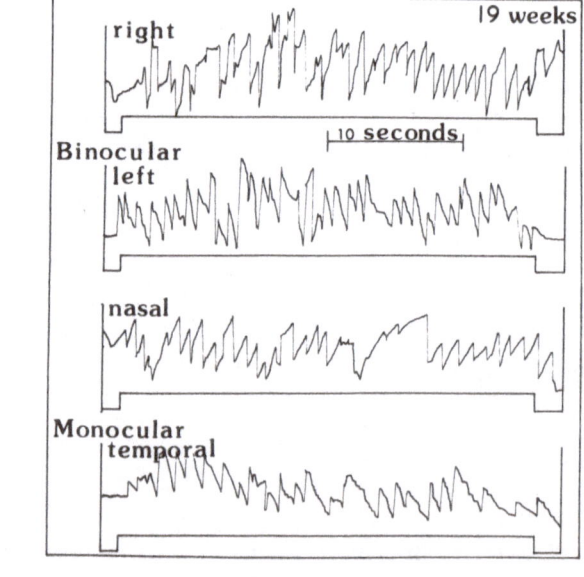

(d)

Fig. 9-2 (continued)

the direction of stimulus motion are generated as a result of excessive retinal error signal.

Sample EOG traces from a 17-week-old are presented in Figure 9-2c. The frequency of nystagmus is approximately equal for both binocular leftward and rightward directions of movement and for the temporonasal monocular condition. When monocular responses to nasotemporal motion were tested, there was a reduction in both the frequency and the slope of the slow-phase movements.

Sample traces from a 19-week-old are presented in Figure 9-2d. The frequency with binocular stimulation is slightly greater than the frequency of monocular OKN. However, when the monocular responses to each direction of motion are compared, the frequently is approximately equal.

In Figures 9-3a through c displacement curves are shown that were obtained by removing the saccades from the EOG traces of the four infants whose eye movement records are shown in Figures 9-2a through d. These curves show the monocular slow-phase displacements over a 30-second trial (i.e., slow-phase velocities). Cumulative slow-phase displacement curves from the 8-week-old (see Figure 9-2a) are shown in Figure 9-3a. The five upper curves were obtained during trials with temporonasal stripe motion; the five lower curves were obtained during trials with nasotemporal stripe motion. From this family of curves, it can be seen that with temporonasal motion, the slope is relatively constant both within and across trials. However, the slope is consistently lower and more variable for trials with nasotemporal stripe motion.

Sample trials from the three older infants are shown in Figures 9-3b through d. For these infants, the trial of maximal beat frequency under temporonasal and nasotemporal directions of stripe movement was chosen for the comparisons. For the 14-week-old infant, nasotemporal and temporonasal conditions elicited different slow-phase slopes, with the steeper slopes (greater velocity) occurring when the stimulus moved in the temporonasal direction (Figure 9-3b). For the 17-week-old infant, the slopes under the monocular conditions are unequal; again, the slow-phase slopes obtained with temporonasal motion exceeded the slow-phase slopes obtained with nasotemporal motion (Figure 9-3c). For the 19-week-old, the two curves coincide, indicating equal velocities under the two monocular conditions (Figure 9-3d).

Within-infant comparisons of gain show that young infants have lower gains for stimuli moving in the nasotemporal direction than for stimuli moving in the opposite direction. By approximately 20 weeks, the gains equalize. These data do not provide quantitative information about age-related increases in gain, but only about within-infant changes in gain relative to the direction of stimulus motion on the retina.

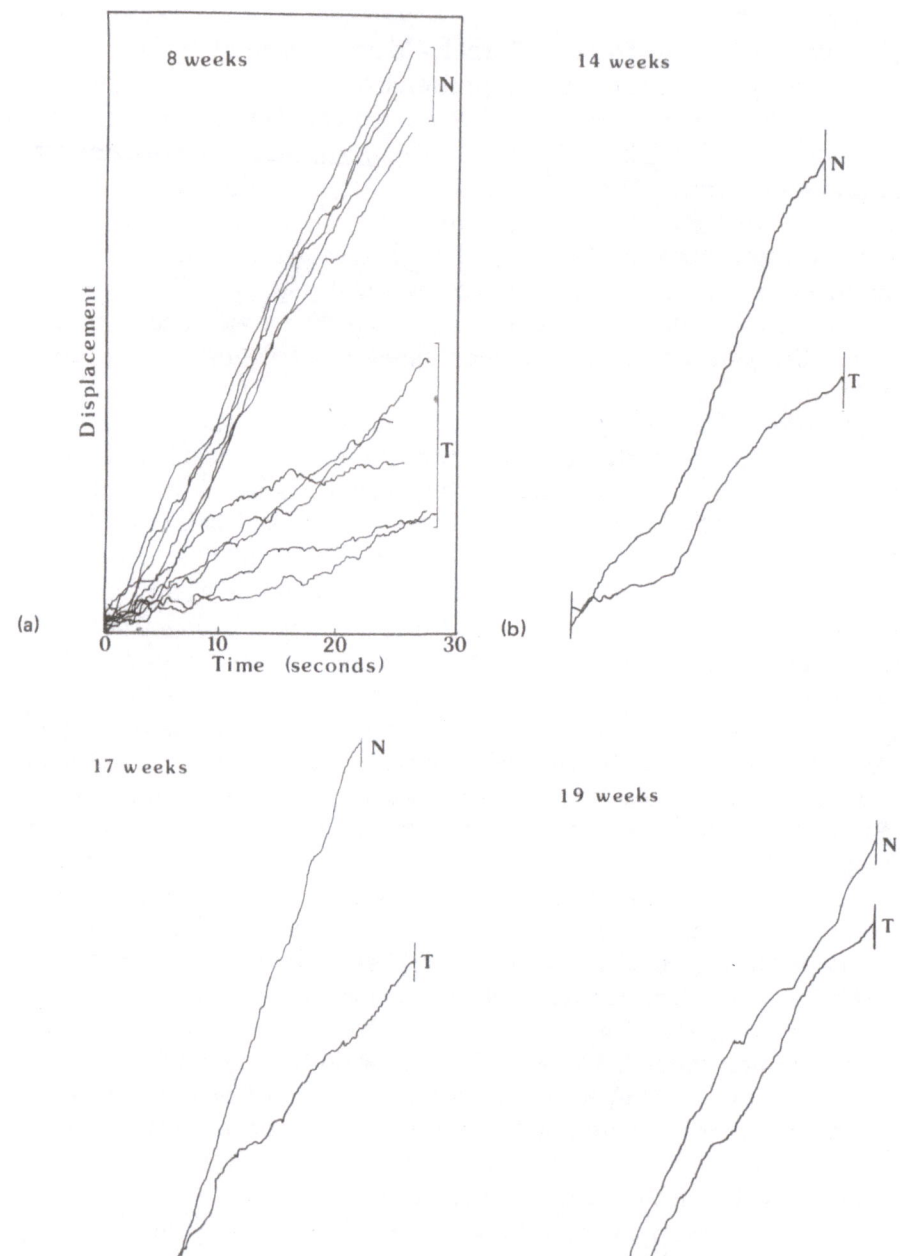

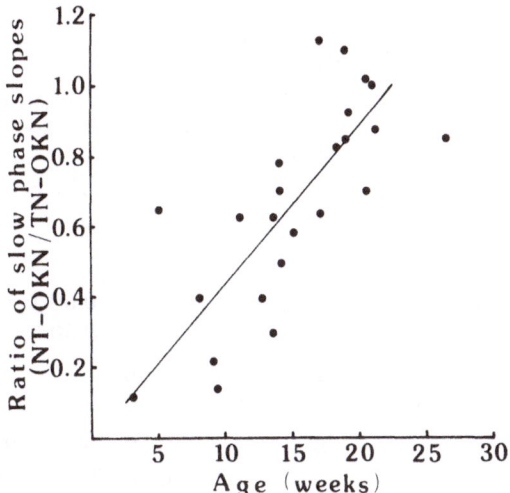

Fig. 9-4. Ratio of the slopes of the slow-phase cumulative displacement curves (the slow-phase slope with nasotemporal motion divided by the slow-phase slope with temporonasal motion) as a function of age. *Dots*, ratio for each of 24 infants, *line*, linear regression line ($r = 0.60$, slope $= 1.1$).

The scatterplot in Figure 9-4 shows symmetry ratios for each of 24 infants categorized by age. The symmetry ratios are smallest in the youngest infants and approach a value of 1 by 20 weeks. Occasionally, the symmetry ratio in an older infant exceeds a value of 1. These data support the general findings of Atkinson (1979) and Atkinson and French (1979) by showing that the slow-phase velocities (shown here by the ratios of the slow-phase slopes) of monocular OKN in response to nasotemporal directions of motion are less than the slow-phase velocities exhibited with temporonasal motion. The direction-dependent disparity between the slow-phase velocities diminishes with increasing age until approximately 20 weeks. We found a longer time course for the development of symmetrical OKN than did Atkinson (1979); our use of a more sensitive recording technique may account for this.

Fig. 9-3. Cumulative displacement curves from infants 8 (a), 14 (b), 17 (c), and 19 (d) weeks of age. Sample monocular trials with the longest succession of optokinetic eye movements were selected for each infant; the fast phases were removed from the records and the remaining slow phases were aligned cumulatively. The family of displacement curves obtained when all monocular trials from an 8-week-old infant are reconstructed this way and superimposed on the same set of axes is shown in (a). The upper curves correspond to trials with temporonasal (T–N) stimulus motion. The lower curves correspond to trials with nasotemporal (N–T) stimulus motion.

Table 9-1: Clinical Data for Infants and Children with Strabismus or Cataract

Child	Clinical diagnosis	Age (weeks)	Cycloplegic refraction	P-L Acuity
SM	Alternating esotropia; fixates with right eye	32	RE: +2.50 −3.00 × 45 LE: +2.50 −5.00 × 35	RE: 20/130 LE: 20/400
JS	Alternating esotropia	42	RE: +2.25 LE: +2.25	RE: 20/200 LE: 20/130
AR	Constant left esotropia	107.5	RE: +3.75 −0.50 × 135 LE: +3.75 −0.50 × 55	RE: 20/65 LE: 20/50
AN	Right cataract	64.5	RE: −8.50 LE: +0.50	RE: —— LE: ——
MC	Right exotropia	58	RE: −0.50 −0.75 × 90 LE: −0.50 −0.75 × 75	RE: —— LE: ——

Abbreviations: P-L, Preferential Looking Method; RE, right eye; LE, left eye.

Clinical Data

In a second study, infants and toddlers with strabismus or unilateral cataract were tested in a clinic. The testing procedure was similar. We recorded eye movements using EOG techniques; however, shorter, 15-second trials were used and all children were tested with a large, random-noise pattern projected onto a flat tangent screen at a velocity of 25°/sec. In order to maintain the attention of the children, the experimenter made sounds from behind the projection screen and animated cartoons were projected between each trial. Monocular testing of acuity occurred within 1 week of the date on which eye movements were recorded. The "preferential looking" method was used (Dobson & Teller, 1978; Gwiazda, Brill, Mohindra, & Held, 1980). The diagnosis, age, cycloplegic refraction, and acuity of five of the infants tested are shown in Table 9-1. The EOG traces in these five patients are shown in Figure 9-5.

In Figure 9-5a, the EOG traces of a 32-week-old infant (SM) are shown. The eyes of this child alternated with an inward deviation (strabismic esotropia); more commonly, the right eye was used for fixation and the left eye deviated. When the acuity of the left eye was

Fig. 9-5. EOG records from five infants and children with strabismus or cataract (see Table 9-1). In each case, several monocular trials with temporonasal (*T-N*) and nasotemporal (*N-T*) motion are shown. (a) Right eye of 32-week-old with strabismus with alternating fixation. (b) Right eye of 42-week-old with esotropia with alternating fixation and a fixation preference for the left eye. (c) Right (*above*) and left (*below*) eyes of 107.5-week-old with left esotropia. (d) Right (*above*) and left (*below*) eyes of 64.5-week-old with right cataract. (e) Left eye of 58-week-old with right exotropia.

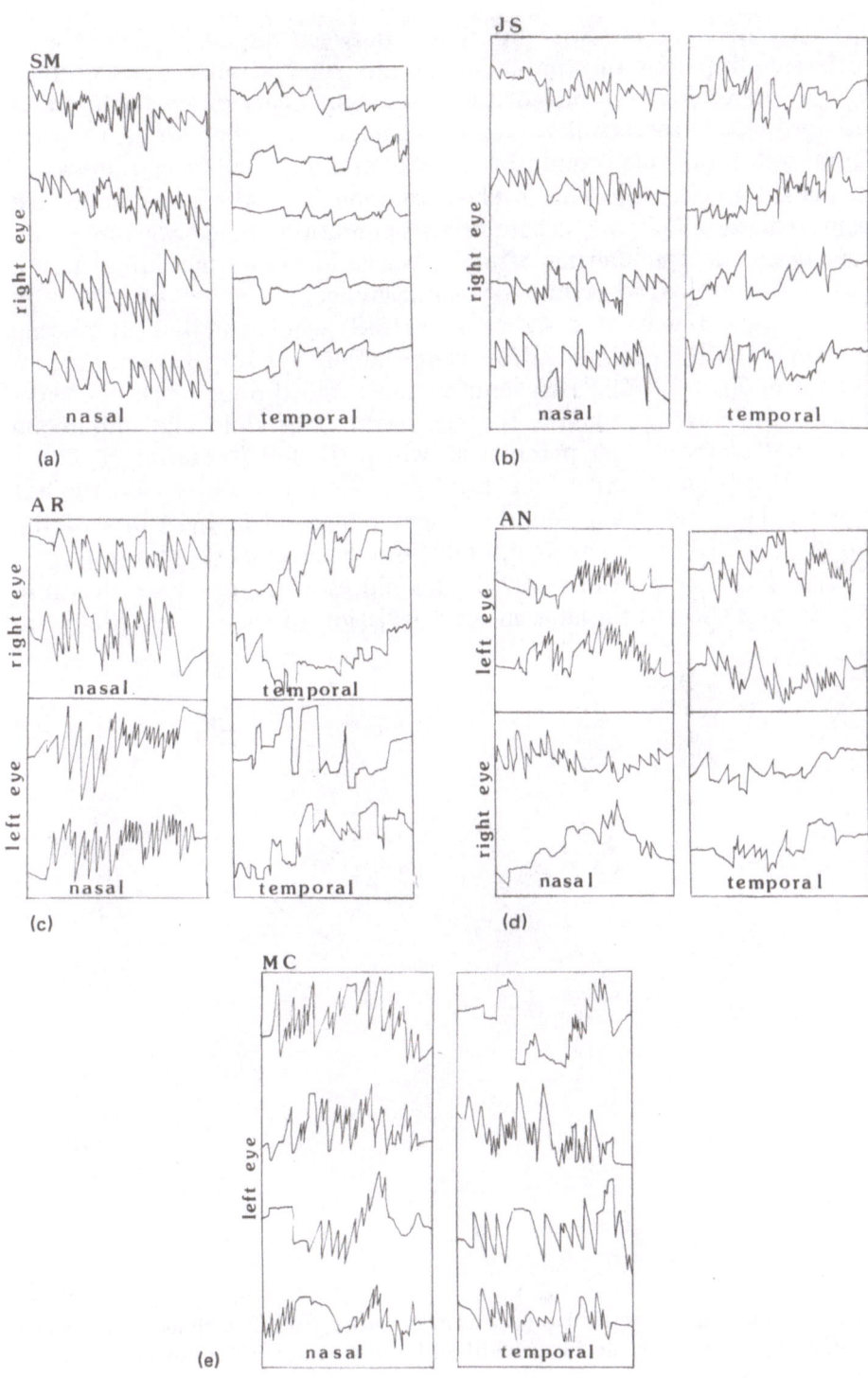

tested, performance was impaired. However, normal acuity measurements for this age were obtained when the right eye was tested. The EOG traces reveal obvious differences between the OKN elicited with different directions of stimulation. Temporonasal OKN was vigorous, yet nasotemporal motion elicited infrequent beats of nystagmus and primarily fast, saccadelike eye movements. Two representative trials from each monocular condition are shown in Figure 9-6a as cumulative displacement records. The marked asymmetry in the slopes reflects a gain reduction for the nasotemporal condition. Since we rarely observed similar asymmetries after 20 weeks in the normal infant population, this prolonged asymmetry is intriguing.

The EOG traces of a 42-week-old (JS) who exhibited alternating esotropia, with a preference for using the left eye for fixation, are presented in Figure 9-5b. The acuities, measured through each eye separately, were 1–2 SD below the age norm. This child had undergone 2 hr/day of monocular patching in which the left (preferred) eye was covered to promote use of the right eye. We tested OKN with the left eye patched and again found a slight reduction in the slope of the slow-phase OKN with nasotemporal stimulus motion (Figure 9-6b).

The EOG traces from a 107.5-week-old esotrope (AR) are shown in Figure 9-5c. She had a large inward deviation of the left eye (35 prism

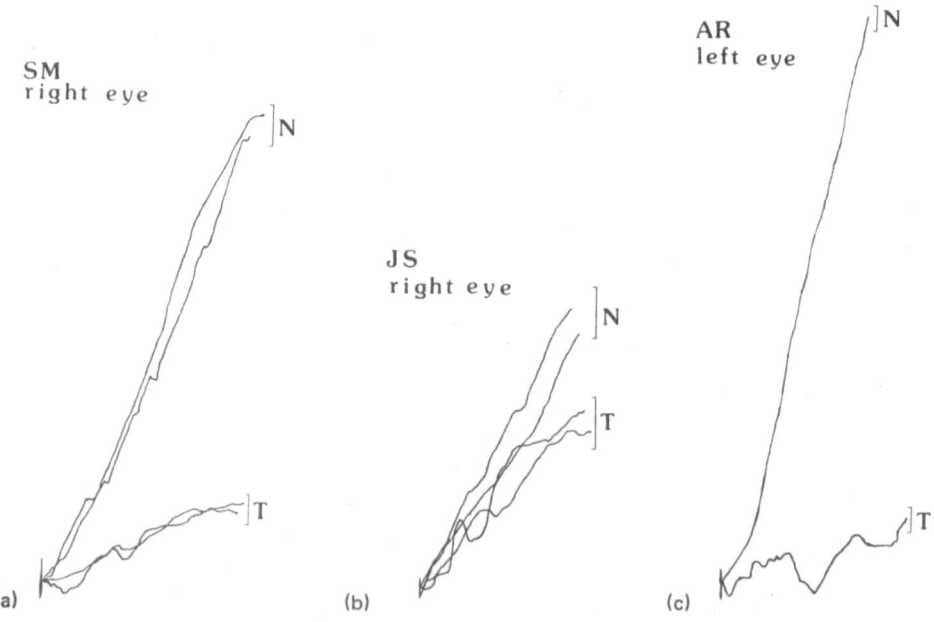

Fig. 9-6. Slow-phase cumulative displacement curves from five clinical infants and children (see Table 9-1 and Figure 9-5) with temporonasal (*T-N*) or nasotemporal (*N-T*) stimulus motion.

diopters) from birth, and had been treated from 10 to 18 months of age with atropine drops in the right eye (atropine for 5 days, none for 2 days). This "penalization therapy" relaxes the accommodation of the atropinized eye, with the result that images upon the retina are blurred. A second treatment, also to encourage use of the left eye, was initiated from 18 to 22 months of age, and consisted of patching the right eye for 3 days of every 4. This child's acuity, tested through each eye, was within the normal range. The EOG traces show highly asymmetrical monocular OKN. Cumulative displacement records, shown in Figure 9-6c, demonstrate that the velocity of the eye movements with nasotemporal motion was highly variable, only occasionally matching the velocity of the slow phase with temporonasal stimulus motion. Furthermore, both the preferred eye and the normally deviating eye displayed monocular OKN slow-phase asymmetry.

In these children with esotropia, the slow-phase asymmetries persisted long after the normal 20-week period for reduction of monocular OKN asymmetry. Interestingly, the asymmetries did not correlate with poor acuities, and the nondeviating eye also showed OKN asymmetries. OKN asymmetries in both the deviating and nondeviating eyes of adults with a previous history of strabismus were found by Schor and Levi (1980). Our data, in addition to that of Schor and Levi (1980), led us to question which conditions correlate with arrested development of monocular OKN, such as interocular misalignment (stabismus) or form deprivation (cataract, monocular patching).

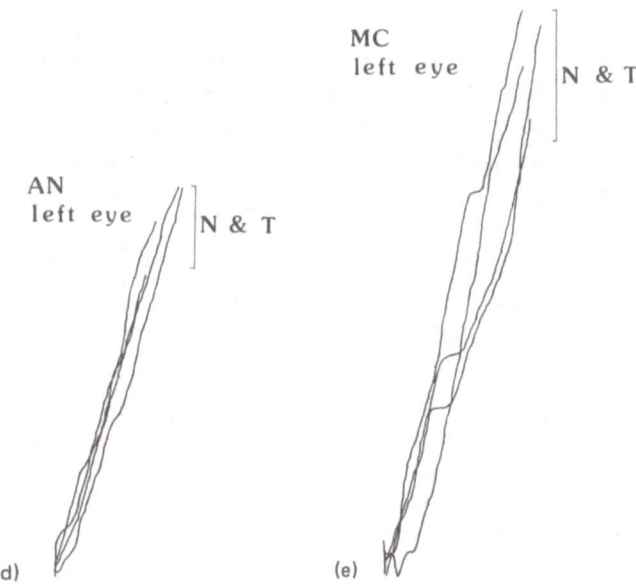

Fig. 9-6. (continued)

A naturally occurring condition for distinguishing between these is unilateral, congenital cataract. In Figure 9-5d the EOG traces of OKN in a child (AN) with this anomaly are shown. When the right eye (with cataract) was tested for monocular OKN, only sporadic beats of nystagmus were recorded, suggesting residual vision in that eye. There were no other indications that vision was mediated through the eye with the cataract. The left eye exhibited normal OKN. Some cumulative displacement curves for the left eye of this child are shown in Figure 9-6d. These curves show identical slopes for both temporonasal and nasotemporal stimulus motion. If one assumes that monocular lid suture is analogous to naturally occurring monocular cataract, a finding of OKN symmetry in the noncataractous eye of this child apparently contradicts findings from monocularly deprived kittens, in which the nondeprived eye, as well as the deprived eye, shows slow-phase gain deficits (Cynader & Harris, 1980). Since AN appeared to retain some visual function through the eye containing the cataract, it is possible that the cataract developed slowly over the first year and some visual experience was acquired early on.

In Figure 9-5, the EOG traces are shown for a child who, at 13.5 months, had a right exotropia that was not exhibited when she accommodated to near targets. There were no significant differences between nasotemporal and temporonasal OKN frequencies or slow-phase slopes (Figure 9-6e).

Conclusions

Three of the five clinical subjects exhibited asymmetrical monocular OKN in which the slow-phase cumulative displacement slopes (velocities) with nasotemporal motion were reduced when compared with the cumulative slow-phase slopes obtained with temporonasal motion. These gain reductions were seen in strabismic esotropia only. Exotropia and monocular cataract, each seen in one infant, were not correlated with asymmetrical slow-phase displacements.

These data on normal and clinical monocular OKN development in these groups of human infants and children show that (a) slopes of cumulative slow-phase displacements in the nasotemporal direction are shallower than slopes of the cumulative temporonasal slow-phase displacements until approximately 20 weeks of age, (b) when the ratios of the slopes are compared across ages, younger infants have a greater magnitude of asymmetry than do older infants and there is a gradual increase in the symmetry ratio with increasing age, (c) the achievement of symmetrical slow-phase velocities may be impaired by some forms of early visual anomaly such as esotropia, and (d) in cases in which deficits do occur following early visual anomalies, both eyes may exhibit monocular OKN impairments long after the time when the asymmetries normally disappear. Unfortunately, these clinical cases were confounded with the treatments that were prescribed (monocular patching and atropine) and, hence, do not allow us to distinguish among these causative agents.

Other Studies

Neural Substrates and OKN

The lack of information on the physiology and anatomy of pathways that mediate monocular OKN in the primate allow only speculation concerning the underlying source of OKN asymmetry in both the normally developing human infant and the infant with early impairments of vision. Recently, a great deal of interest has been generated by findings that relate monocular OKN asymmetry to retinal specialization, that is, the existence of a fovea, visual streak, or ganglion cell subtypes. Anatomical and physiological studies have shown more explicitly that OKN asymmetry in the rabbit can be accounted for by the selective excitation of subpopulations of retinal ganglion cells by temporonasal motion across the receptive field of the cell (Hoffman & Schoppman, 1975; Oyster, Takanashi, & Collewijn, 1972). In the rabbit, direction-selective retinal ganglion cells with a preference for temporonasal stimulus motion also respond to the range of velocities that elicit OKN (Oyster et al., 1972). Several investigators have proposed that the foveal region of the retina is important for the production of symmetrical monocular OKN in higher species, since afoveate animals and humans with maldeveloped foveas show temporonasal OKN biases (Baloh, Yee, & Honrubia, 1980; Tauber & Atkin, 1968).

Despite a great deal of information concerning brainstem circuitry involved in the production of optokinetic eye movements (reviewed by Raphan & Cohen, 1978), there is very little information available on the contributions of cortical and/or pretectal neurons in the mediation of OKN in primates. Possibly, postnatal development of the binocular properties of either cortical or pretectal neurons may account for the changes we observed in normal infant optokinetic eye movements, but many other possibilities exist. For instance, postnatal anatomical changes have been described in the human retina (foveal and parafoveal region) until at least 20–25 weeks postnatally (Abramov, Gordon, Hendrickson, Hainline, Dobson, & LaBossiere, 1981; Spira & Hollenberg, 1973). Given the parallel developmental time courses for OKN symmetry and the maturation of the human central retina, it is possible that retinal development could be an important factor limiting processing of optokinetic stimuli. In the human lateral geniculate nucleus the parvo- and magnocellular laminae show the most rapid

postnatal cell growth in the first 6 months and 1 year, respectively
(Guillery, 1972; Hickey, 1977). This growth continues at a slower rate
for 1–2 years postnatally. In other species, such as the cat and monkey,
monocular deprivation has been shown to stunt the growth of neuronal
size in lateral geniculate laminae that receive input from the deprived
eye (Vital-Durand, Garey, & Blakemore, 1978), with parallel deficits
in the visual acuity of the animal (Von Noorden & Crawford, 1978).
Based on his observations of rapid periods of lateral geniculate cell
growth and the accumulating knowledge about behavioral effects of
early abnormal visual experience, Hickey (1981) has suggested that
there exists a close relationship between the beginning of the critical
period and the period of most rapid cell growth in the dorsal lateral
geniculate nucleus.

Infant Psychophysics: Behavioral Measurement
of Binocularity and OKN

Recent psychophysical studies of infant vision have revealed some interesting parallels between the developmental time courses for monocular
OKN and binocular vision. Stereopsis in the human infant, as measured
by a modified preferential looking technique, shows a rapid rise in de-

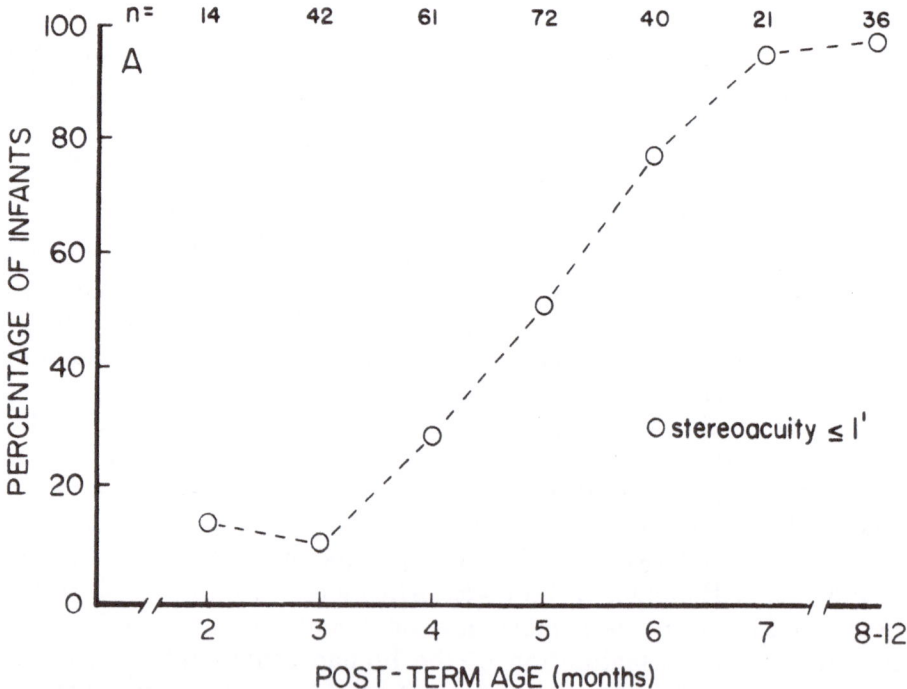

Fig. 9-7. Percentage of infants showing stereoacuity of 1 min arc as a function of
age. *n*: number of infants in each age group.

tection of fine disparities to an adultlike level between 20 and 30 weeks of age (Birch, Gwiazda, & Held, 1982; Held, Birch, & Gwiazda, 1980). This growth of stereopsis in young infants is shown in Figure 9-7. Both stereoscopic vision and the degree of interocular transfer of the tilt after effect are grossly reduced in human observers who have a history of early childhood strabismus (Movshon, Chambers, & Blakemore, 1972). A retrospective study of binocularity in humans with periods of abnormal interocular alignment occurring at various times after birth implied that the critical period of susceptibility begins within 6 months of birth and may extend for several years (Banks, Aslin, & Letson, 1975). The critical period in which visual acuity in the deviating eye may be impaired by strabismus also begins shortly after birth and appears to last through the first year of life (Jacobson, Mohindra, & Held, in press). Together, these results suggest that early interocular alignment is necessary for the normal development of binocular vision and acuity. Our data from three strabismic infants, showing impairment of slow-phase optokinetic following for nasotemporal stimulus motion, indicate that early interocular alignment is important for symmetrical opto-kinetic responses as well.

Future Research

Future studies of infant visual development should examine the relationship between developing binocularity and OKN symmetry. Is cortical binocularity a prerequisite for OKN symmetry? If so, tests of OKN symmetry may be a reasonable test of cortical binocularity in human infants. Careful documentation of the treatment history and onset of strabismus may clarify which aspects of visual experience (such as interocular alignment and stimulation of corresponding retinal points) are necessary for the normal development of symmetrical OKN. Animal models that parallel early human visual anomalies will help to elucidate physiological and anatomical characteristics of optokinetic pathways and the functional consequences of early postnatal visual experience.

Acknowledgments

This work was supported by Grants EY-01191, EY-02649, and EY-02621 from the National Institute of Health. J. R. Naegele was supported by grants from the Spencer Foundation and by NIH Pre-doctoral Training Grant T-32-GMO7484. We are grateful to the following people, who have provided useful feedback during the course of this study and preparation of this manuscript: Joseph A. Bauer, Jr., Eileen Birch, Janet Conway, Max Cynader, Kenneth Ciuffreda, Allan Doyle, Jane Gwiazda, David Israel, Samuel Jacobson, Roberto Lent, Indra Mohindra, Jan Nagle, Alex Pentland, Whitman Richards, Sean True, and Jeremy Wolfe.

References

Abramov, I., Gordon, J., Hendrickson, A., Hainline, L., Dobson, V., & LaBossiere, E. Postnatal development of the infant retina. *Investigative Ophthalmology and Visual Science* (Suppl.) 1981, *20*(3), 46. (Abstract)

Aslin, R. N. Development of smooth pursuit in human infants. In D. F. Fisher, R. A. Monty, & J. W. Senders (Eds.), *Eye movements: Cognition and visual perception.* Hillsdale, N. J.: Erlbaum Press, 1981.

Atkinson, J. Development of optokinetic nystagmus in the human infant and monkey infant: An analogue to development in kittens. In R. D. Freeman (Ed.), *Developmental neurobiology of vision.* New York: Plenum Press, 1979, pp. 277–287.

Atkinson, J., & French, J. Development of symmetrical optokinetic nystagmus in infants. *Proceedings of the Society for Research in Child Development Biennial Convention,* 1979, *2*, 216. (Abstract)

Baloh, R. W., Yee, R. D., & Honrubia, V. Optokinetic asymmetry in patients with maldeveloped foveas. *Brain Research,* 1980, *186*, 208–210.

Banks, M. S., Aslin, R. N., & Letson, R. D. Sensitive period for the development of human binocular vision. *Science,* 1975, *190*, 675–677.

Birch, E. E., Gwiazda, J., & Held, R. Stereoacuity development for crossed and uncrossed disparities in human infants. *Vision Research,* 1982, *22*, 507–513.

Collins, W. E., Schroeder, D. J., & Elam, G. W. Effects of D-amphetamine and of secobarbital on optokinetic and rotation induced nystagmus. *Aviation, Space and Environmental Medicine,* 1975, *46*, 357–364.

Crone, R. A. Amblyopia: The pathology of motor disorders in amblyopic eyes. *Documenta Ophthalmologica,* 1977, *11*, 9–17.

Cynader, M., & Harris, L. Eye movement in strabismic cats. *Nature,* 1980, *286*, 64–65.

Dobson, V., & Teller, D. Y. Assessment of visual acuity in human infants. In J. Armington, J. Krauskopf & B. Wooten (Eds.), *Visual psychophysics: Its physiological basis.* New York: Academic Press, 1978.

Fantz, R. L., Ordy, J. M., & Udelf, M. S. Maturation of pattern vision in infants during the first six months. *Journal of Comparative and Physiological Psychology,* 1962, *55*, 907–917.

Fukuda, V. T., & Tokita, T. Uber die beziehung der richtung der optischen reize zu den reflextypen der augen und skelettmuskeln. *Acta Oto-Laryngologica,* 1957, *48*, 415–424.

Gorman, J. J., Cogan, D. G., & Gellis, S. S. An apparatus for grading visual acuity of infants on the basis of optokinetic nystagmus. *Pediatrics,* 1957, *19*, 1088–1092.

Guillery, R. W. Binocular competition in the control of geniculate cell growth. *Journal of Comparative Neurology,* 1972, *144*, 117–130.

Gwiazda, J., Brill, S., Mohindra, I., & Held, R. Preferential looking acuity in infants from two to fifty-eight weeks of age. *American Journal of Optometry and Physiological Optics,* 1980, *57*, 428–432.

Harris, L. R., Lepore, F., Guillemot, J. -P. & Cynader, M. Abolition of optokinetic nystagmus in the cat. *Science,* 1980, *210*, 91–92.

Held, R., Birch, E. E., & Gwiazda, J. Stereoacuity of human infants. *Proceedings of the National Academy of Sciences, U.S.A.,* 1980, *77*, 5572–5574.

Hickey, T. L. Postnatal development of the human lateral geniculate nucleus: Relationship to a critical period for the visual system. *Science*, 1977, *198*, 836-838.

Hickey, T. L. The developing visual system. *Trends in Neurosciences*, 1981, *4*(2), 41-44.

Hoffman, K.-P., Behrend, K., & Schoppmann, A. Visual responses of neurons in the nucleus of the optic tract of visually deprived cats. *Society for Neuroscience Abstracts*, 1977, *3*, 563.

Hoffman, K.-P., & Schoppman, A. Retinal input to direction selective cells in the neucleus tractus opticus of the cat. *Brain Research*, 1975, *99*, 359-366.

Hubel, D. H., & Wiesel, T. N. Binocular interaction in striate cortex of kittens reared with artificial squint. *Journal of Neurophysiology*, 1965, *28*, 1041-1059.

Huizinga, E., & Meulen, P. Vestibular rotatory and optokinetic reactions in the pigeon. *Annals of Oto-Rhinolaryngology*, 1951, *60*, 927-947.

Jacobson, S. G., Mohindra, I., & Held, R. Monocular visual form deprivation in human infants. *Documenta Ophthalmologica*, in press.

Montarolo, P. G., Precht, W., & Strata, P. Functional organization of the mechanisms subserving the optokinetic nystagmus in the cat. *Neuroscience*, 1981, *6*, 231-246.

Movshon, J. A., Chambers, B. E. I., & Blakemore, C. Interocular transfer in normal humans and those who lack stereopsis. *Perception*, 1972, *1*, 483-490.

Naegele, J. R., & Held, R. Optokinetic nystagmus shows asymmetry in human infants. *Investigative Ophthalmology and Visual Science* (Suppl.), 1980, *19*, 210.

Naegele, J. R., & Held, R. The postnatal development of monocular optokinetic nystagmus in infants. *Vision Research*, 1982, *22*, 341-346.

Oyster, C. W., Takanashi, E., & Collewijn, H. Direction-selective retinal ganglion cells and control of optokinetic nystagmus in the rabbit. *Vision Research*, 1972, *12*, 183-193.

Rademaker, G. G. J., & Ter Braak, J. W. G. On the central mechanism of some optic reactions. *Brain*, 1948, *71*, 48-76.

Raphan, T., & Cohen, B. Brainstem mechanisms for rapid and slow eye movements. *Annual Review of Physiology*, 1978, *40*, 527-552.

Schor, C. M., & Levi, D. M. Disturbances of small field horizontal and vertical optokinetic nystagmus in amblyopia. *Investigative Ophthalmology*, 1980, *19*, 668-683.

Spira, A. W., & Hollenberg, M. J. Human retinal development: Ultrastructure of the inner retinal layers. *Developmental Biology*, 1973, *31*, 1-21.

Tauber, E., & Atkin, A. Optomotor responses to monocular stimulation: Relation to visual system organization. *Science*, 1968, *160*, 1365-1367.

van Hof-van Duin, J. Early and permanent effects of monocular deprivation on pattern discrimination and visuomotor behavior in cats. *Brain Research*, 1976, *111*, 261-276.

van Hof-van Duin, J. Direction preference of optokinetic responses in monocularly tested normal kittens and light deprived cats. *Archives Italiennes de Biologie*, 1978, *116*, 471-477.

Vital-Durand, F., Garey, L. J., & Blakemore, C. Monocular and binocular deprivation in the monkey: Morphological effects and reversibility. *Brain Research*, 1978, *158*, 45-64.

Von Noorden, G. K., & Crawford, M. L. J. Morphological and physiological changes in the monkey visual system after short-term lid suture. *Investigative Ophthal-mology*, 1978, *17*, 762–768.

Wood, C. C., Spear, P. D., & Braun, J. J. Direction-specific deficits in horizontal optokinetic nystagmus following removal of visual cortex in the cat. *Brain Research*, 1973, *60*, 231–237.

Chapter 10

Spatially Determined Visual Activity in Early Infancy

Marshall M. Haith

Early infancy is a period of dramatic growth in learning how to negotiate through the environment. Within approximately 14 months, the immobile newborn becomes the racing, exploratory, pan-toppling toddler, accomplishing, along the way, the skills of precision reaching, sitting, creeping, standing, and walking. These postural accomplishments depend more on visuospatial anchoring than we might suspect. For example, infants who have recently learned to walk will fall when the room they are in appears to move forward or backward even though they are on a stationary surface (Lee & Aronson, 1974). A similar dependence of postural stability on visual cues can be found in prewalking babies who are tested in a sitting position (Butterworth & Hicks, 1977).

In considering infants' acquisition of the ability to negotiate space and to interpret the meaning of changing relations between themselves and, for example, the walls and ceilings of a room, one must ponder how spatial knowledge emerges. Newborn infants are pitifully incapable of self-induced motion through space or purposeful manipulation of items in it, so one might suppose them to be largely undiscriminating of the spatial characteristics of their world. Certainly, classic pieces of research have demonstrated the contribution of self-initiated movement to the acquisition of skill in moving one's body through a spatial frame of objects (Held & Hein, 1963). Nevertheless, human infants do possess action schemes of a sort, at birth, which could provide them with cues concerning the direction and position of objects. Such action schemes include eye and head motion. Although substantial development occurs over the first few months of life in both the neuromuscular and neurosensory components of the human visual system, infant eye movements are more similar to those of the adult than for any other action system. As Bullinger (1979) has noted, the operation of these action schemes provides the infant an opportunity to learn how the body's instruments of exploration work, as well as the properties of the world and the intersection between the two.

Although our research on the interface between visual perception and action in the human newborn and infant has not focused on questions about spatial knowledge, our study of infant eye movement activity has revealed several relevant phenomena, and our findings are presented from this perspective.

Appreciation of Spatial Relations

An important step for the infant, in organizing the visual world, involves the appreciation of spatial relations among visual elements. Elements that go together must cohere, perceptually, to enable the infant to fractionate the visual field into separable whole objects. We used subjective contour arrays (Figure 10-1) to inquire about the infant's ability to appreciate spatial relations among identical elements (Bertenthal, Campos, & Haith, 1980). Subjective contour stimuli are quite useful because it is possible to modify only the relation in the spatial orientation among elements to destroy the illusion. Thus, one can avoid confounding factors that often weaken the interpretation of studies that purport to manipulate spatial relations among visual elements; such factors include changes in the total area that elements occupy, contour density, the elements themselves, their orientation, and so on.

For one nonillusion array, NI_1, in Figure 10-1, the elements along one diagonal in the original illusion array have been exchanged; for NI_2, the diagonals have been exchanged. Thus, there is as much difference, in terms of element change or rotation, between NI_1 and NI_2, as there is between array I and array NI_1, and twice as much change as between I and NI_2. Through a comparison of the effects of these various arrays on infant looking, we asked whether the illusion array has any special status; if so, we are in a position to argue that babies are sensitive to the spatial relations among elements that comprise a "stimulus."

We showed combinations of these patterns to babies 5–7 months of age. The design of the study is shown in Table 10-1. At each age, there were three subgroups who were first familiarized with the I, the NI_1, or the NI_2 stimulus. The familiarization stimulus was first shown for a series of trials that were terminated when the baby looked away from the stimulus. Babies normally look at new stimuli for a relatively longer time and look less as the stimuli become more familiar (a process termed *habituation*). The familiarization series was continued until the baby reached a habituation criterion, defined as two successive trials for which the average looking time was 50% of the average for the first three trials. Then one-half of the babies in each subgroup, the change subjects, were shown a changed stimulus for two postcriterion trials.

The remaining babies in each subgroup continued to see the familiarization stimulus during two postcriterion trials; these babies served as a control for spontaneous recovery of looking, whereas the change babies would presumably demonstrate, by recovery of looking activity, how different the changed stimulus looked from the familiarization stimulus.

Fig. 10-1. Subjective and nonsubjective contour arrays; original format consisted of blue corner elements on a yellow ground. Stimulus I was the illusion stimulus; it is usually seen by adults as containing subjective contours between the corner elements formed, perhaps, by an occluding square that overlaps circular corner elements and the ground. The NI₁ and NI₂ stimuli contain exactly the same elements, but elements along one diagonal are interchanged in NI₁ and elements along both diagonals are interchanged in NI₂ to produce nonsubjective arrays. (Adapted from Bertenthal, B. I., Campos, I. & Haith, M. M. Development of visual organization: The perception of subjective contours. *Child Development*, 1980, *51*, 1072-1080.)

Table 10-1. Design of Spatial Relations Study.

Group	Habituation stimulus	Novel stimulus
1a	I	NI_1
1b[a]	I	I
2a	NI_1	NI_2
2b[a]	NI_1	NI_1
3a	NI_2	I
3b[a]	NI_2	NI_2

[a]Control group.

Groups 1a and 3a experienced an illusory shift, whereas Group 2a experienced a nonillusory shift; Groups 1b, 2b, and 3b experienced no shift at all.

Using the two last precriterion trials as a baseline, we see that babies at 7 months of age recovered more when the change was away from stimulas I (Group 1) or to stimulus I (Group 3) than when it was from one NI stimulus to another (Group 2) (Figure 10-2). Of course, the performance of the no-change subgroups across criterion and postcriterion trials serves as the comparison for chance recovery in each case. Thus, at 7 months, babies consider array I as somewhat special. Babies at 5 months showed differential recovery only when they had been familiarized with stimulus I (Group 1); therefore, for them, the outcome was ambiguous. We first speculated that it may take more time for the relations among elements to be appreciated by 5-month-olds than by 7-month-olds, and Group 3 babies, who received only two postcriterion presentations of stimulus I, may not have had sufficient time to organize it. However, further work has led us to believe that 5-month-olds simply do not organize the subjective contour array (Bertenthal, Haith, Campos, & Tucker, 1980). We have, then, some evidence for a development of ability to appreciate at least a class of spatial relations in babies between 5 and 7 months of age. To date, we have not tied this development to changes in scanning activity; future studies will address that issue.

The subjective contour array, while providing a useful stimulus for experimental purposes, may place a fairly heavy demand on the baby. Intuitively, some effort seems to be required to see the illusion; it does not appear immediately. Moreover, the baby may need prolonged visual experience to learn about occlusion of a ground by a figure stimulus; many observers see the illusion as a central foreground square that partially occludes a quadrant of each of four circles. In short, the subjective contour study tells us something about age-related responsivity to

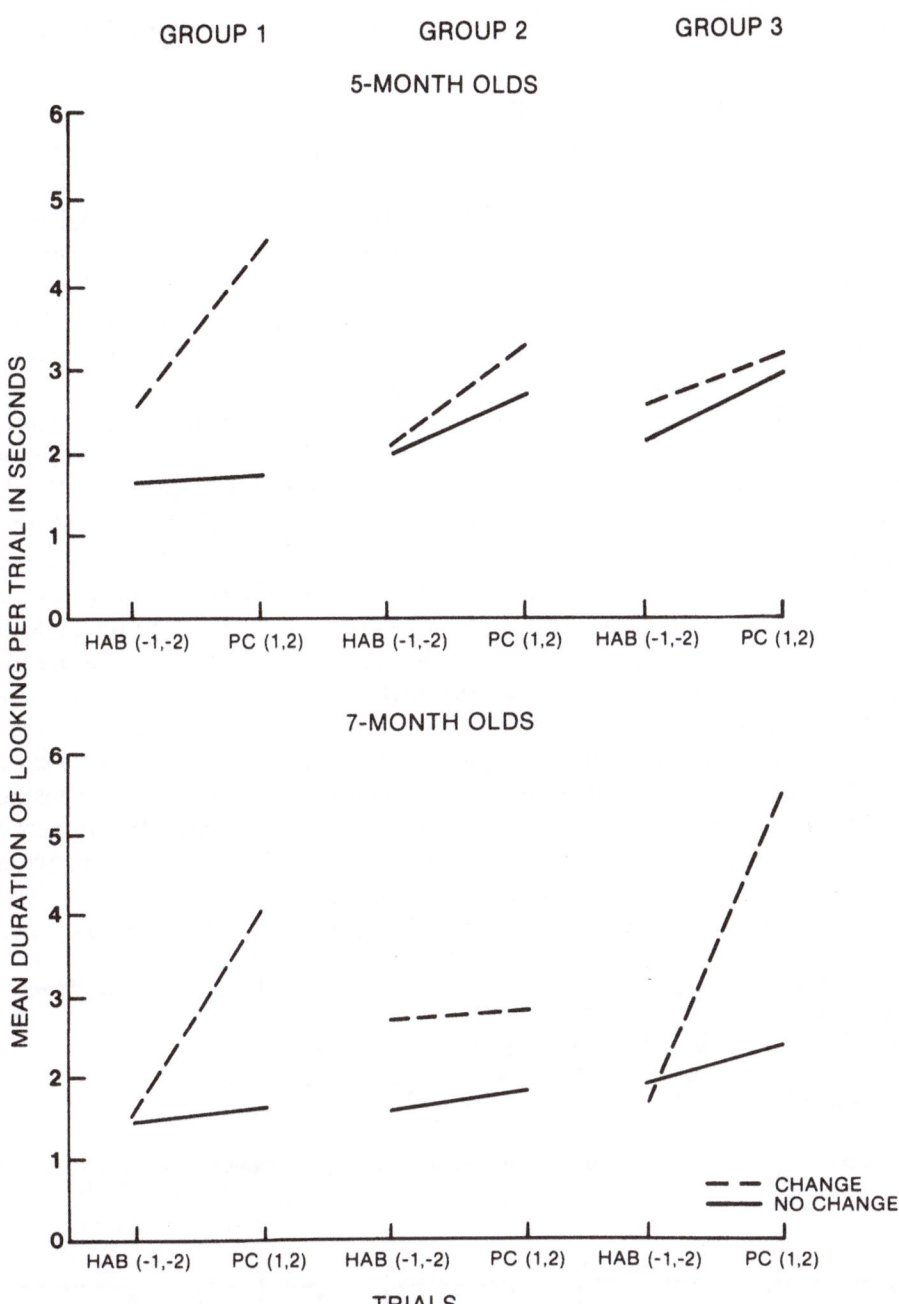

Fig. 10-2. Average duration of looking for 5- and 7-month-old infants as a function of subgroup membership and whether or not a stimulus change occurred after the habituation criterion was reached. Average looking duration for the two trials prior to criterion is shown [*HAB (−1, −2)*], as well as that for the two postcriterion trials [*PC (1, 2)*] for subjects who experienced a changed stimulus (*broken line*) and for no-change controls (*solid line*). (Adapted from Bertenthal, B. I., Campos, J., & Haith, M. M. Development of visual organization: The perception of subjective contours. *Child Development*, 1980, *51*, 1072–1080.)

spatial relations, but an uncritical acceptance of those findings might lead us to underestimate the age of appreciation of any type of spatial relation.

Van Giffen (1980) speculated that early appreciation of spatial relations might obey the same kind of Gestalt laws that are operative in adults. She created "good Gestalt" arrays that essentially "drive" visual organization. By distorting one element of each array (Figure 10-3) and looking at whether infants noticed it, she asked whether a more basic sensitivity to spatial relations might occur at an earlier age.

Van Giffen presented these stimuli to 1- and 3-month-old babies and reconstructed their scanning patterns from measurements of fixations recorded each 0.5 seconds. Whereas the Bertenthal, Campos, and Haith (1980) study required the recording only of looking duration at the whole stimulus, the Van Giffen study required the detailed analysis of fixation points. The apparatus employed in this and several other studies is shown in Figure 10-4. A baby lies horizontally on its back in an infant seat with its head in a cloth sling (not shown). Above the infant is an aluminum panel that supports two television cameras. The top camera is aligned so that the image of the baby's right eye is recorded after transmission through mirror Y and reflection by mirror X. The lower television camera records the image of the infant's visual field, reflected to the baby's eye via mirror Y. Six invisible infrared light sources are mounted directly above the baby's head; these provide illumination for the eye camera, which contains an infrared tube. The point reflections of these lights from the surface of the infant's cornea

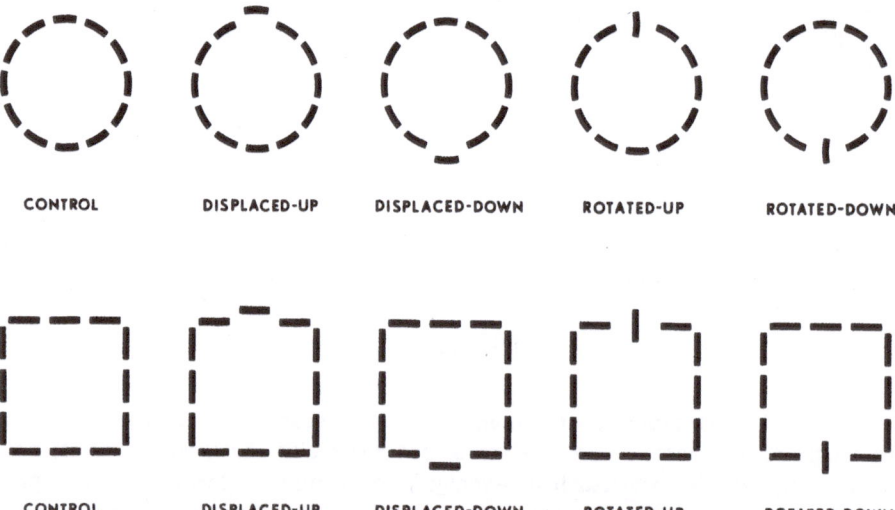

Fig. 10-3. Experimental stimuli differ from the control "good-Gestalt" prototype only in the change in the top or bottom element. (From Van Giffen, K. *The emergence of form appreciation in infancy.* Unpublished doctoral dissertation, University of Denver, 1980.)

are used to determine where the infant is looking at a particular moment. This determination is possible, because the location of the lights with respect to points in the virtual visual field is known. Because mirror Y contains a special coating that reflects most of the visible light and transmits most of the infrared light, the baby may scan the reflected visual field undistracted while the eye camera records the image of the eye with the reflected infrared marker lights. The eye image and the visual field of the infant are videotaped for later processing.

The baby's scan pattern is reconstructed at a later time. First, the videotape is replayed and up to 500 video images are recorded onto a videodisc at a rate of 1/.5 second. Each image can then be displayed

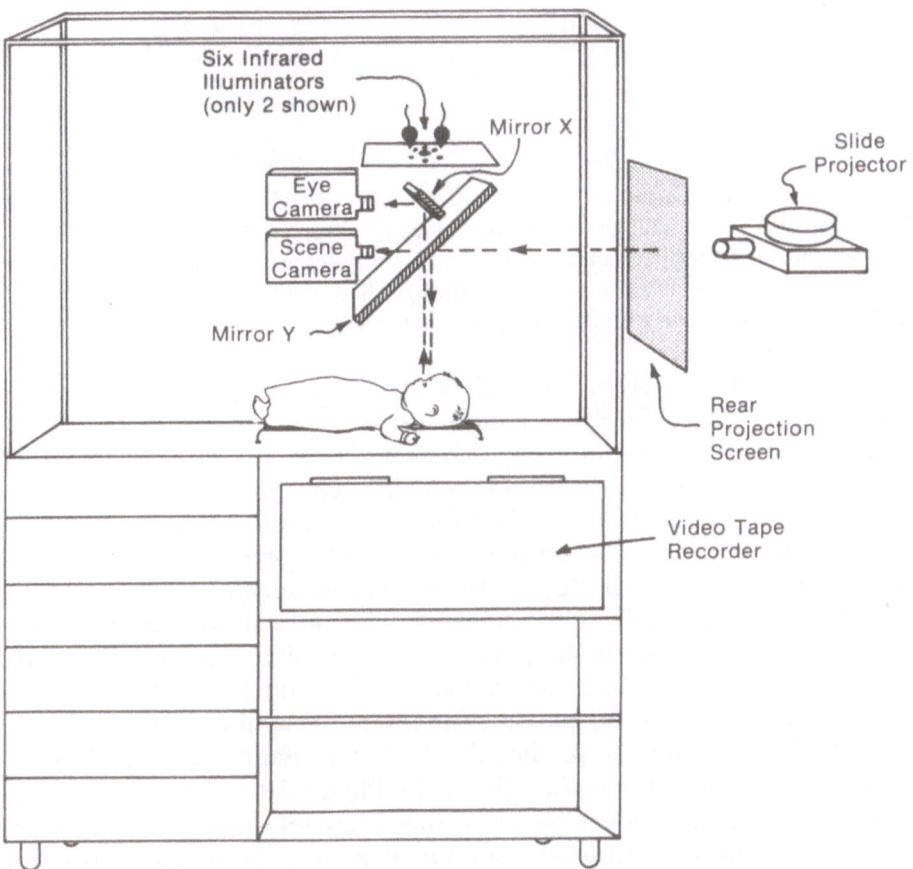

Fig. 10-4. Apparatus for the Van Giffen study and for the eye movement studies on newborns. The supine baby looks at the stimuli by reflection from mirror Y of the projection screen. Mirror Y reflects visible light and transmits infrared light. A television camera (*eye camera*) records the image of the infant's eye by reflection from mirror X and transmission of mirror Y. Infrared light sources illuminate the eye and four small reflections on the corneal surface.

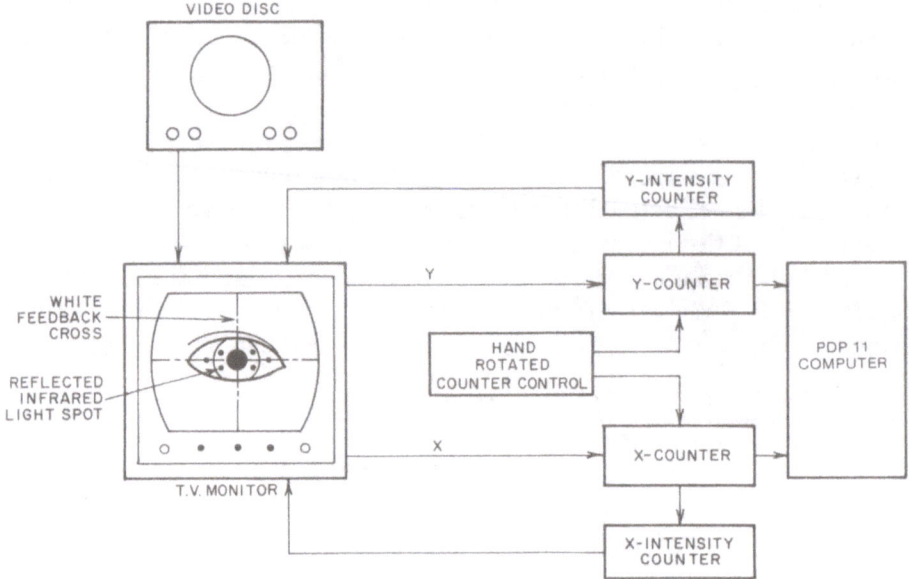

Fig. 10-5. Apparatus used to measure eye position. A videodisc displays one video frame of the eye while an electronic interface displays X, Y crosshairs. Hand-rotated knobs are adjusted until the crosshairs intersect at the point to be measured (center of the pupil or an infrared reflection). These coordinates are fed to a computer, and then a sequence of fixations over a display can be reconstructed.

indefinitely for analysis. The data reduction sequence is shown in Figure 10-5. The Cartesian coordinate position of the center of the pupil and the closest infrared light reflection are measured and recorded. Vertical and horizontal crosshairs that appear on the television monitor may be moved by rotating knobs so that they intersect at the point to be measured. Electronic counters keep track of the number of television lines from the top of the television screen to the horizontal line and from the left of the screen to the vertical line, providing the Cartesian coordinates of the point. These coordinates are recorded into a computer by a button press. Later, computer programs make the necessary calculations for plotting the infant's scanning sequence.

Van Giffen found the predicted effects on infant scanning. When 3-month-olds were shown the stimuli in Figure 10-4 that had the distorted element in a down position, they scanned lower on the stimulus than when the nondistorted stimulus array was presented. When the distorted element was presented in the up position, they scanned significantly higher than when it was in the down position. These results are displayed in Figure 10-6. The 1-month-olds provided only a tentative suggestion that they might have sensed the distortion, whereas the 3-month-olds gave strong evidence. Since the misplaced element was identical to the comparable element in the control array and varied

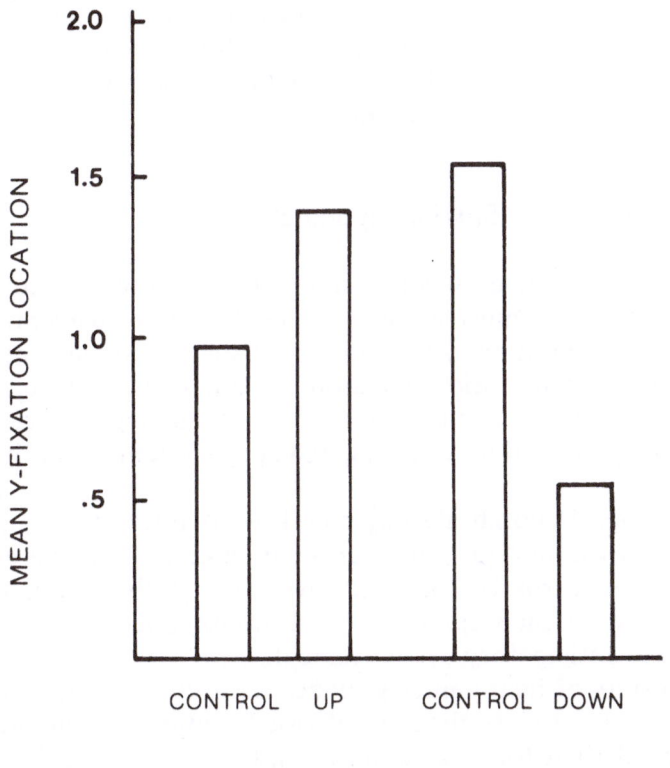

Fig. 10-6. Locatin of fixation on *Y* axis when control stimuli were presented and when the up-distorted and down-distorted stimuli (see Figure 10-3) were presented to 3-month-olds. (From Van Giffen, K. *The emergence of form appreciation in infancy.* Unpublished doctoral dissertation, University of Denver, 1980.)

only in its spatial relation to neighboring elements, it should have attracted fixations only if all the other elements were seen as somehow belonging together, or spatially concordant. Van Giffen's results provide good evidence of infant sensitivity to spatial relations of stimulus elements by 2 months of age.

The 1- to 3-month age period is the same period that we reported earlier as being pivotal in the baby's scanning of the components of the mother's face, such as the eyes (Haith, Bergman, & Moore, 1977). We speculated that the onset of internal scanning might reflect the baby's emerging ability to see the face as a whole; the age correspondence between Van Giffen's findings and the face finding of Haith et al. (1977) suggests that, in fact, the sensitivity to spatial relations has some generality.

Consideration of these studies provides a lesson that we had probably known at some level but often overlooked. The development of sensitivity to spatial relations need not be thought as an all-or-none affair, a skill that the baby possesses or does not. Probably, sensitivity to various relational parameters develops throughout life. It would be an interesting task to specify those parameters and the developmental functions that correspond to them.

Visual Anticipation of Spatiotemporal Events

Negotiating visual space involves more than simply responding to the here and now. One must anticipate the short-term perceptual future, certainly in locomoting, and, most probably, in organizing dynamic events in the visual world. A reasonable argument can be made that perceptual anticipation plays a role in stabilizing the visual world whether eye and head movements, object movement, or both are under consideration.

Sandy Pipp, Deborah Porter, and I were interested in the age at which babies can develop short-term anticipation of visual events that occupy different spatial locations. We studied the rudiments of the baby's ability to map spatiotemporal events, and our index of such mapping was anticipatory eye movements.

Two clusters of light-emitting diodes were used, one about 9° to the left and one about 9° to the right of visual center. For one presentation series, they flashed for .5 second in simple alternation with a .5-second interflash interval. We were interested in whether the baby, after a certain number of exposures to this alternating sequence, would show interflash anticipation of the next light flash. We also used a second flash sequence for which the timing was the same, but the spatial position of the flash was randomly determined. We used the random sequence to test the possibility that babies might tend to switch sides in the interflash interval after a certain number of exposures to the sequence even though there was no basis for developing an anticipation. An examination of preliminary data from this study will help to communicate the potential of the paradigm and the questions at issue. In Figure 10-7 the tracking data for one 3-month-old infant are shown for the regular sequence (Figure 10-7a) and for the random sequence (Figure 10-7b). Light onsets are represented by the squares in their appropriate left/right position with time proceeding vertically from top to bottom. Each sequence is divided into two 30-second periods. The solid

---

Fig. 10-7. Display of events in the anticipation study for two periods of a regular alternation (a) and for two periods of an irregular alternation (b). Only two lights were used, one on the left and one on the right. Time is represented from top to bottom. *Squares*, illumination of a light; *solid line*, position of the eye. Light flashes were always .5 second in duration, separated by a .5-second interval. Periods were 30 seconds in duration, separated by a 15-second interval.

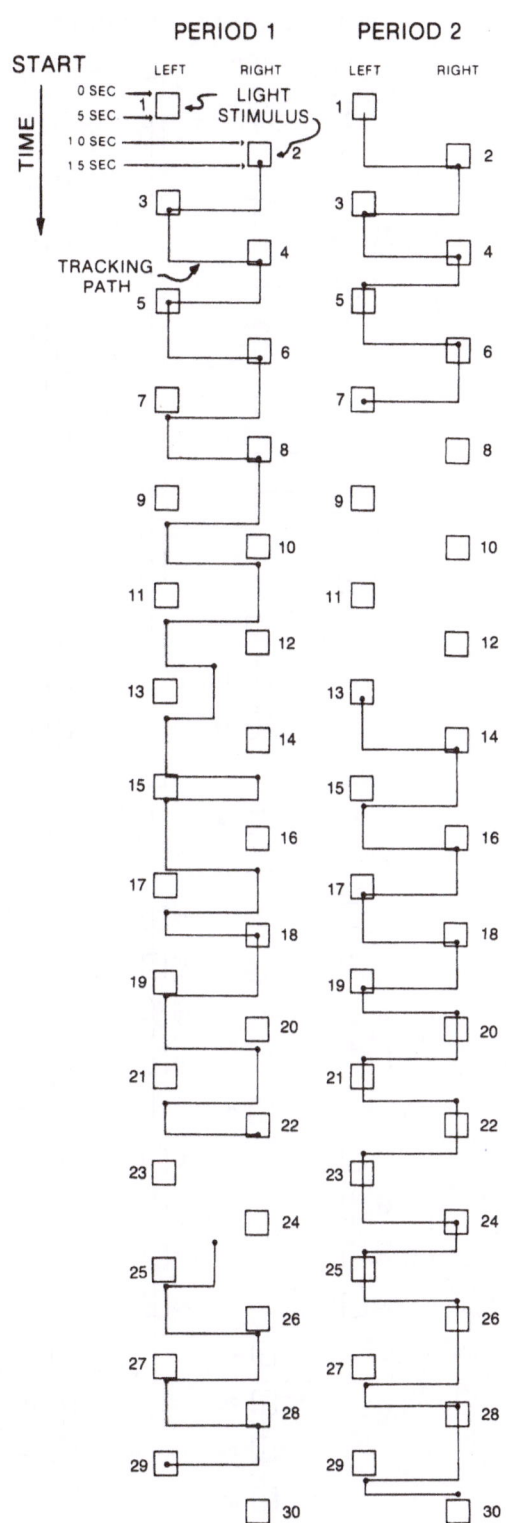

(a)

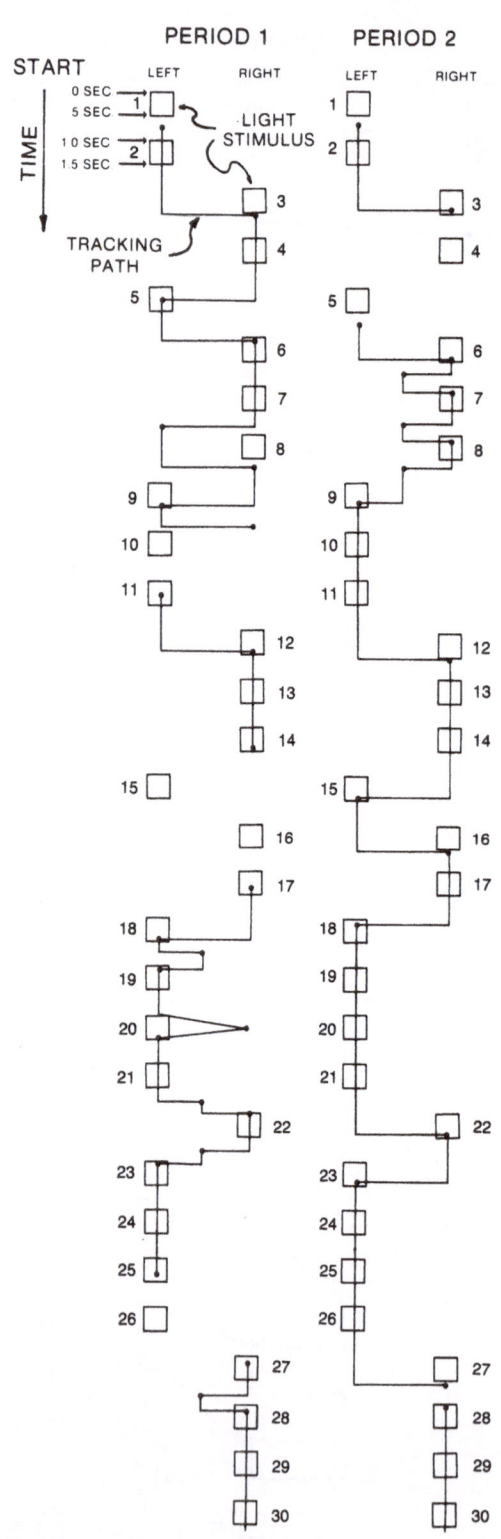

(b)

Fig. 10-7 (continued)

line represents the scan pattern of the infant based on our sampling rate of 8/sec.

Starting at the top of Period 1 in Figure 10-7a, one sees that the infant fixated flashes 2-5 well after their onset and just before their offset. For the remainder of Period 1, the infant directed his gaze to the relevant light either just before offset (7 times) or after it was extinguished (13 times).

Period 2 in Figure 10-7a followed Period 1 after a 15-second rest interval; here we see a different picture. The baby fixated the first 7 lights reasonably early during the onset period in 3 cases and with one anticipation a fourth. After 8-12 flashes, for which data are missing (head movement and eye closure), we see a sequence of late fixations and one miss from flash 13-19. Then, the baby generated a sequence of 7 anticipations and one early fixation between flash 20-26. Notice that these anticipations were accomplished after approximately 45 seconds of exposure to the light sequence.

Did the baby simply alternate fixation from one side to the other because he was prone to do so after looking at light flashes for 45 seconds? Evidence concerning this possibility is provided in Figure 10-7b. If anything, alternate fixation diminished with time in this sequence. The infant engaged in some interesting episodes that add support to the notion that his visual activity was governed by appreciation of the spatiotemporal structure of the sequence to which he had been exposed. Period 1 of the random sequence was presented 15 seconds after the completion of the alternating phase. In the early parts of Period 1 in Figure 10-7b, the infant moved away from the activated light in the interflash interval two times (between 7-8 and 8-9) and made a very early alternation fixation (flash 6), presumably based on the exposure to the preceding alternating sequence. Generally, after that he adopted an interesting ad hoc response strategy. Apparently, the baby dealt with the uncertainty in a number of cases by selecting an intermediate station point between the two light positions in the interflash interval (e.g., between 18-19, 21-22, 22-23, and 27-28). This is an optimal position from which a flash on either side can be responded to with an eye movement. This tack was used seven times for the random sequence but appeared only twice in the alternating sequence.

This work is still in progress. However, we expect to demonstrate that the 3-month-old baby is capable of mastering simple rules of time-space correspondence. The role of his own activity in learning such rules is open to question; it will occupy our attention as soon as we have established an optimal paradigm. Obviously, this paradigm holds the potential for examining the infant's ability to deal with visual sequences that obey more complicated rules than simple alternation. It could provide an opportunity to examine the development of the infant's ability to master various levels of spatiotemporal regularity at different ages.

Spatial Sensitivity at Birth

At this point in our presentation, we have demonstrated that sensitivity to at least some spatial relations is present by 3 months of age. Are babies sensitive to spatial cues before then? The evidence says "yes," even at birth, but that evidence is based on phenomena in a different realm than those we have been considering. The evidence with newborn babies takes four forms: (a) scanning activity that is affected by the spatial locus of auditory input; (b) localization of eye movements to a transient peripheral visual stimulus; (c) visual localization that does not depend on transient peripheral cues; and (d) coupling of stimulus orientation and scanning activity.

Localized Auditory Input and Visual Scanning

Interest in sound localization in newborn infants has been evident for some time, and the pertinent literature has been reviewed often (e.g., Kessen, Haith, Salapatek, 1970; Mendelson & Haith, 1976). In addition to the important question about whether sounds can be localized without experience, the use of eye and head movements to index that ability led to the question of whether newborns possess a supramodal or amodal map of space; such a map might serve to coordinate sensory modes.

A popular paradigm asks whether newborns move their eyes and/or head toward the locus of a sound, addressing whether the visual and auditory modes share a common spatial map. Investigators typically use clicks and other transient inputs for auditory stimuli with no visual input, the direction of the first eye movement being used as the indicator response. Controversy abounds in the literature regarding the results of studies using this paradigm, with reports of significant probability of eye movements toward the stimulus, significant probability of eye movements away from the stimulus, and no stable probability of either (Kessen et al. 1970; Mendelson & Haith, 1976; see also Dodwell, Chapter 11, this volume).

We have argued that the type of map a newborn possesses might be one by which the activity in one system biases the activity in a second—specifically, the activity of listening to a localized sound might bias the scanning of a localized visual stimulus (Mendelson & Haith, 1976). This notion replaces the concept of a trigger in which the sound produces a reflexive movement of the eyes toward it. Instead of using transient clicks, we used a continuing natural stimulus, an audio-recorded voice. The voice emanated from a speaker to the left or right of the baby's head. We also provided the baby with something to look at, another deviation from prior procedures. We presented the newborn two vertical bars, one on the right and one on the left. Finally, we examined the baby's scanning activity rather than measuring only the first eye movement.

We found that babies first scanned the bar that was spatially coherent with the sound source more often than the other bar. Bar approach was not reflexive or stereotyped; rather, the centroid of visual fixations simply shifted toward the bar location on the sound-coherent side. We concluded that, in fact, the activity of listening to sound emanating from a given direction does affect the directional activity of scanning. It seemed to us that there was a loose spatial bias built in that would optimize the likelihood of the baby finding a sound-producing object if it were there; however, we supposed that experience with particular sound–sight combinations would permit sharpening or, perhaps, even overriding of the spatial bias. The possibility of such flexibility was indicated in our finding that babies would search the opposite side of the field after looking at the sound-coherent side. Such plasticity may underlie visual capture of sound sources.

Localization of Transient Peripheral Stimuli

The second type of evidence for some type of spatial sensitivity in young infants comes from work on eye movements toward flashing peripheral stimuli. Using EOG, Aslin and Salapatek (1975) carried out the most complete study in 1- and 2-month-old babies and found significant tendencies of babies to make their first eye movement toward the appropriate side when a flashing light was 10°–30° from a starting fixation point. Harris and MacFarlane (1974) carried out a somewhat similar, observational, study with newborns and reported similar findings. Response times were much longer for infants than for adults. For the Aslin and Salapatek study, response times ranged up to 2 seconds for 2-month-olds and longer for 1-month-olds; under the most optimal conditions, 1-month-olds averaged around a 1-second latency. The observational procedure used in the Harris and MacFarlane study did not permit precise calculation of response times; however, it is worth noting that they found it necessary to accept directionally appropriate responses that occurred as long as 5 second after stimulation.

Visual Scanning Affected by Stable Peripheral Inputs in Free-Scan Situations

The third line of evidence comes from our own work on newborn scanning of edges (Kessen, Salapatek, & Haith, 1972). In this study, we presented newborns off-center, black–white edges. Vertical and horizontal edges were presented for 60 seconds each, preceded and followed by homogeneous black or white fields that were presented for 30 seconds each. Comparisons of control and experimental periods established that vertical edges did influence scanning activity. The interesting finding was that the probability of the baby making an eye movement toward the edge increased with the proximity of the originating fixation. Comparison probabilities were obtained from control periods during which homogeneous light fields were presented. These data suggest that peripheral stimuli affect free-scan behavior of newborns with stable stimuli in a spatially appropriate way; that is, the newborn is capable of more than a directionally appropriate reflexive response to a transient peripheral stimulus.

A further set of experiments demonstrated that the newborn is capable of still more complex visual activity (Haith, 1980). The Kessen et al. (1972) study established that the newborn possesses whatever localizing tendencies are required to keep black–white edges near the center of the retina while engaging in continual refixations. A trial-and-error model could account for this activity with little need to appeal to skills for preprograming of eye movements; however, data from other experiments require us to consider the preprograming possibility. These experiments were similar to the previous one with the exception that a variety of stimuli were used and the stimuli were located near the resting position of the infant's gaze. In one study, we used vertical and horizontal edges; in a second, we used vertical and horizontal bars; in a third, we used angles. These stimuli are shown in Figure 10-8.

Two relevant findings were obtained in these studies. The first was that eye movements were longer when they crossed a contour than

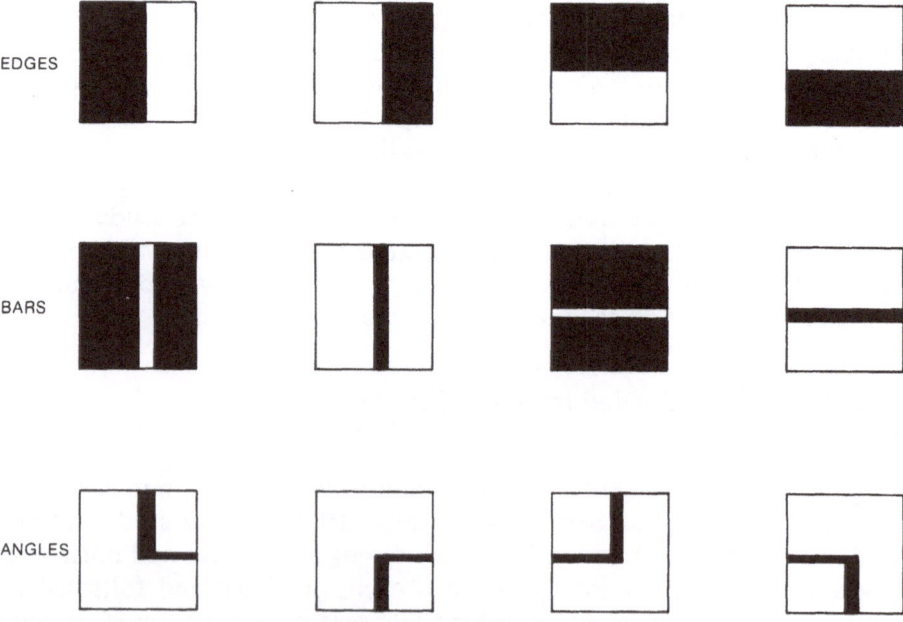

Fig. 10-8. Stimuli used in studies of newborn visual scanning.

Table 10-2. Size of Cross and No-Cross Eye Movements (in Centimeters) for Vertical Stimuli

Condition	Cross	No cross
Edges		
Stimulus	6.10	3.02
Control	4.88	3.18
Bars		
Stimulus	4.22	2.08
Control	3.73	2.18

when they did not. This was true for horizontal and vertical edges, horizontal and vertical bars, and angles; again, the important comparison is with newborn activity in control, homogenous fields. Some relevant data for the vertical stimuli in the edge and bar studies are shown in Table 10-2. Under stimulus conditions the eye movements that crossed the location of an edge were substantially larger than those that did not; this difference was much smaller for the nonstimulus, control condition. The size of the difference was 25% greater for the stimulus condition in the edge study and 13% larger for the bar study. These statistics were based on hundreds of eye movements, and the differences were quite stable.

Since the terminus of an eye movement cannot be reestablished "in flight," it could not have been the case that the trajectory was extended as a consequence of contour motion across the fovea; rather, the path length must have been determined prior to movement. This is not the place for a full discussion of speculations for why newborns locate and cross contours; briefly, this phenomenon might be accounted for by assuming that the newborn is programed to maximize cortical firing activity (see Haith, 1980). The important point for the present argument is not only that the newborn makes directionally appropriate eye movements, but that the characteristics of those eye movements, characteristics that must be preprogramed, are modified to coincide with spatial events. We should not consider the human newborn a trial-and-error organism.

Stimulus Orientation and Scanning Activity

The second finding in the abovementioned study concerned the newborn's sensitivity to stimulus orientation. Vertically oriented bars and edges were scanned differently from horizontally oriented ones. Scan dispersion along the horizontal axis doubled for horizontal bars and edges (over control periods), whereas it decreased for vertical bars and edges. In addition, eye movement characteristics were different. Vertical edges and bars produced an increase in the size of the X component of eye movements. Although horizontal edges and bars also produced an increase in the X component of eye movements, we have reasonable evidence that the Y component increased also, especially for horizontal bars (see Haith, 1980, for details).

Pipp (1978) provided even better evidence. She presented newborns, 1-month-olds, and 2-month-olds several types of line patterns, oriented either horizontally or vertically, and found that the ratio of the size of the X to the Y component of eye movement was significantly larger for vertically oriented than for horizontally oriented stimuli. This difference did not change with age. Thus, by their scanning activity, even newborns demonstrated sensitivity to orientation.

Conclusions

Newborns apparently enter the world with some relatively loose constraints on behavior that make sense in terms of spatial reality. Their visual activity is generally appropriate when localized sounds and visual stimuli are presented, and even, in stable situations, when the localized presentation is a consequence of their own scanning activity. While honoring their skills, we also need to acknowledge the lack of determination of their behavior. Auditory stimuli do not produce stereotypical, reflexive eye movements toward an appropriate location. We found only a tendency for the baby to scan more on the side from which sound emanated. In fact, babies looked at the stimulus on the side opposite the sound after a while. Thus, there is opportunity for plasticity and for experience to refine a spatial map; still, the baby is biased toward discovering expectable concordances in a reasonable world.

Similarly, peripheral fixations in newborns are not the precise, sudden type of movement recorded in adults. Statistics are often required even to reveal the appropriateness of first eye movements. Response latencies are quite long. The bias is there, but there is room for refinement.

I suggest that in the early weeks of life babies further develop these biased correspondences and learn about how the eye orientation system works. One could easily make the case that an appreciation of stimulus orientation and location results from infants' appreciation of the type of eye movements they have to make to localize or scan stimuli. For example, a discrimination of horizontal from vertical lines may depend on a discrimination of the eye movements that are used to cross them. A similar case for the precedence of action over spatial knowledge can be made for sound localization. Newborn eye and head movements biased toward a sound source may, in fact, reflect a sound-balancing routine in the two ears from which sound–visuospatial maps are generated.

To coordinate the visuomotor behavior, the baby must learn the sensory consequences of motor output. Given, for example, that a bar is located to the left of the fovea, the baby must learn that a command to move the eyes left or right will produce a certain perceptual consequence. The original biasing tendency to move the eyes toward a peripheral stimulus must be calibrated and augmented by appropriate anticipation of a particular command–percept outcome. Thus, the baby must come to appreciate something about how the eyes operate, must coordinate that knowledge with current sensory input, and must anticipate the consequences of action. It can be argued that this is partially what newborns are doing in the first weeks of life. Until they possess the ability to look at a figure and anticipate the perceptual consequences of moving the eyes to all portions of it, they cannot see it as a whole. The reader may recognize Hebb's influence on these thoughts; Hebb (1949), however, gave little attention to the process by which newborns learn about what their eyes do, a point that has been emphasized by Bullinger (1979; see also Bullinger, Chapter 12, this volume).

Once babies have developed some knowledge of their oculomotor capacities and have enough experience with stimuli to extract information from the afferent–efferent loop, they can process the type of form Van Giffen (1980) presented. Still later, their ability to anticipate the perceptual consequences of their own eye movements through a stationary spatial array can be transferred to anticipation of a sequential spatial array, as in our light alternation study. Finally, experience with object occlusion may permit the type of perceptual inference required for integration of the subjective contours we used in our infant studies.

I have not thought through the specifics of a model of development of spatially determined behavior, but it seems to me that similar processes must be involved in developing a general conception of space and its contents; that is, instruments of inquiry are applied, and through their application reciprocal knowledge is acquired about both the instrument and the world. Campos, Svejda, Campos, and Bertenthal (1982), for example, found a correspondence between the development of locomotor skill in infants and their development of fear on the visual cliff. They speculated that interpretation of depth cues and judgments of size constancy and absolute distance depend heavily on the development of self-directed activity.

Although somewhat removed from this discussion, we might note analogies to the loop formed by instrument application and mastery and the acquisition of space knowledge throughout adult development and even over the stages of scientific inquiry. For example, advanced techniques of distance monitoring and spatial framing in the fields of celestial mechanics and radioastronomy involve such a loop; but, obviously, care must be given to analyses of such parallels.

The research findings discussed in this chapter lead to the conclusion that visual behavior is spatially determined from the earliest age. What the spatial determination of behavior implies about spatial knowledge is, however, still uncertain. At the least, spatial determination of activity tutors the acquisition of such knowledge. Natural biases exist that couple different sensory modes and sensorimotor systems, and these biases make sense in terms of the way the body and the world are organized. However, there is room for experience to contribute, and it does so through a continuing interplay between self-directed action and its perceived effect in the external world.

Acknowledgments

This work was supported by National Institute of Mental Health Research Grant MH-23412 and by Research Scientist Award MH-00367 to Marshall M. Haith. Thanks are due Betty Richardson and Denise Hall for assistance in preparing the manuscript, and to more graduate students and research assistants than it is possible to mention here for contributions of ideas and labor to the research reported.

References

Aslin, R. N., & Salapatek, P. Saccadic localization of peripheral targets by the very young human infant. *Perception and Psychophysics*, 1975, *17*, 293-302.

Bertenthal, B. I., Campos, J., & Haith, M. M. Development of visual organization: The perception of subjective contours. *Child Development*, 1980, *51*, 1072-1080.

Bertenthal, B., Haith, M. M., Campos, J., & Tucker, P. *Infants' sensitivity to subjective contours*. Paper presented at the International Conference on Infant Studies, New Haven, April 1980.

Bullinger, A. *Posture et oculomotricite*. Paper presented at the International Congress of Child Psychology, Universite Rene Descartes, Paris, July 1979.

Butterworth, G., & Hicks, L. Visual proprioception and postural stability in infancy: A developmental study. *Perception*, 1977, *6*, 255-262.

Campos, J. J., Svejda, M. J., Campos, R. G. & Bertenthal, B. The emergence of self-produced locomotion: Its importance for psychological development in infancy. In D. Bricker (Ed.), *Intervention with at-risk and handicapped infants: From research to application*. Baltimore: University Park Press, 1982.

Haith, M. *Rules that babies look by: The organization of newborn visual activity*. Hillsdale, N.J.: Erlbaum, 1980.

Haith, M. M., Bergman, T., & Moore, M. J. Eye contact and face scanning in early infancy. *Science*, 1977, *198*, 853-855.

Harris, P., & MacFarlane, A. The growth of the effective visual field from birth to seven weeks. *Journal of Experimental Child Psychology*, 1974, *18*, 340-348.

Hebb, D. O. *The organization of behavior*. New York: Wiley, 1949.

Held, R., & Hein, A. Movement-produced stimulation in the development of visually guided behavior. *Journal of Comparative and Physiological Psychology*,
1963, *56*, 872–876.

Kessen, W., Haith, M. M., & Salapatek, P. Human infancy: A bibliography and
guide. In P. H. Mussen (Ed.), *Carmichael's manual of child psychology*. New
York: Wiley, 1970.

Kessen, W., Salapatek, P., & Haith, M. M. The visual response of the human newborn to linear contour. *Journal of Experimental Child Psychology*, 1972, *13*,
9–20.

Lee, D. N., & Aronson, E. Visual proprioceptive control of standing in human infants. *Perception and Psychophysics*, 1974, *15*, 529–532.

Mendelson, M. J., & Haith, M. M. The relation between audition and vision in the
. human newborn. *Monographs of the Society for Research in Child Development*,
1976, *41*(4, Serial No. 167).

Pipp, S. *A test of theories of infant visual perception in the first two months of life.*
Unpublished doctoral dissertation, University of Denver, 1978.

Van Giffen, K. *The emergence of form appreciation in infancy.* Unpublished doctoral dissertation, University of Denver, 1980.

Chapter 11

Spatial Sense of the Human Infant

Peter C. Dodwell

Understanding of the perceptual capabilities of infants has undergone a vast transformation in the past two decades. Before 1960 there had been a strong tradition, at least among English-speaking psychologists, that regarded infancy as a period in which very little of interest was going on, or, more exactly stated, that there was little that could be studied in the perceptual activities of the human infant. All this changed when Berlyne (1958) and Fantz (1958) reported almost simultaneously that infants are sensitive to visual patterns and respond to them differentially. Berlyne investigated infants that were several months old, and Fantz concentrated his attention on the newborn. Both used the so-called "differential looking" technique, in which the infant is presented with two displays and the experimenter simply observes which display is looked at. In the intervening years information has accumulated showing that the infant in the first year of life is sensitive to many visual factors including contrast, contour orientation, complexity, contour density, color, and so on (see Dodwell, Humphrey, & Muir, in press; also see Haith, Chapter 10, this volume). Complementary to these developments has been a growing interest in the infant's ability to orient itself in space and to coordinate spatial information that is gathered through the different senses.

Concurrent with this increased interest in the perceptual behavior of the infant there has been a considerable movement to a more nativistic interpretation of that behavior. Several reasons for this can be discerned: First, there is the suggestion, already mentioned, of at least rudimentary pattern perception in the infant (Berlyne, 1958; Fantz, 1958). In addition, there have been advances in the electrophysiological investigation of the visual nervous system, particularly the now classical findings of Hubel and Weisel (1962). Since then, many investigators have been at work demonstrating that the visual systems of vertebrates, far from being randomly organized, have a remarkable degree of struc-

ture that appears to be present at a very early age and before experience
has had a chance to mold it (see Blakemore, 1978). Moreover, ethologi-
cal studies have had a powerful influence on psychology (e.g., Eibl-
Eibesfeldt, 1970; Ingle, 1978). This has been a factor in the move away
from the abstract theorizing, characterized, for example, by the work
of Hull (1943), toward a realization that psychological functioning is
subject to strong biological constraints. Perceptual functioning is subject to evolutionary pressure no less than other behavior systems. On
the theoretical side, the arguments of Gibson (1950, 1966) have strongly
affected the way psychologists view the perceiving organism. While
Gibson claimed not to be a nativist as such, it is clear that his later
theoretical work is based on an ethological-evolutionary approach to
perception and is antithetical to much that is central to the empiricist's
point of view.

Finally, developments in artificial intelligence have also had their influence on perceptual theorizing. Early attempts to simulate perceptual
behavior with randomly connected networks of the type postulated in
Hebb's (1949) theorizing were not successful (Rosenblatt, 1960). It
soon became evident that for a machine to display perceptuallike behavior of any complexity, it was desirable, if not necessary, to have a
considerable degree of structure built in. Pattern recognition by the
computer, now known as computer vision (Winston, 1975), continues
to demonstrate the necessity for something like a nativistic approach to
perceptual processing in the sense of a well-specified set of preprocess-
ing routines.

Despite this change in perspective, new information available on the
details of single-cell activity and the cytoarchitecture of the brain, etc.,
it is still true that we know little about some of the general properties
of the visual system. In particular, our knowledge of how feature information extracted at the single-cell level is integrated into representations of the world, and how the spatial orientation of the organism first
occurs and is developed through interaction with the environment is
incomplete. It is generally accepted that no simple or extreme form of
nativism or empiricism approximates the facts of perception (Dodwell,
1978; Dodwell et al., in press), but there is still little agreement about
what the overall structure is, how it functions, and how our experience affects pattern and object recognition or spatial orientation and
guidance.

The movement toward a nativistic view of infant competence has, in
certain respects, gone too far. There are numerous instances in which a
claim has been advanced for remarkably elaborate or high-level responding to external stimulation in the neonate or young infant. It has been
claimed, for example, that the very young infant can respond "defen-
sively" to approaching objects (Ball & Tronick, 1971), that the neonate
can grasp objects but does not attempt to grasp their two-dimensional

representations (Bower, 1972), and that the very young child can imitate more or less elaborate facial gestures (Meltzoff & Moore, 1977). The investigations on which such claims have been based were motivated by one strand of the neonativistic argument, asserting that the infant has evolved to respond appropriately to distal properties in its environment. The argument, in this instance following Gibson, is that a perceptual system responds initially not to isolated elements such as patches of color and randomly oriented lines or angles, but to certain aspects of the inherent structure of a stimulating environment. These structural properties, which are rooted in the nature of the physical world, Gibson referred to as *invariances*. This point of view probably has considerable validity, but it is often taken to mean—it seems quite erroneously—that in responding to such structure the neonate is displaying significant cognitive ability. The biologically appropriate response to some external event might just as well be of the nature of what ethologists used to call "innate releasing mechanisms," as a response based on real understanding of what the environmental stimulus signifies, be it an object, fact, or other event. The latter interpretation seems quite implausible, unless one has other strong grounds for believing that cognitive activity is within the repertoire of the very young baby. Such evidence, however, seems to be conspicuous by its absence. On the other hand, the evolutionary argument itself appears plausible, given the natural biological constraints under which the organism has evolved.

How does one assess which is the better interpretation? We have conducted experiments that were designed to address questions about the perceptual world of the infant. We have been especially concerned with how perceptual organization is manifested in the infant's ability to orient itself in space and the extent to which stimulation of its different senses elicits coordinated activity. We have studied both intersensory as well as sensorimotor coordination.

Experiments in Reaching

Bower's (1972) claim that neonates can distinguish between a solid object and its two-dimensional representation, evidenced by grasping for the former but not the latter, first brought me to research on the nature of the infant's spatial sense. If his claim were true, it would contradict a long tradition in infant research concerning the development of the object concept (e.g., Piaget, 1954; White, Castle, & Held, 1964) and would undermine the empiricist's argument that development of the infant's perceptual world is a result of experience and interaction with its environment. Bower made a distinction between what one can infer if the neonate simply shows a discriminative response between object and picture and what one can infer if the response is a naturally useful one, or, as we say, has "ecological validity." His argument was that a simple discrimination shown, for example, by the preferential looking technique, merely establishes that there are cues by which the infant differentiates object from picture. However, attempting to grasp a ball, but ignoring a picture of it, shows that some objectlike properties of the ball have been apprehended.

Our first attempt (Dodwell, Muir, & DiFranco, 1976) to substantiate Bower's claim failed despite our strenuous efforts to replicate precisely the conditions of his experiment. The essential differences between our results and his are shown in Table 11-1. Whereas Bower claimed a very high rate of reaching for his infants and a strong difference in grasping for objects and pictures, we were unable to find either a great deal of reaching or a differential response. The stimuli used in these experiments were an orange sphere viewed against a blue background and a colored photographic reproduction of it. Our results, subsequently replicated in essence by others (Ruff & Halton, 1978), showed that neonates had a keen visual interest in such stimuli. They spent, for example, about 60% of a 4-minute trial period inspecting the stimulus, but directed reaching was infrequently observed and did not occur selectively in response to objects and pictures.

Although we were unable to substantiate the claim that reaching serves to differentiate objects from nonobjects, we did find properties of the infant reaching response itself to be of considerable interest. In a subsequent publication we reported that the same babies had a repertoire of arm movements at 1–3 weeks that included some well-defined common components (DiFranco, Muir, & Dodwell, 1978). These components seemed to be antecedents of the truly directed and intentional grasping that appears later in infancy (Field, 1976; Halverson, 1931; van Hofsten, 1981). We found, for example, a variety of reaching styles

Table 11-1. Contrasting Results Concerning Infants' Reaching for Objects (O) and Pictures (P)

	Contacts		Reaches	
Experiment	O	P	O	P
Bower, 1972	12.0	0.0	53.0	0.5
Dodwell et al., 1976	0.46	0.39	2.2	1.9

Note. From Dodwell, P. C., Muir, D. W., & DiFranco, D. Responses of infants to visually presented objects. *Science*, 1976, *194*, 209–211.

Infants 7–23 days old were exposed for 4 minutes each to either an object or a picture. Mean number of responses per infant in each observation period for each target is shown.

Table 11-2. Reaching Style Preferences in Infants

Reaching style cluster[a]	Infants[b] (n = 20)	Reaches	
		Preferred[c]	Total
Full-arm shaped raise	3	56	121
Full-arm unshaped swipe	3	21	65
Full-arm shaped swipe	2	16	44
Forearm shaped swipe	2	14	44
Forearm unshaped swipe	1	12	25
Full-arm unshaped raise	1	7	29

Note. From DiFranco, D., Muir, D. W., & Dodwell, P. C. Reaching in very young infants. *Perception*, 1978, 7, 385–392.
[a] "Shaped" and "unshaped" refer to whether the fingers of the hand were appropriately extended or not.
[b] Number of infants showing a majority of reaches of a particular style.
[c] Number of reaches in the preferred style.

in the babies, and large differences in the frequency with which different babies attempted to reach, whether for objects or pictures. The major findings are summarized in Tables 11-2 and 11-3. Our general impression was that, whereas many of the components of directed reaching are present very early, it is relatively rare to see them well coordinated in a coherent gesture. Of particular interest are the data shown in Table 11-3. The three infants who reached most frequently also showed a very consistent pattern, one which may be described as "mature" reaching. They raised the arm, extended the fingers, and approached the visual target with a smooth action. Even these infants, however, did not show the differential form of responding to objects that Bower had claimed for all of his babies. Our studies suggest that neonates have spatial organization within the visual system, in the sense that they scan a visual target. In addition, they possess a degree of organization in the motor system, in the sense that components of directed activity toward a visual target can be seen at an early age. These components seem to proceed, at least in the first 2–3 weeks of life, in relative independence.

Experiments in Intermodal Spatial Coordination

The fact that neonates do not selectively grasp for objects does not necessarily mean that they cannot discriminate them, but it does reveal that the ability to discriminate cannot be evaluated with this response. Our observation that neonates reach toward visual targets, whether objects or not, and spend a great deal of time inspecting them visually is

Table 11-3. Percentage of Reaches in Each Style Cluster

No. of infants	No. of reaches	Full arm				Forearm			
		Unshaped		Shaped		Unshaped		Shaped	
		swipe	raise	swipe	raise	swipe	raise	swipe	raise
3[a]	121	5	3	13	48*	2	3	7	19
15[b]	201	17	12	10	13	16	8	15	9

Note. From DiFranco, D., Muir, D. W., & Dodwell, P. C. Reaching in very young infants. *Perception*, 1978, 7, 385–392.

[a] Three most frequent reachers.

[b] Remainder of infants. The preponderance of full-arm, shaped-hand raises for the frequent reachers is highly reliable; other categories are within the expected range of variation.

*$P < .01$.

in general agreement with the findings that infants and young children respond to some depth and object properties (Field, 1976; Piaget, 1954; White et al., 1964). Another way of asking whether infants are sensitive to objectlike properties in the environment is to ask whether they can detect relations between stimuli in different modalities. For example, one could ask whether infants can detect whether a sight and a sound come from the same location in space; this is one of the most general objectlike properties one can imagine, as an object is commonly a source of stimulation for more than one modality. This possibility was examined in experiments that determined the sensitivity to the common position in space of such stimulation. We used orienting as a dependent measure. Infants, as well as older children, turn the head and/or eyes toward a source of stimulation in the visual modality. Our experiments have demonstrated that this is also true for hearing and touch. Responsivity of the neonate to sound and touch has been a source of some disagreement in the literature, disagreement which our experiments have begun to resolve.

Coordination of Sight and Sound

Our first experiment was an attempt to discover whether infants are sensitive to correlated sources of vision and hearing. The intention was to use the approach first with older infants and, if positive results were obtained, then to find how much earlier in life this sensitivity could be demonstrated. In the first experiment we studied 2½-month-old infants (Field, DiFranco, Dodwell, & Muir, 1979). Lying comfortably supported, the infant faced a black semicircular screen. A visual stimulus was presented immediately in front of the infant; while the infant was observing it steadily that stimulus was extinguished and another was presented either on the right or left side, 60° from the midline. The second stimulus was one of three possible types: visual (condition V), a female face suddenly appearing over the screen; auditory (condition A), a female voice speaking poetry; or a combination of the two. In the combined condition the visual and auditory stimuli either came from the same position (condition A + V) or from opposite locations, one 60° to the left and one 60° to the right (condition A–V). As in all our infant experiments, the subject was videotaped from two positions simultaneously, the two views being displayed on a split-screen monitor. We measured the speed and accuracy of turns toward the visual and auditory targets alone, or toward the two in combination. In the control condition no stimulation was presented.

The results were quite straightforward. Infants at this age oriented with nearly 100% accuracy toward a visual target presented at 60° from

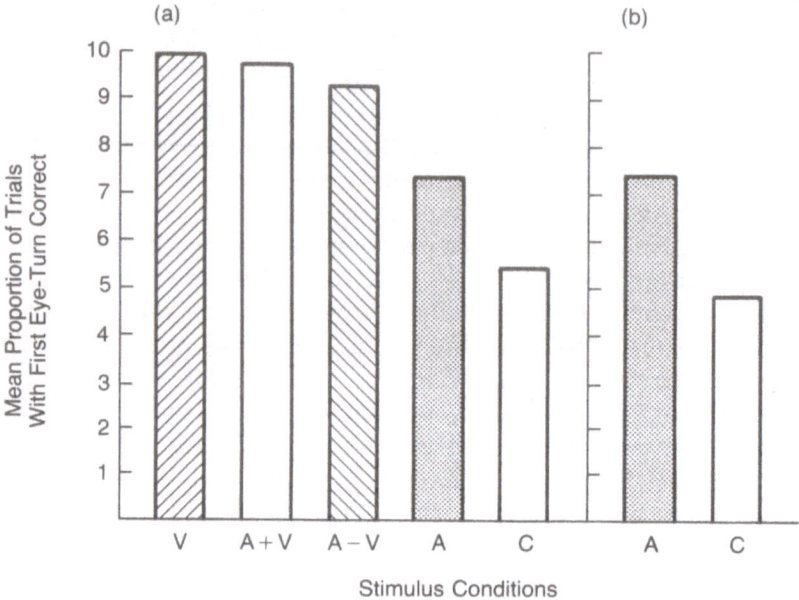

Fig. 11-1. Responses to visual and auditory stimuli in 2½-month-old infants. Mean proportion of trials for each stimulus condition in which subjects' first eye turns were in the correct direction. V, visual; A, auditory, C, control; A + V, "same location" combined condition; A-V, "different location" combined condition. (b) Replication of auditory and control conditions. (From Field, J., DiFranco, D., Dodwell, P. C., & Muir, D. W. Auditory-visual coordination in 2½-month-old infants. *Infant Behavior and Development*, 1979, 2, 113-122.)

the midline, whether or not it was accompanied by a sound from the same location. Auditory stimulation from another location (condition A-V) had little affect on orienting to visual targets. Infants did show a tendency to orient toward the auditory target when presented alone, but accuracy was lower (around 70%) and the latency considerably longer. These results are summarized in Figure 11-1. The difference in latency to orient to visual and auditory targets (Figure 11-1) could explain why the visual target always dominated when it was present. This domination of sound by sight is not really surprising, although it had not been so clearly demonstrated before. What was more unusual was the clear indication that infants at 2½ months orient toward a sound alone. Earlier evidence on this point has been somewhat conflicting, so we decided to replicate the auditory only condition (A). There was almost perfect replication; we concluded that infants at this age orient with moderate accuracy, although slowly, toward an appropriate source of auditory stimulation.

The emphasis of our research had now shifted from questions of object recognition to an attempt to understand sensory mapping, and the extent to which common maps for the different modalities can be identified very early in life (see Chapter 10, this volume). The fact that an infant orients head and eyes toward a source of stimulation is evidence of spatiomotor coordination; this is obviously so if the stimulus is visual. The same orientation response to nonvisual targets is evidence of some preprogrammed ability for intermodal spatial coordination.

Development of Orienting to Sound

Muir and Field (1979) demonstrated convincingly that orienting to a sound source can be elicited in very young infants. They used a steady, rhythmic, broadband noise source and showed that the neonate (only a few minutes old or even premature) turns its head and eyes toward a source of sound 90° to the left or right. It appears that the earlier uncertainties and conflicting findings were due to the use of inadequate auditory stimuli, typically, brief buzzes or clicks. Muir and Field's findings showed that the response, while reliably present, is of relatively long latency, and in the form of a slow and deliberate turning of the head. About 70% of neonates show the response, and those that respond do so quite reliably.

It is instructive here to consider the developmental history, which we have followed in some detail (Field, Muir, Pilon, Sinclair, & Dodwell, 1980; Muir, Abraham, Forbes, & Harris, 1979). Although most neonates respond to the auditory stimulus, we found that responsiveness declines with age, and there is an interval, approximately during the third month of life, when it disappears almost completely. (From this point of view we were perhaps lucky to obtain consistent auditory responding in our earlier experiment with 2½-month-olds.) Auditory orienting reappears, in essentially all infants, by the fourth month of life. This may be the origin of the conventional pediatric view that auditory orienting does not occur before 4 months of age. Muir et al. (1979) followed four infants over the first 4 months of life in great detail. They were able to track not only the decline in auditory orienting (which, incidentally, cannot be attributed to mere boredom with the stimulus—see Figure 11-2d), but also its rapid increase by the end of the fourth month, with significant changes in the latency of responding. These results are shown in Figure 11-2. It is probably not true that the reappearance of auditory orienting in the fourth month is simply a return of the earlier behavior; it has a different character at this stage. Rather than being "drawn to" the sound, the older infant appears to search with a rapid glance for the source of stimulation. On the average, the latency of these these later turns is only about one-half that of the earlier movements.

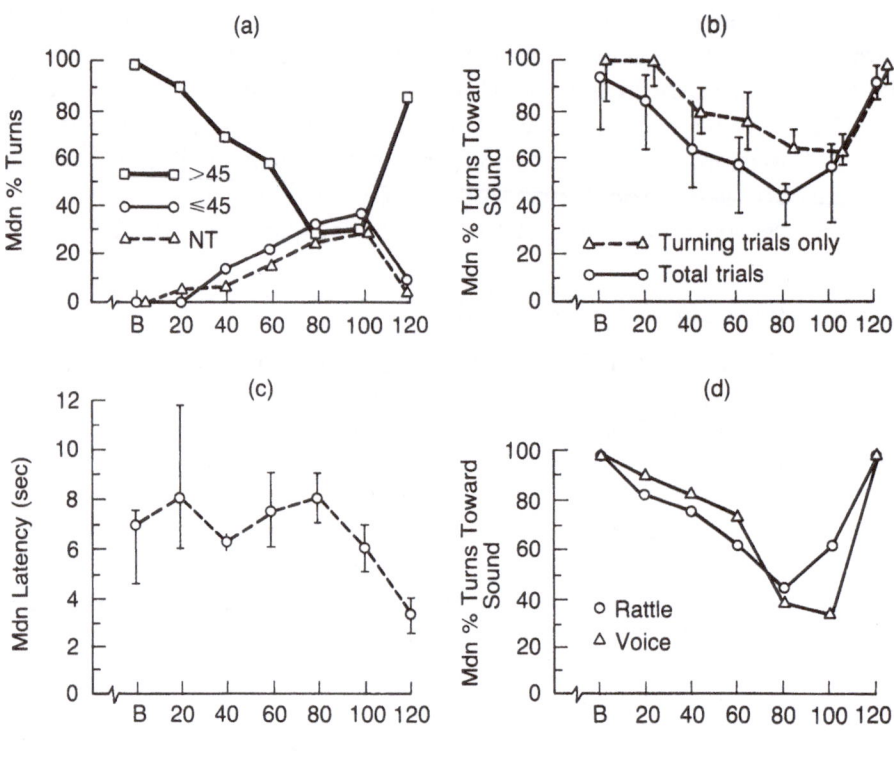

Fig. 11-2. Responses to sound as a function of age; median data for four infants. (a) Percentage of trials on which large, small, and no turns (*NT*) occurred. (b) Results of turning trials compared with results of all trials. (c) Latency to complete a head turn. (d) Percentage of turns one infant made toward two types of sound pesentations. (From Muir, D. W., Abraham, W., Forbes, B., Harris, L. S. The ontogenesis of an auditory localization response from birth to four months of age. *Canadian Journal of Psychology*, 1979, *33*, 320–333.)

Orienting to Touch

D. DiFranco, D. W. Muir, & I recently completed a parallel series of experiments on sensitivity to tactile stimulation in the newborn. It is well known that the neonate has a highly developed reflex response to perioral stimulation (the so-called rooting response). While this has obvious biological utility, it is not the form of response to tactile stimulation that is our primary interest. Our question is as follows: Can the neonate respond with an appropriate orientation of the head, eyes, or both to tactile stimulation of the more distant parts of the body?

We have investigated orienting responses to tactile stimulation of the ulnar surface of the forearm, either by gently stroking with a finger, or by light air puffs. While we obtained evidence of general orientation toward the source of tactile stimulation, this is not by any means as

clear as in the case of hearing. In three experiments we found that newborns consistently detected the side of the body stimulated. This is
signaled by turning of the head or eyes in the appropriate direction.
Our head-turning data are summarized in Figure 11-3. Somewhat differ-

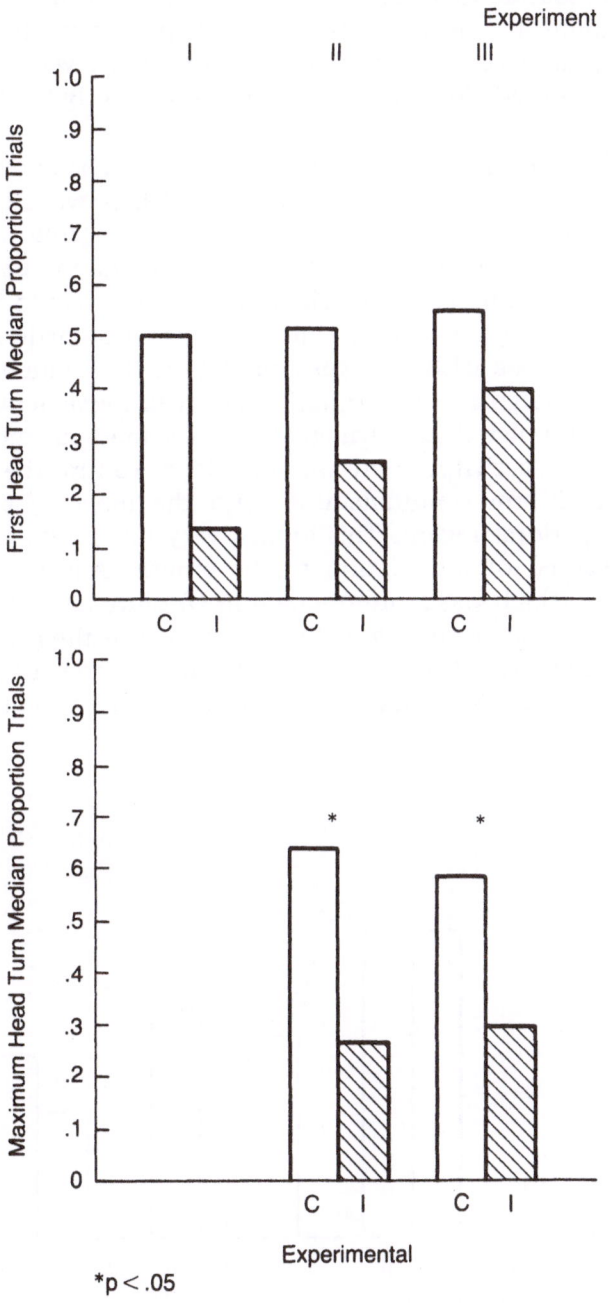

Fig. 11-3. Head turns to tactile stimulation in the newborn: results of three separate experiments. Turns in control trials were essentially random. C, correct side;
I, incorrect side.

ent measures were taken over the three experiments, depending on the particular question being asked; it is clear, however, that in all cases sensitivity to the place of stimulation on the body was manifested. The strength of these results lies in their consistency over replications. Compared to the sound-orienting results, the extent of head turning was less (typically about 45°, in contrast to the 90° turn found in our auditory experimental situation) and the latency of response was greater. Nevertheless, typically 70–80% of infants oriented correctly to the stimulated side.

Of considerable interest was the pattern of responding to tactile stimulation shown in Figures 11-4 and 11-5. There was a strong suggestion that the limb stimulated was the first to move, followed by a head turn, and then a turning of the eyes. It is as if the infant was initially alerted to the stimulation at its site, and slowly oriented the head and eyes to inspect that site. This contrasts with the startle response one might expect to see with stronger stimulation. The data in Figure 11-5 demonstrate a degree of coordination between movement of the head and of the stimulated arm which suggests something more elaborate than mere reflex activity. There was a tendency to turn the head toward the stimulated arm simultaneously with the movement of the arm toward the perioral region. This looked very much like an attempt to bring the source of stimulation to the mouth. Another tendency in coordination, which was more frequent in absolute terms but somewhat less reliable, as shown in Figure 11-5, was to turn the head toward the source of stimulation when the arm also moved toward the stimulus. This type of response looked more like what the infant does in the

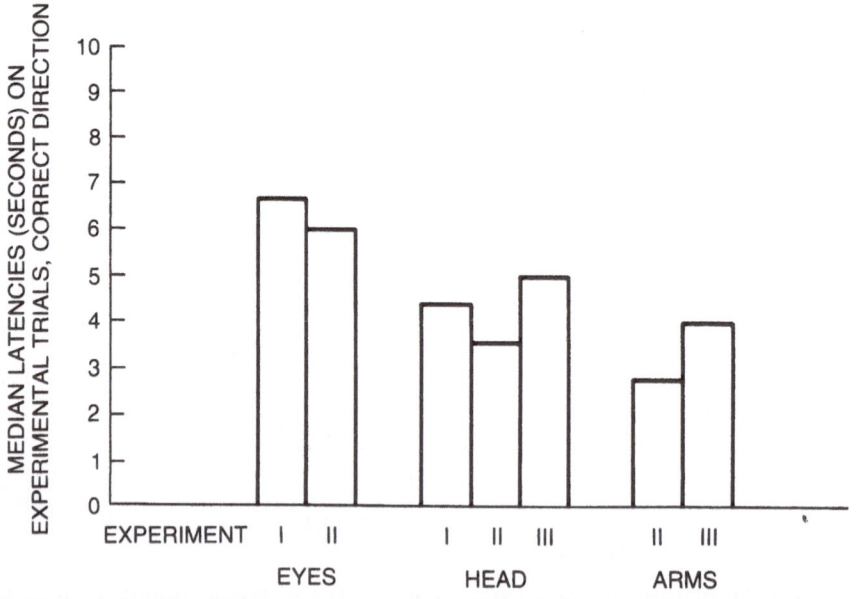

Fig. 11-4. Latencies of response to tactile stimulation in the newborn: results of three separate experiments.

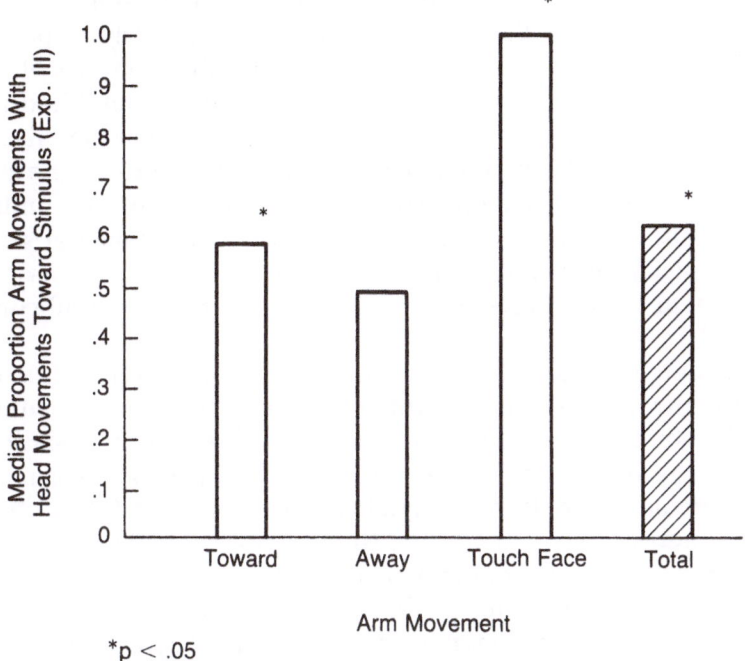

*p < .05

Fig. 11-5. Pattern of head and limb responses to tactile stimulation in the newborn.

presence of a visual target, although it was clearly not visually guided since the arm movement usually preceded the head movement. Once again, as in the case of our earlier study of reaching to visual targets, we see the appearance of somewhat organized components of directed movement, which at a later stage is coordinated into the purposive behavior of the older infant.

We have not yet followed the developmental history of responses to touch; for example, we do not know whether it shows the same decline and reappearance as is found in hearing. What we can say is that the early responses to touch have the slow, deliberate, and almost tropistic character that is found in the early responses to hearing. However they may subsequently develop, at least the early responses to touch share some of the features of early responding to hearing, although differences are also evident.

Conclusions

Taken together, the findings on the coordination of vision, hearing, and touch in the young infant do suggest strongly that there is a built-in and biologically adaptive tendency to orient toward external sources of stimulation. Although our evidence is still quite incomplete, it shows that these orientational tendencies are modality specific, and that the neonate differentiates between seeing, hearing, and feeling by touch; its typical motor responses to different modalities are dissimilar. Evidence in the infant research literature has suggested that early responding is controlled by subcortical mechanisms (e.g., Bronson, 1974, 1982; Maurer & Lewis, 1979), and that these responses are inhibited around the third month of life, perhaps by emergent cortical activity, to be supplanted by cortically controlled behavior thereafter. The clearest case in which such an argument might be applied is to our findings in audition. Although we have not followed the developmental sequence for responsivity to touch, certainly the salient tactile responses in the neonate seem to be consistent with the view that they are subcortically controlled.

There is physiological evidence that supports this notion of early subcortical control. In particular, the mammalian superior colliculus contains cells that are selectively sensitive to a particular modality, and some that are multimodal. These cells could be part of the general system subserving orientation and guidance, as Schneider (1967) first suggested. Moreover, it is known that the superior colliculus in the cat and monkey contains visual, auditory, and tactile sensory maps that directly overlay each other (Chalupa & Rhoades, 1977; Stein, Lobos, & Kruger, 1973). These could well embody the mechanism controlling the early visuomotor tropistic responses of the neonate to hearing and touch. Our experimental findings are certainly compatible with this idea and support Bronson's (1974) conception of the substrate of early behavior.

To return to the general position of the neonativist, we might argue that our results support a form of nativism, in that we find built-in coordinated behavior that is apparently modality specific, and biologically useful. However, our results are far from indicating any elaborate form of cognitive processing. In view of what we know of the undeveloped state of the neocortex at birth (Bronson, 1982), we would be surprised if evidence were to suggest that the very young infant has more than minimal cognitive capacity. However, our findings are consistent with the view that the early orientational behavior provides the base from which perceptual and cognitive abilities later develop in somewhat the same ways that were proposed in general terms by Piaget (1954). These remarks do not, of course, solve any of the problems of cognitive psychology, but they do serve to remind us that, however much we may have learned of the infant's competence in many areas over the past two decades, the claims should not be overstated.

There have been disputes in the literature not only about the interpretation of results, but also about the facts themselves, so it is worth mentioning some of the difficulties in obtaining reliable results in the type of infant research described here. First, many infants are not suitable as experimental subjects; they tend to fuss, go to sleep, or otherwise prove intractable. It is also important to know that the infant's behavior is heavily dependent on its state of arousal. According to

Brazelton's (1972) scheme, the infant can be in one of six different states, from deep sleep to overarousal, which is accompanied by screaming and other motor activity. To obtain useful results on the responsiveness of infants to external stimulation, it is best to have the baby in a quiet but alert state, which in Brazelton's scheme is State 4. To some degree it is possible to manipulate an infant into this state, and in most of our experiments this was done. We also found that the way the infant is held and handled during an experiment is important. Generally an infant held in the hands or on the lap responds better than one that is in a chair or lying supported by cushions. There is an obvious possibility of introducing bias into the experimental results if the infant handler knows what the experimental conditions are, and we normally take precautions to make sure that this cannot happen. One cannot be certain that results obtained with infants that can be kept in a suitable state for a considerable period of time will be typical for all infants. However, this is a problem that is common to virtually all types of behavioral research with infants.

Our experiments demonstrated that sensitivity to different modalities is present essentially at birth. Different forms of response are found to sight, hearing, and touch, but in our experiments they were all orientational responses, and ones that could form the basis for later intentional activity. Similarly, we demonstrated components of motor control and sensorimotor coordination in the first days of life—again types of activity that can form the basis for later, more complex behavior. The early spatial sense of the infant is based on these abilities, rather than on some more elaborate form of cognitive processing. The infant is competent in many respects, but not to the extent claimed by some neonativist researchers.

References

Ball, W., & Tronick, E. Infant responses to impending collision: Optical and real. *Science*, 1971, *171*, 817–820.

Berlyne, D. E. The influence of albedo and complexity of stimuli on visual fixation in the human infant. *British Journal of Psychology*, 1958, *49*, 315–318.

Blakemore, C. Maturation and modification in the developing visual system. In R. Held, H. W. Leibowitz, & H. L. Teuber (Eds.), *Handbook of sensory physiology* (Vol. VIII): *Perception.* New York: Springer-Verlag, 1978.

Bower, T. G. R. Object perception in infants: *Perception*, 1972, *1*, 15–30.

Brazelton, T. B. *Neonatal behavioral scale.* Philadelphia: Lippencott, 1972.

Bronson, G. The postnatal growth of visual capacity. *Child Development*, 1974, *45*, 873–890.

Bronson, G. W. Structure, states, and characteristics of the nervous system at birth. In P. M. Stratton (Ed.), *The psychobiology of the human newborn.* New York: Wiley, 1982.

Chalupa, L. M., & Rhoades, R. W. Response of visual somatosensory and auditory neurones in the golden hamster's superior colliculus. *Journal of Physiology* (London), 1977, *270*, 595-626.

DiFranco, D., Humphrey, G. D., Dodwell, P. C., & Muir, D. W. *Touch localization in newborns.* Unpublished manuscript, Queen's University.

DiFranco, D., Muir, D. W., & Dodwell, P. C. Reaching in very young infants. *Perception*, 1978, 7, 385-392.

Dodwell, P. C. Human perception of patterns and objects. In R. Held, H. W. Leibowitz, & H. L. Teuber (Eds.), *Handbook of sensory physiology* (Vol. VIII): *Perception.* New York: Springer-Verlag, 1978.

Dodwell, P. C., Humphrey, G. K., & Muir, D. W. Shape and pattern perception. In L. B. Cohen & P. Salapatek (Eds.), *Handbook of infant perception.* New York: Academic Press, in press.

Dodwell, P. C., Muir, D. W., & DiFranco, D. Responses of infants to visually presented objects. *Science*, 1976, *194*, 209-211.

Eibl-Eibesfeldt, I. *Ethology, biology of behavior.* New York: Holt, Rinehart, & Winston, 1970.

Fantz, R. L. Pattern vision in young infants. *Psychological Record*, 1958, *8*, 43-47.

Field, J. Relation of young infants' reaching behavior to stimulus distance and solidity. *Developmental Psychology*, 1976, *12*, 444-448.

Field, J., DiFranco, D., Dodwell, P. C., & Muir, D. W. Auditory-visual coordination in 2½-month-old infants. *Infant Behavior and Development*, 1979, *2*, 113-122.

Field, J., Muir, D. W., Pilon, R., Sinclair, M., & Dodwell, P. C. Infants' orientation to lateral sounds from birth to three months. *Child Development*, 1980, *51*, 295-298.

Gibson, J. J. *The perception of the visual world.* Boston: Houghton-Mifflin, 1950.

Gibson, J. J. *The senses considered as perceptual systems.* Boston: Houghton-Mifflin, 1966.

Halverson, H. M. An experimental study of prehension in infants by means of systematic cinema records. *Genetic Psychology Monographs*, 1931, *10*, 110-286.

Hebb, D. O. *The organization of behavior.* New York: Wiley, 1949.

Hubel, D. H., & Wiesel, T. N. Receptive fields, binocular interaction and functional architecture in the cat's visual cortex. *Journal of Physiology* (London) 1962, *160*, 106-154.

Hull, C. L. *Principles of behavior.* New York: Appleton-Century-Crofts, 1943.

Ingle, D. Mechanisms of shape-recognition among vertebrates. In R. Held, H. Leibowitz, & H. L. Teuber (Eds.), *Handbook of Sensory Physiology* (Vol. VIII): *Perception.* New York: Springer-Verlag, 1978.

Maurer, D., & Lewis, T. L. A physiological explanation of infants' early visual development. *Canadian Journal of Psychology*, 1979, *33*, 232-252.

Meltzoff, A. N., & Moore, M. K. Imitation of facial and manual gestures by human neonates. *Science*, 1977, *198*, 75-78.

Muir, D. W., Abraham, W., Forbes, B., & Harris, L. S. The ontogenesis of an auditory localization response from birth to four months of age. *Canadian Journal of Psychology*, 1979, *33*, 320-333.

Muir, D. W., & Field, J. Newborn infants orient to sounds. *Child Development*, 1979, *50*, 431-436.

Piaget, J. *The construction of reality in the child.* New York: Basic Books, 1954.

Rosenblatt, F. Perception simulation experiments. *Proceedings of the Institute of Radio Engineers*, 1960, *48*, 301–309.

Ruff, H., & Halton, A. Is there directed reaching in the human neonate? *Developmental Psychology*, 1978, *14*, 425–426.

Schneider, G. F. Contrasting visuomotor function of tectum and cortex in the golden hamster. *Psychologische Forschung*, 1967, *31*, 52–62.

Stein, B. E., Lobos, E., & Kruger, L. Sequence of changes in properties of neurons of superior colliculus of the kitten during maturation. *Journal of Neurophysiology*, 1973, *36*, 667–679.

van Hofsten, C. *Eye-hand coordination in the newborn.* Unpublished manuscript, Uppsala University, 1981.

White, B. L., Castle, P., & Held, R. Observations in the development of visually-guided directed reaching. *Child Development*, 1964, *35*, 349–364.

Winston, P. H. *The psychology of computer vision.* New York: McGraw-Hill, 1975.

Chapter 12

Space, the Organism and Objects, Their Cognitive Elaboration in the Infant

André Bullinger

This chapter is concerned with the development of spatially oriented behavior in the human infant during the first few months following birth. We have chosen to concentrate on visual tracking of a moving object as this activity requires at least minimal sensorimotor coordination, especially with movements of large amplitudes, where both head and the eyes must be involved. The research to be described will also be discussed within a particular developmental perspective that could provide a new interpretation of some existing studies. We will argue that the kind of sensorimotor coordination necessary for adequate tracking of a moving object is not complete at birth but has to be constructed, along with the cognitive elaboration of an object which implies properties that are independent of its state of motion and its location.

We would like to make a distinction between what may be called "global" or "undifferentiated" behavior and so-called "pre-wired" or "pre-coordinated" behavior. In the former case, the whole organism participates in an ongoing activity and as such certain behavioral acts will be correlated or co-temporal, but not necessarily coordinated. In the latter case, an innate coordination is presupposed, which, we feel, may be more apparent than real. We feel that much confusion arises when it is assumed that the individual body segments involved in a given activity such as reaching, have an *a priori* autonomy of control and thus can be studied without reference to the organism as a functioning totality.

From our point of view, the newborn infant *is* just such a functioning totality, unable to discriminate between "self" and "other" or between "body" and "object." External (and internal) realities interact with biological properties of the organism, ensuring that the organism functions. It is through this interactive functioning that invariant properties of objects and of the sensorimotor systems themselves may be extracted, leading to differentiation and eventually to coordination of body parts with the potential for dissociating an object from a particular action. When this state is reached it becomes possible to speak of an object or a sensorimotor system as being "represented" at a cognitive level. These representations would then be the basis for the sensori-motor coordinations involved in complex actions such as tracking or reaching. A developmental approach to the study of spatial behavior thus considers the transformations in this behavior as being essential steps towards more automatized, goal-directed performance which are evidence not only of physiological maturational but also of cognitive changes with more general implications for the organism.

The Development of Organized Movements for Tracking a Moving Visual Object

By 3 days of age infants sitting in an upright position are able to orient towards a visual target placed at eye level. They are also able to track a moving target with rotations of the head (Bullinger, 1977). For target displacements of 60° (lasting 4 to 6 seconds), it may take up to 15 seconds for the head to stabilize in a new position. The head seems to be suddenly released from its starting position, accelerating quickly to a position generally oriented towards the target (cf. Fig. 1). After a period of oscillation, the head stabilizes in a new position which reflects the target location. This kind of movement seems is not predicted by the feed-forward mechanism described by Bizzi, Kalil, and Tagliasco (1971) in the adult monkey. The impression is more one of feed-back, permitting calibration and thus paving the way for feed-forward at a later stage in development.

It should be noted that this tracking activity is accompanied by variations in heart-rate concomitant with the appearance and disappearance of the target in the visual field (Bullinger, 1979). This is an important observation because it suggests the engagement and total mobilization of the baby by the task.

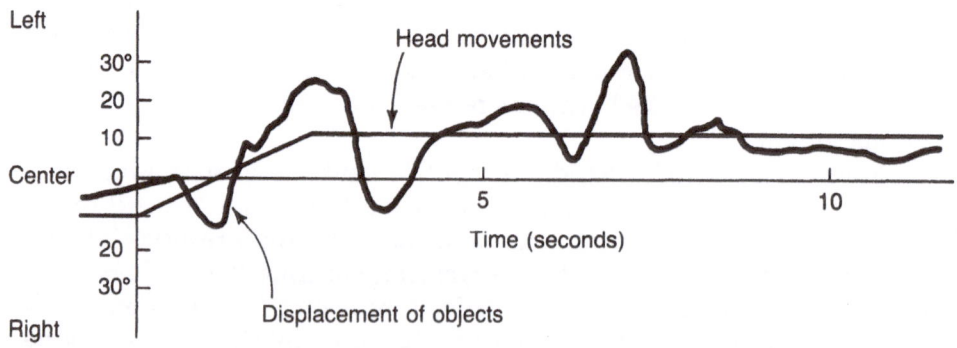

Fig. 12-1. Plot of head movements with displacement of the object.

A simple analysis of the baby's movements leads us to believe that the postural support for this tracking activity is unusual: there is generally an asymetrical posture at the onset of tracking, rather like the Tonic Neck Reflex (TNR). The laterality of this asymetrical posture changes with the trajectory of the moving object. It thus seems as if tracking activates a dynamic form of the TNR, causing the baby to pass from a left-facing asymetrical posture to a right-facing asymetrical posture, or vice versa, depending on the direction of target movement. These observations clarify the biologically-determined foundations of the organism capacity for orientation and for maintainence of its posture relative to a source of stimulation.

Experiment 1

In order to analyze this postural evolution in more detail, we constructed another experimental situation, similar to the one from which the above observation were obtained. Our aim was to discover which kinds of transformation of the global activity described above would provide evidence for the emergence of a cognitive organization which would produce better-adapted, more rapid tracking performance, in which the head alone would move. Subjects in this study belonged to 3 age groups: 15–40 days, 40–80 days and 80–120 days. The experimental situation was as follows: The baby was seated in a chair that was adapted to his or her size that gently restrained the pelvis, supported the back and allowed free movement of the arms and head. A mobile target consisting of a red woolen pom-pom subtending a visual angle of $6°$-$8°$ was suspended in front of the infant at a distance of 70 cm. The object was moved through a semicircle according to a pre-determined program. The total trajectory was $170°$ ($85°$ to the left and $85°$ to the right of the midline). The angular velocity was $10°$/second and thus the time to complete the movement in one direction was 17 seconds. A 16mm movie motion picture camera positioned before the baby and a mirror placed above it provided simultaneous front and top views. A digital chronometer and a counter indicating the angular position of the mobile were also recorded on the film.

An analysis of a variety of parameters has provided some interesting results (Table 1).

Table 12-1 Results of Tracking Experiments in Infants

Age (days)	Rate of tracking (%)	Complete tracking (%)	Rate of segmentation (%)	Mean amplitude (degrees)
15–40	50	13	34	87
40–80	76	47	14	114
80–120	75	12	61	113

In the youngest infants tracking with the head over the entire trajectory was rare, the mean amplitude being close to one half the field. These babies were able to track the target object to the midline, rarely crossing it. The possibility of tracking appeared largely to be determined by the posture at the onset. The rate of tracking was higher for infants in the next age group for whom tracking was rarely interrupted and seemed to be slaved to the trajectory of the target. If the baby did stop tracking he or she generally returned the head to the object's starting place, producing a characteristic (TNR type) posture. For the oldest group of infants the task seemed to present no problems; they tracked with few interruptions. When these occurred they were followed by reorienting movements which appeared to take the trajectory of the object into account.

Postural analysis based upon the relative positions of body parts (position of eyes, nose and wrists) enabled us to chart the evolution of tracking behavior over the 3 age groups (Bullinger, 1981). For the youngest group the possibilities for tracking were restricted by the typically asymmetrical initial posture (TNR) and involved a rotation of the whole trunk. Observation of the arm movements from the sequence uncovers a form of 'early reaching' described by other investigators (Bower, Broughton, & Moore, 1970). The intermediate group revealed a strict association between arm position and spatial location of the mobile: in the course of tracking the mobile, the posture changed from one asymmetrical form of the tonic neck reflex to another. The trunk was involved to a much lesser extent. In *the oldest group* arm movements were no longer related to the location of the mobile object in space and the head alone was mobilized by the task of tracking. Thus it appears that, starting with a global and biologically-determined engagement of the organism in the task, tracking behavior evolved with the developing independence of body segments, leading to greater precision and economy of movement. Ultimately, only the body segment involved actually moves. This evolution is summarized in Figure 2.

The white star shows the region of intersection of imaginary lines connecting the infant's head and the mobile object at different moments during tracking. For the youngest group the center of rotation was situated between the target and the head reflecting displacement of the body axis and mobilization of the entire trunk. In the intermediate group, the arms and shoulders participated in the activity of tracking but the torso did not and so the center of rotation was nearer to the body axis. For the oldest group, the center of rotation corresponded to the axis of rotation of the head. The task provoked only head movements, with the torso serving as a stable reference and the arms no longer participating in tracking. At this level of development performance was well coordinated with the spatial properties of the trajectory. These new coordinations are genuinely spatial inasmuch as they concern the movements of an object in space defined with respect to the body.

A problem is raised by these results concerning the nature of underlying oculomotor activity and a second experiment was carried out to clarify this issue.

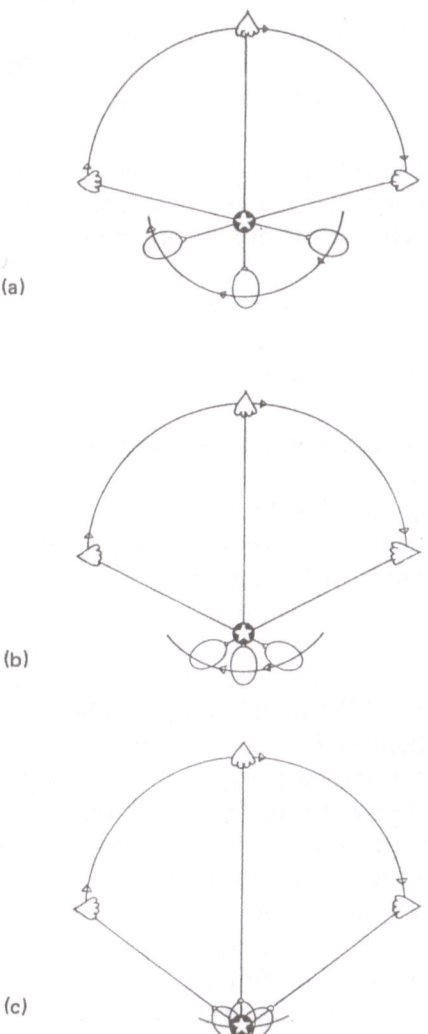

Fig. 12-2. Evolution of the position of the center of rotation (⊗) of the pair formed by the head and mobile. This position progressively approaches the axis of rotation of the head. The global participation of the organism thus diminishes ultimately to provide a fulcrum for the rotation of the head centered on the cervical axis. (a) Infants 15–40 days old. (b) Infants 40–80 days old. (c) Infants 80–120 days old.

Experiment 2

Currently available techniques unfortunately preclude the recording of eye movements in infants while leaving the head free to move. We did, however, wish to carry out a study of 9–11 week-old infants during a tracking task and decided to accept the inevitable consequences of restricting head movements.[1] The task required exploration of a three-dimensional object (a construction of Lego pieces) subtending a visual angle of 9° at a distance of 80 cm from the infant's head. The object could be moved 20° to the left or right of the midline. Five situations were presented to the infant, as summarized below:

1. Stable object → movement → stable object with change of location
2. Stable object → movement → stable object with same location
3. Stable object → occlusion of movement → stable object with change of location
4. Stable object → occlusion of movement → stable object with same location
5. Continuous presentation of stable object in same location

In situations 1–4 the stationary phases lasted 20 seconds and the movement or occlusion phases lasted 5 seconds. In situation 5 the object was presented continuously for 45 seconds.

The results showed that the three-dimensional object held the infants' gaze much longer than is usual with a two-dimensional stimulus. In terms of the location of fixations, when the infant looked toward the object his or her eye movements remained within the object zone for more than 85% of fixations. The modes of oculomotor functioning described by Haith's (1978) "rules" were seen here particularly clearly.

As the object was displaced, its movement was tracked with the eyes. When the object was stationary or came to rest it was explored intensely. One result, observed only when the object's trajectory of displacement was visible, was unanticipated. After the infant had seen the stationary object in its second location, the gaze tended to return to the first location, which now was empty.

This phenomenon is difficult to demonstrate statistically since, following displacement, there were few fixations in the empty zone relative to the large number of fixations on the object. Nevertheless, if we compare the frequency of fixations in the zone previously occupied by the object (and now empty) with the frequency of fixations in other empty zones in the stimulus field, the difference is significant. In order to interpret these results it is necessary to recall that stable objects

---

[1] Research carried out in the laboratory of Professor M. Haith at the University of Denver, Colorado, with his collaboration and that of K. Purcell.

elicit the saccadic eye movement system, whereas objects in motion elicit both the slow pursuit and the saccadic systems.

If the properties of an object are to be elaborated by means of the oculomotor system (and *mutatis mutandis* by any other sensorimotor system) it follows that such intrinsically different modes of functioning of this system as slow pursuit and saccadic eye movements should be coordinated at a psychological level. If this were not the case, one would have to explain why the moving object (eliciting the pursuit system) and the stationary object (eliciting the saccadic system) should be perceived as one and the same. Although this coordination may exist on a biological level, it has to be reflected on a psychological level, with consequences not only for the perceived object, but also for the oculomotor system, since both objects and eye movements come to be represented within cognitive space. We suggest that these results would be difficult to explain from a purely biological perspective. As a consequence they might constitute the kind of evidence for the cognitive elaboration discussed above. From a developmental point of view, we should recall that development of this behavior coincides with the emergence of capacities for analyzing the spatial properties of both an object and its trajectory (possibility of leaving the object and finding it again) as shown in Experiment 1, above.

A similar kind of coordination may be seen in the monkey (Paillard, 1971) between the activities of seizing and manipulating an object. These two activities may be carried out by the visual system alone or in coordination with manual activity. Both activities are involved in the same act and their coordination on a cognitive level ensures the perceived unity of the object. Developmentally, it is the initial lack of coordination that becomes evident from our results. The newborn's postural activities toward the object would be interpreted, from our point of view, as related to an orientation of the visual system and not to any manipulatory activity ("pre-reaching," etc.). Such manipulatory activity exists, but it is dissociated from orientation even though they may be contiguous or cotemporal.

The head and visual system have to be become tools (analogous to the way tools are manufactured to carry out specific jobs) before the moving and stationary object have the perceived unity that permits ballistic capturing with the goal of manipulation. Seen as a whole, this kind of activity would be evidence of a yet more complex spatial coordination.

The experiments described here may not be seen simply as the development of sensorimotor coordination adequate to the task of tracking a moving object. We argue that this coordination is not simply sensorimotor but that it has a simultaneous cognitive component (perception of a single object within a spatial framework) that is in fact necessary for the establishment of such sensorimotor coordinations.

Acknowledgments

The author would like to thank Marlyse Hoegen for secretarial assistance and Peter Coles for translating this paper and for commenting on an earlier manuscript. This work was supported by the Fonds National Suisse de la Recherche Scientifique, Grant 1.828-0.78.

References

Bizzi, E., Kalil, R. E., & Tagliasco, V. Eye-head coordination in monkeys: Evidence for central patterned organization. *Science*, 1971, *173*, 452–454.

Bower, T. G. R., Broughton, J. M., Moore, M. K. The coordination of visual and tactual input in infants. *Perception and Psychophysics*, 1970, *8*(1), 51–53.

Bullinger, A., Orientation de la tête du nouveau-né en présence d'un stimulus visuel. *L'Année Psychologique*, 1977, *2*, 357–364.

Bullinger, A. La réponse cardiaque comme indice de la sensibilité du nouveau-né à un spectacle visuel. *Cahiers de Psychologie*, 1979, *22*, 195–208.

Bullinger, A. Cognitive elaboration of sensorimotor behaviour. In G. Butterworth (Ed.), *Infancy and epistemology*. London: Harvester Press, 1981, pp. 173–199.

Haith, M. M. Visual competence in early infancy. In R. Held, H. W. Leibowitz, & H. L. Teuber (Eds.), *Handbook of sensory physiology* (Vol. 8): *Perception*. Berlin: Springer, 1978, pp. 311–356.

Paillard, J. Les déterminants moteurs de l'organisation de l'espace. *Cahiers de Psychologie*, 1971, *14*(4), 261–316.

Chapter 13

Motion Parallax Sensitivity and Space Perception

Ken Nakayama

Spatially coordinated behavior is critically dependent on a visual appreciation of the surrounding three-dimensional environment. Many cues are available, including binocular disparity, motion parallax, perspective, shading, and interposition. Binocular disparity has received the most attention, perhaps justly so, because it admits to a neurophysiological analysis. It appears to be processed by binocular cortical neurons at a relatively early level in the visual pathway (Barlow, Blakemore, & Pettigrew, 1967; Poggio & Fischer, 1977). Less appreciated, but possibly as primary as stereopsis, is the contribution of optical velocities for monocular encoding of depth. This view is supported by the spatially adaptive behavior of monocular individuals as well as those animals having little or no binocular overlap. Recent evidence also has indicated that motion parallax alone can provide an unambiguous and compelling sensation of depth, equal to that obtained with binocular cues (Rogers & Graham, 1979).

Space Perception

Optical Velocity Vector Field

Before a description is given of a possible neural mechanism that could efficiently code the monocular optical velocity field, this vector field is reviewed in simplified mathematical terms as previously outlined (Nakayama & Loomis, 1974). We considered an observer, free to move his eyes, as he moves through a rigid environment. First, it was recognized that the combination of eye and head motions can be intricate. The head can translate through space with either rectilinear or curvilinear motion, and it can also rotate simultaneously. Furthermore, the eye can also rotate simultaneously about a different axis other than the axis of the head. Thus, the corresponding kinematic equations of motion can be complex (Goldstein, 1950). Surprisingly, however, the optical velocity flow field for even the most complex kinematic case is simpler. Given a rigid environment, the instantaneous optical velocity vector field is the vector sum of only two elementary vector fields: a *rotational* and a *translational* field.

The rotational component is determined by the vector sum of head and eye angular velocity with respect to a rigid environmental coordinate reference frame. It generates a field of optical velocities that is solely dependent on the angular velocity of the eye with respect to space. If we adopt a two-dimensional spherical optical array to describe the monocular field, the rotational vector field consists of velocity vectors tangent to latitude lines, with the poles defined by the axis of rotation. The magnitude of the velocity rotational vector can be described as follows:

$$\omega_r = \omega \sin \phi \tag{13-1}$$

Thus the angular velocity of any optical point due to rotation (ω_r) is proportional to the eye angular velocity (ω) and the sine of the angle ϕ between the resultant axis of rotation and the vector formed by the optical point and the center of rotation of the eye. The velocity of this rotational component is uniquely determined by the observer's rotation and not by the distances of points in the environment. Therefore, in a rigid and stationary environment, the rotational component is purely *proprioceptive*, and only provides information as to the angular velocity of the eye with respect to space.

As stated above, there is also a translational component of the optical velocity field accompanying the normal translational movements of an observer. Its appearance is familiar to those acquainted with the well-known work of Gibson (1950, 1966), showing an optical expansion in the direction of the translational motion and a contraction in the opposite direction. It consists of velocity vectors parallel to longitude lines, where the origin of the vectors is the point in space toward which the observer is moving. In contrast to the rotational field, the translational field carries information regarding the distances of environmental points. The optical velocity of this translational field can be described by the equation:

$$\omega_t = V \sin \Theta / s \tag{13-2}$$

where ω_t is the optical velocity vector magnitude accompanying pure translational motion, V is the observer's velocity, s is the distance of the environmental point in question, and Θ is the angle that this point forms with the observer's direction of motion. It should be clear from Equation (13-2) that the angular velocity of the translational component is inversely related to the distance of the point in space. Thus the translational field carries environmental distance information, and if Equation (13-2) could be evaluated, all information regarding the distances of objects would be available.

However, the evaluation of Equation (13-2) requires knowledge regarding the velocity of the observer (V). Without such an estimate, the translational velocity field only supplies information regarding the relative distances of objects, not absolute distances. Although this may seem to be a severe limitation, it need not be. Independent *a priori* information regarding the distance of only one environmental point contains implicit information regarding the observers velocity, by Equation (13-2), and thus suffices to determine the absolute distances of all objects in space. Points on the ground at different angles with respect to straight ahead, for example, always have the same approximate distance from the eye, at least for terrestrial animals that locomote on the ground. These points and their fixed distances comprise such a priori knowledge. Thus, the absolute distances can be theoretically determined from the optical velocity field without the need to make a direct measurement of observer velocity, as might be suggested by Equation (13-2). An illustration of how points of known distances can calibrate the velocity-based measurement of unknown distances is given by the equation in Figure 13-1.

The component optical velocity fields accompanying pure translation

$$S_i = S_0 \left(\frac{\dfrac{d\theta_i}{dt}}{\dfrac{d\theta_0}{dt}} \cdot \frac{\operatorname{Sin} \theta_0}{\operatorname{Sin} \theta_i} \right)$$

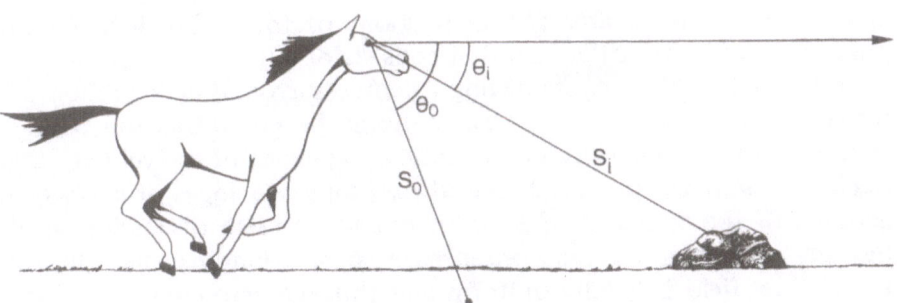

Fig. 13-1. Demonstration that the optical velocity field plus the known distance of only one environmental point (S_0) provides sufficient information to obtain the absolute distances of all other environmental points without requiring direct knowledge of one's own translational velocity. S_0, a known and ordinarily fixed distance on the ground; S_i, an unknown distance. From the equation in the inset it can be seen that S_i is fully defined by the velocity field.

and pure rotation are the only ones that can be generated even by the most complex forms of observer movement in a rigid environment. There are no others. Any optical velocity field generated in a rigid environment is simply the vector sum of each of these two fields:

$$\omega = \omega_r + \omega_t. \tag{13-3}$$

A more complete theoretical treatment of this subject is reported by Nakayama and Loomis (1974) and subsequent contributions have been made by Koenderink and Van Doorn (1976), Longuet-Higgins and Prazdny (1980), and Prazdny (1980).

In summary, each component of the velocity field contains different types of functional information. The rotational component is proprioceptive, providing information as to the angular velocity of the eye. The translational field is both proprioceptive and exteroreceptive, depending on eye translational velocity as well as on the distances of environmental points. Given this distinction, it would seem to be of biological advantage to keep these component vector fields separate.

Two alternative operations seem available to perform this suggested segregation. In one case, the components could be separated in the visual system itself by a process of pure visual decomposition. In the other case, they might be better extracted with the help of the oculomotor system, by oculomotor decomposition.

Neural Theory of Visual Decomposition

As one example of the first alternative, Nakayama and Loomis (1974) proposed a neural model for coding optical flow so as to organize the visual field into distinct surfaces. We postulated a "convexity" function that would measure the difference of velocity between a center region and the surrounding area for all portions of the visual field. We proposed a neural wiring diagram that consists of a neuron receiving inputs from a number of subunits having a center–surround organization with respect to velocity (Figure 13-2). It should be noted that this formulation is close to taking the mathematical Laplacian of the velocity field. Such a system has the advantage of outlining the edges of surfaces regardless of the direction of motion. For the present discussion, it also has the advantage of being insensitive to rotations because the rotational flow field is locally uniform and thus has zero convexity. Figure 13-3 shows how such a mechanism could delineate the outline of an object regardless of the direction of motion. The model is physiologically attractive because it involves only local computations. It requires only local pooling of excitation and inhibitory influences, a well-recognized principle of neural connectivity (Hubel & Wiesel, 1965).

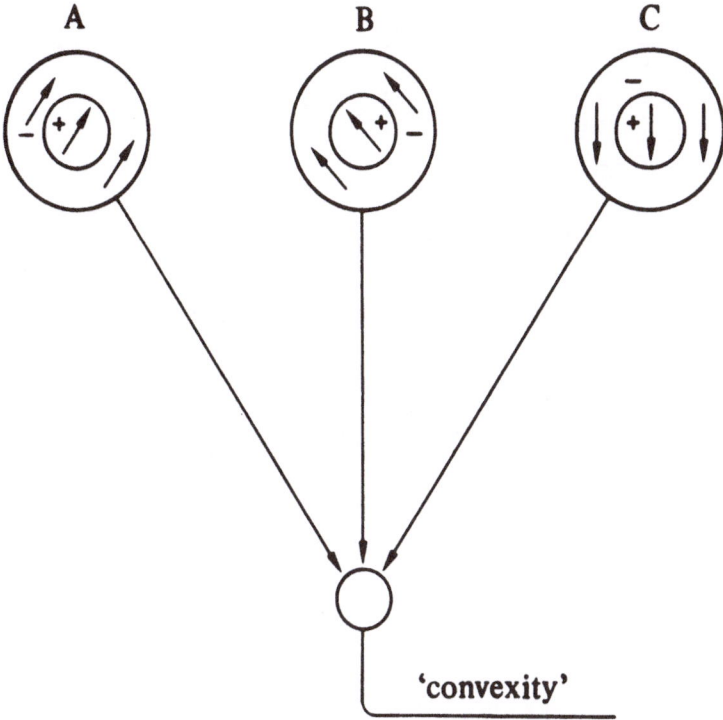

Fig. 13-2. Neural wiring diagram of a hypothetical receptive field type that would emphasize the edges of real objects in three-dimensional space. The convexity detector sums the output of subunits, each of which has a center–surround opponent organization with respect to velocity. (From Nakayama, K., & Loomis, J. Optical velocity patterns, velocity sensitive neurons, and space perception: A hypothesis. *Perception*, 1974, *3*, 63–80.)

Support for our hypothetical mechanism has just emerged from recent neurophysiological research on the receptive field properties of velocity-sensitive neurons in pigeon tectum. Frost, and Nakayama (1983) demonstrated the existence of receptive fields having close similarities to that proposed in Figure 13-2, showing separate center and surround antagonistic mechanisms with respect to velocity. Although such cells are nearly omnidirectional, a striking relativity linking center to surround can be seen. For example, when test spots were moved upward in the receptive field center, downward movement of surround patterns produced facilitation, whereas upward patterns produced inhibition. For downward movement of a test spot, it was exactly the opposite: upward movement in the surround produced facilitation and downward movement produced inhibition. Thus, the directional tuning curve of the surround of a single tectal cell could change as a function of the direction of motion in the excitatory receptive field. As such, these tectal cells share some important properties with the convexity

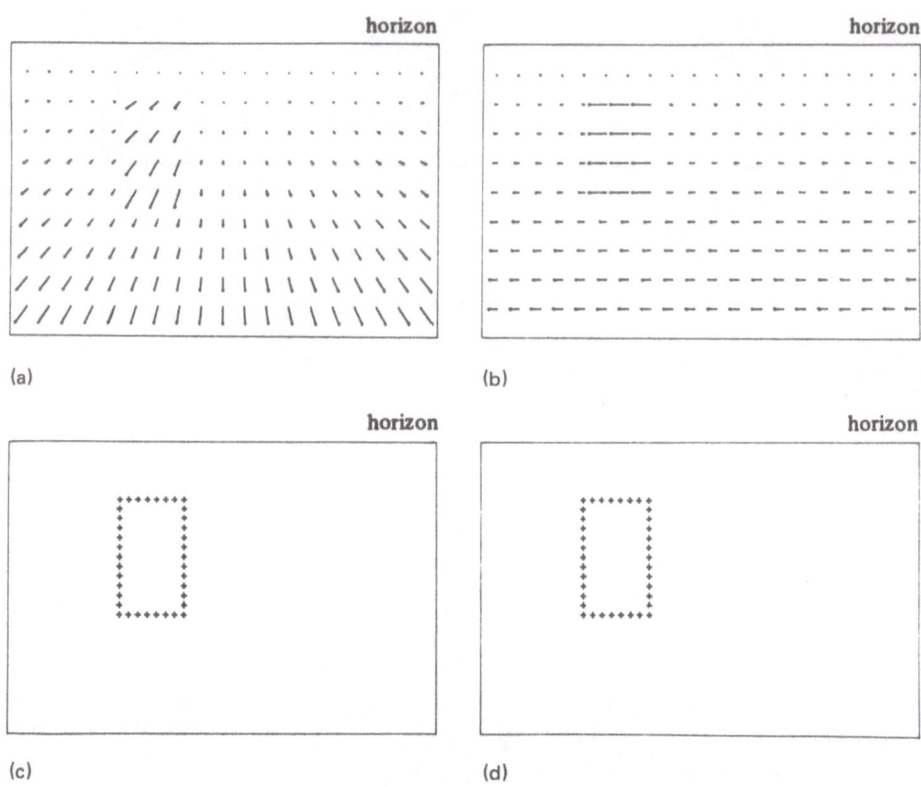

Fig. 13-3. Demonstration of how a population of motion-sensitive cells (depicted in Figure 13-2) could delineate the edge of a vertical surface suspended over the ground plane. (a, b) The optical velocity fields associated with forward and lateral motion of an observer. (c, d) The corresponding output of a two-dimensional array of convexity detectors operating on each of these vector velocity fields. Note that despite the large difference in the raw velocity vector field, the edges of the surface are similarly enhanced in each case. (From Nakayama, K., & Loomis, J. Optical velocity patterns, velocity sensitive neurons, and space perception: A hypothesis. *Perception*, 1974, *3*, 63–80.)

detectors postulated by Nakayama and Loomis (1974). With appropriate eye movements, they can sense relative motion independent of the direction of translational motion and are hypothetical neural candidates to aid in the segregation of object from background.

Although the vector field of velocities can be theoretically decomposed by mathematical operations to obtain meaningful correlates of the distal stimulus (Gibson, 1950; Nakayama & Loomis, 1974), and there are plausible physiological substrates to perform these operations (see above), more detailed information would be desirable. Whether or not this theory is reasonable depends in part on the quantitative characteristics of the motion detection system itself. For this reason some of these characteristics are described in some detail in the next section.

Psychophysical Observations in Motion Parallax

In order to examine motion processing, we made some direct measurements of the relative motion detection system. In order to begin, however, we needed to overcome a longstanding problem that has hampered experimental examination of human velocity sensitivity in the past—the contamination problem. Every time an object moves with respect to another, it occupies a new relative position, and this change has contaminated the measurement of motion sensitivity. The movement of the minute hand of a clock provides a familiar example. Do we actually see that the hand is moving, or do we infer that it has moved because we recognize it to be in a new position relative to other points? In other words, are we using a motion-sensitive mechanism or a position-sensitive mechanism?

The confounding of motion and position information has made it extremely difficult to examine motion sensitivity in isolation, to measure its minimum threshold, and to determine its other characteristics (Graham, Baker, Hecht, & Lloyd, 1948; Leibowitz, 1955).

Removal of Contaminating Cues to Position

In order to overcome the confounding of position with velocity information, we developed a moving random dot stimulus that proved to be devoid of familiar position cues (Nakayama & Tyler, 1981). When the display shown in Figure 13-4 is stationary, it appears essentially the same as the monocular half of a random dot stereogram with no features. A horizontal shearing motion is introduced such that each horizontal row of dots oscillates sinusoidally with a velocity proportional to a sinusoidal function of its vertical position (Figure 13-5). This particular movement is that of a transverse standing wave with characteristic nodes and antinodes. The amplitude, the spatial frequency (the reciprocal of distance between two nodes), and the temporal frequency of the motion waveform are under experimental control. There are no recognizable features in this random dot pattern. Any other pattern having similar pixel density appears essentially equivalent. Thus, changes in position would be imperceptible even if different parts of the display were moved differentially over large distances.

In order to demonstrate that this was indeed the case we varied the temporal frequency of the standing wave movement and measured the threshold amplitude required to see motion, concentrating especially

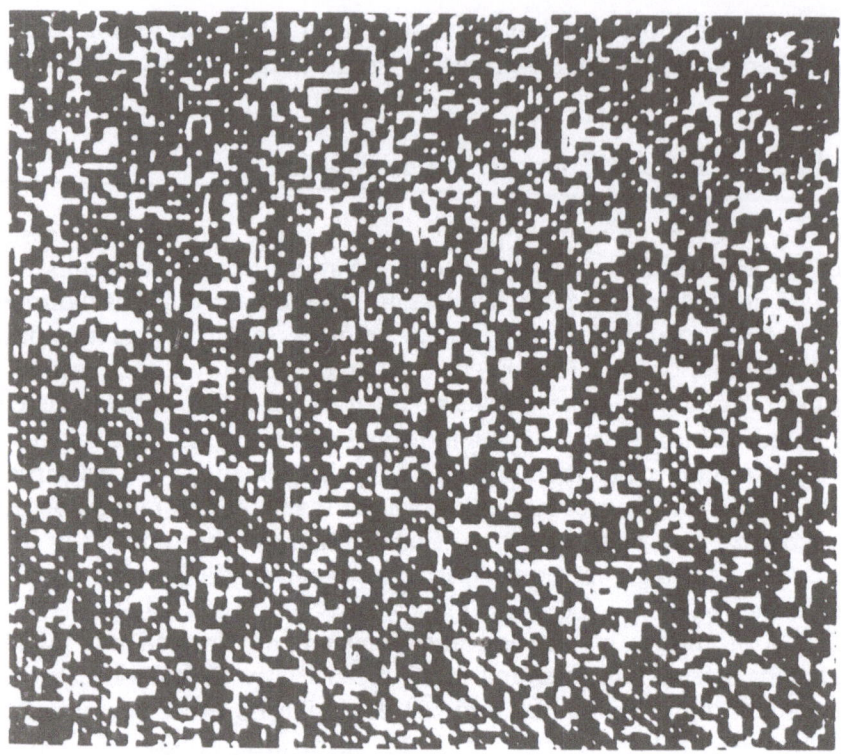

Fig. 13-4. Stimulus that has no recognizable position cues: random field of pixel elements as it appears in its static form on the cathode ray oscilloscope. (From Nakayama, K., & Tyler, C. W. Psychophysical isolation of movement sensitivity by removal of familiar position cues. *Vision Research*, 1981, *21*, 427–433.)

on the characteristics of motion at low temporal frequencies. Our prediction was that, because the peak velocity of sinusoidal movement is proportional to its temporal frequency, a velocity-sensitive system should show a reciprocal relationship between threshold amplitude and temporal frequency, at least over some significant range

$$d(\sin \omega t/dt) = \omega \cos \omega t.$$

We found that the minimum threshold amplitude to see motion in the random dot movement grating was indeed inversely proportional to the movement temporal frequency, showing a slope of −1 in a log–log coordinate representation (Figure 13-6). Note that the threshold displacement for very low temporal frequencies of movement was very high, 10 times the best value seen at 2 Hz. Despite a differential displacement of over 100 sec arc, well above the vernier acuity threshold, the observer could not detect motion at this low temporal frequency. This occurred because the pattern was not moving fast enough to stimulate a velocity-

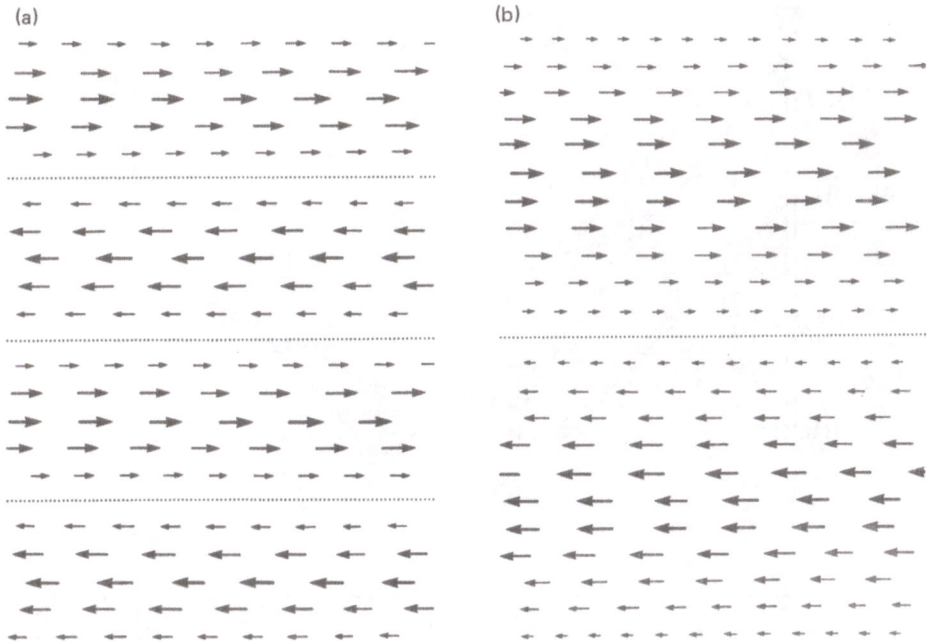

Fig. 13-5. Vector field representation of the instanteous velocity field of the movement of random dots for two different spatial frequencies of movement. Movement spatial frequency in (a) is twice that of (b). Length of each arrow is proportional to velocity. Contrary to the more conventional usage, *spatial frequency* refers to the spatial frequency of the differential movement rather than to the spatial frequency of the luminance distribution (From Nakayama, K., & Tyler, C. W. Psychophysical isolation of movement sensitivity by removal of familiar position cues. *Vision Research*, 1981, *21*, 427–433.)

sensitive system and the observer could not recognize the large positional offset because of the lack of any familiar and codable position cues in the random dots (Attneave, 1954). Thus, the variation in amplitude to see motion for these low movement temporal frequencies is dependent on a minimum threshold velocity and not on a minimum change of position.

In order to highlight this point with a contrasting counterexample, we imposed the same differential standing wave motion on a stimulus *with* codable position cues, a single vertical line. In this case, the observer could see the change in appearance of a line as it moves differentially (Figure 13-7, inset). As a consequence, the reciprocal relationship between threshold amplitude and temporal frequency seen for random dots was not obtained (Figure 13-7). Instead, the relationship between temporal frequency and threshold between .1 and 1 Hz was flat, indicating the dominance of positional information.

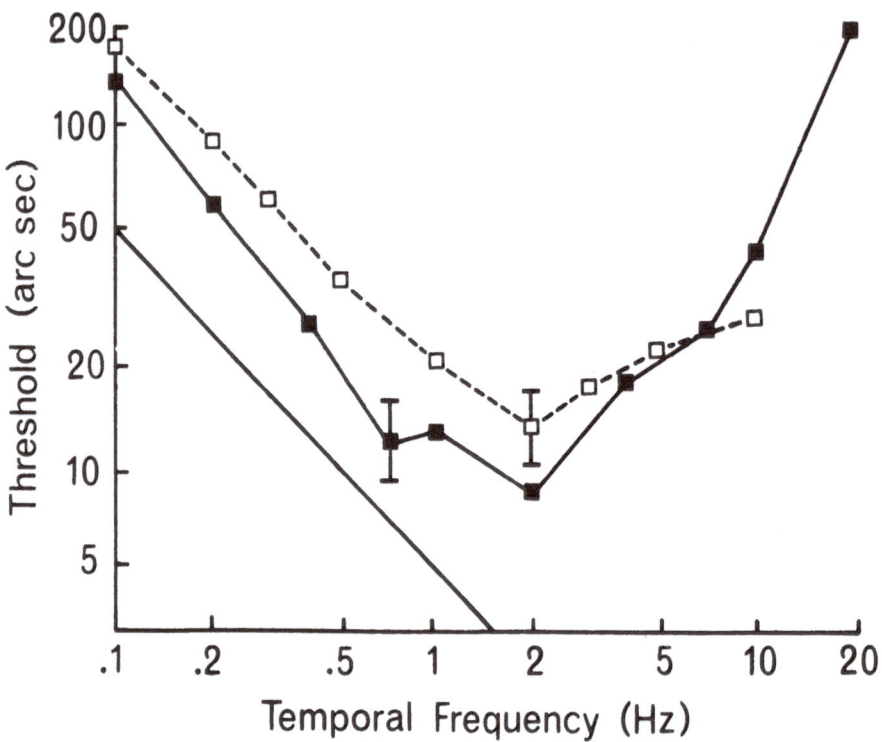

Fig. 13-6. Demonstration that movement in random dots is detected by velocity-sensitive mechanisms. Motion threshold amplitude plotted as a function of temporal frequency for a random dot moving grating having a spatial frequency of 3 sinusoidal cycles/degree. Note that between .1 and 1.0 Hz the function falls as the reciprocal of temporary frequency, indicating that the observer's threshold is determined by relative movement rather than relative position sensitivity. Data are from two separate observers. (From Nakayama, K., & Tyler, C. W. Psychophysical isolation of movement sensitivity by removal of familiar position cues. *Vision Research*, 1981, *21*, 427–433.)

This contamination of position information in the measurement of motion suggests that a single line or small aggregates of point stimuli having codable features are unsuitable as a probe to test the motion system. Observers can notice the change in shape. On the other hand, random dot patterns, having no such recognizable features, can be used to explore additional characteristics of motion sensitivity.

Spatial Properties of Motion Parallax Sensitivity

In order to examine spatial characteristics of motion sensitivity, we varied only the spatial frequency of the movement, leaving the random dot pattern unchanged. A vector representation of two different spatial frequencies of movement is shown in Figure 13-5. Using the temporal frequency having the best sensitivity (2 Hz), we obtained thresholds for

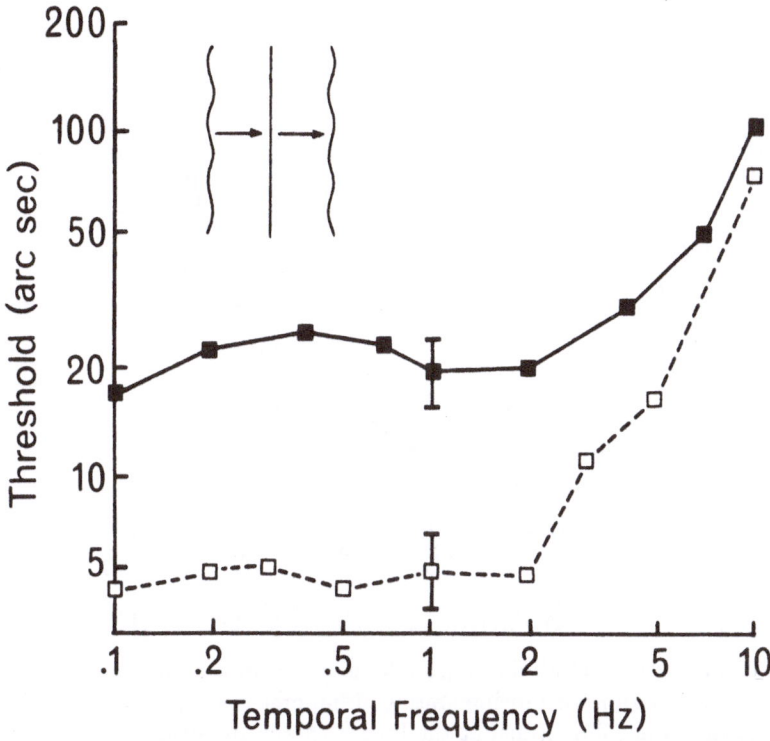

Fig. 13-7. Demonstration that differential movement in a single line is dictated by position-sensitive mechanisms. Motion threshold amplitude for a line undergoing exactly the same standing wave motion as in Figure 13-6. *Inset:* line changes appearance as it moves. Note that instead of showing a threshold rise for the very low frequencies, the function is flat, indicating that a minimum positional offset rather than a minimum velocity determines the psychophysical threshold.

different movement spatial frequencies (Figure 13-8). Several features of this function should be noted. First is the fact that the sensitivity to differential motion was extremely good, requiring only 5 sec arc of differential displacement. This figure was as low as those obtained for hyperacuity tasks such as vernier acuity. Second is the rise in thresholds above the rather low spatial frequency of .75 cycle/deg. Thus, in contrast to previous views (Brown, 1931), greatest proximity of differentially moving points was not advantageous in determining the best differential motion sensitivity. In fact, sensitivity was optimal even at the lowest spatial frequency tested, corresponding to a 2.5° separation of the points undergoing the maximum differential movement.

Before we comment on the significance of these results with differential motion sensitivity, it is instructive to provide a comparison with differential position sensitivity. In order to do this along the same spatial metric, we used the measure of periodic vernier acuity, consisting of a single, static, sinusoidal line stimulus, replicating the results originally reported by Tyler (1973). In order to obtain the position-sensi-

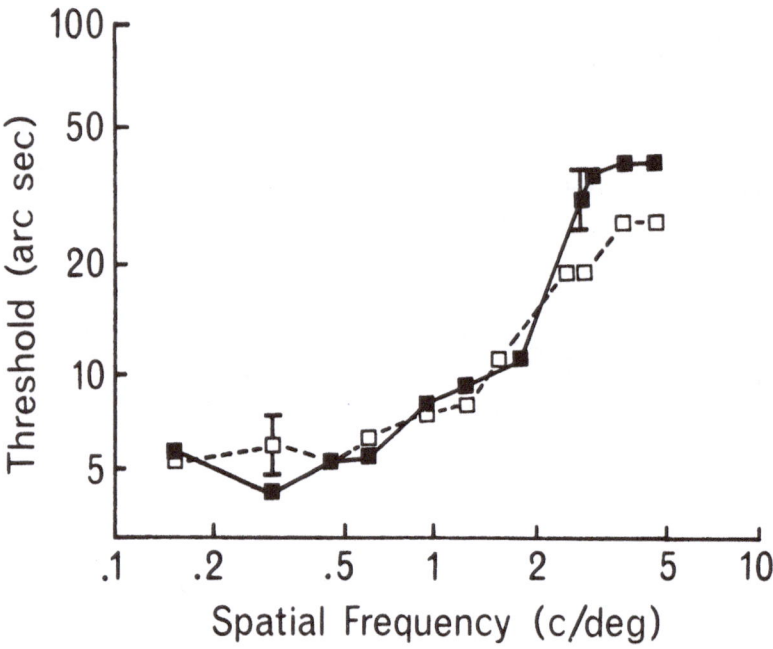

Fig. 13-8. Movement threshold amplitude as a function of movement spatial frequency for a random dot moving grating. Temporal frequency is 2 Hz. Note that the sensitivity is not diminished at the lowest spatial frequencies tested, indicating that increased proximity of differentially moving points does not offer advantages in the detection of relative motion.

tivity threshold we had the observer determine the smallest amplitude that would make the line deviate from perceived collinearity or straightness. The resulting amplitude versus spatial frequency curves are plotted in Figure 13-9. This target had some similarity to that used in the classical vernier acuity task, and the thresholds could be as low, at least for the intermediate spatial frequency of 2 cycles/deg. Two features are of interest in comparing the results obtained for motion sensitivity. Most important is the radically different function relating threshold to the spatial frequency of the undulating sinusoidal line. Instead of the very low thresholds obtained for the detection of movement in random dots, the low-frequency portion of the period vernier acuity function showed increased thresholds as spatial frequency was decreased. For the lowest spatial frequencies it was 10 times worse than the movement threshold. It should also be noted that for high spatial frequencies the sensitivity for position was better than for movement.

The results show that the most prominent difference between position and motion sensitivity is the much greater areal integration of motion information. Maximum sensitivity of the motion system can occur over a distance that is over 20 times more than the distances spanned by the position system. This indicates that the receptive fields mediating motion sensitivity are probably very large, even in the fovea.

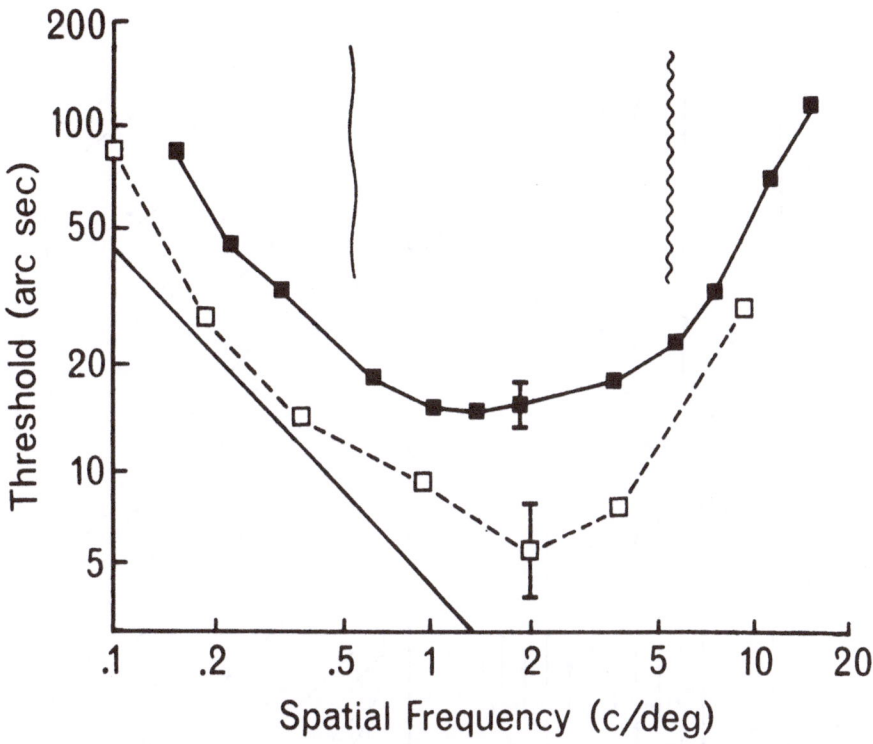

Fig. 13-9. Spatial characteristics of human position sensitivity. A single sinusoidal line having differing spatial frequencies (two examples are shown above the plotted functions) is adjusted so as to be just perceptibly different from collinearity. Note that in contrast to the function derived for motion sensitivity (Figure 13-8), differential position sensitivity is extremely poor at the lowest spatial frequencies, just where movement sensitivity is at its best.

Influence of Common Image Motion

A primary feature of our model to encode the edges of object surfaces in three-dimensional space is the existence of neural elements that are able to sense differences in velocity between neighboring retinal areas (see Figures 13-2 and 13-3). Ideally, such a mechanism should be able to extract the same amount of motion difference over a wide range of common image motion velocity. This would, for example, permit the extraction of the same amount of velocity difference irrespective of an added rotational flow component. In other words, common image motion should have little effect on differential motion sensitivity.

With differential position sensitivity, Westheimer and McKee (1975) showed that observers can detect very small vernier offsets despite surprisingly large amounts of common image motion. Velocities up to 3°/sec have essentially no degrading effect on vernier acuity thresholds; they remain at the same low level of about 5 sec arc (one-fifth of a receptor diameter) even as this tiny offset is sweeping across the retina at over 300 receptors/sec. Because of these findings with position sen-

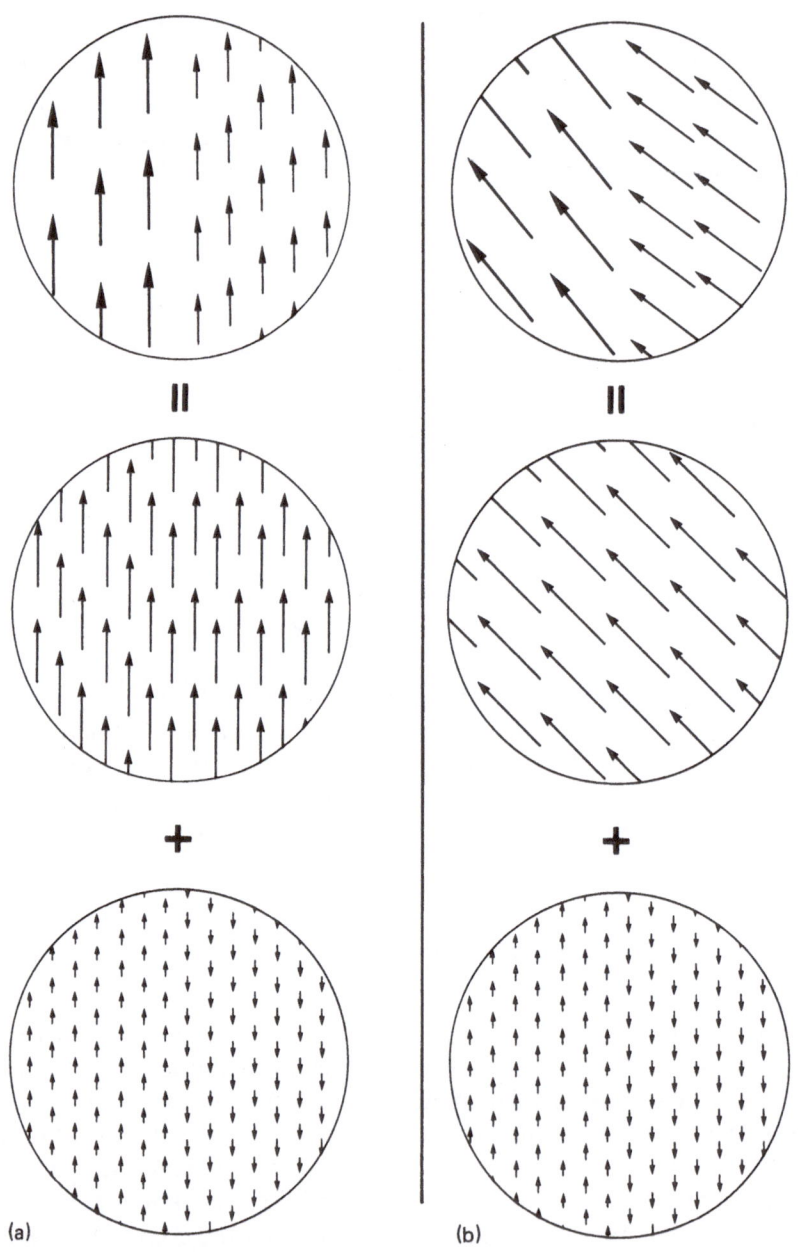

Fig. 13-10. Vector field representation of differential and common image components as they add to form the composite motion field used in psychophysical experiments. (a) Addition of horizontal differential motion to a horizontal common image motion. (b) Addition of horizontal differential motion to an oblique common image motion. The observer's task is to detect the differential motion. (From Nakayama, K. Differential motion hyperacuity under conditions of common image motion. *Vision Research*, 1981, *21*, 1475–1482.)

sitivity, it is of additional interest to see how motion parallax sensitivity is maintained under similar conditions of image motion. Is motion parallax sensitivity also immune to such large amounts of common image motion?

In order to answer this question, two definable component vector fields of motion were imposed on a circular field of random dots (Nakayama, 1981). First, there was a differential component, consisting of opposing horizontal movement on the top and bottom position of the screen. To this was added, vectorially, another common motion whose direction and amplitude could be controlled. Two examples of such a superimposition of common and differential motion are presented in Figure 13-10. The observer's task was to detect the differential motion in the presence of various amplitudes and directions of common motion. The movement duration was kept at 100 msec, lower than the latency of pursuit eye movements. Furthermore, the horizontal directin of the common motion was randomized from trial to trial to prevent anticipatory pursuit eye movements (Kowler & Steinman, 1979).

A plot of differential motion sensitivity as a function of common motion amplitude is presented in Figure 13-11. Above a velocity of approximately .3°/sec (or 2 min arc displacement) there is a progressive increase in motion thresholds, reaching approximately 60 sec arc at 3°/sec. Relative motion sensitivity is severely degraded by the presence of common image motion. This is especially noteworthy in comparison to the lack of any rise in vernier acuity thresholds at an image motion of 3°/sec (Westheimer & McKee, 1975).

Measurements of the ability to see differential motion with other orientations of common motion show that the interference is greatest

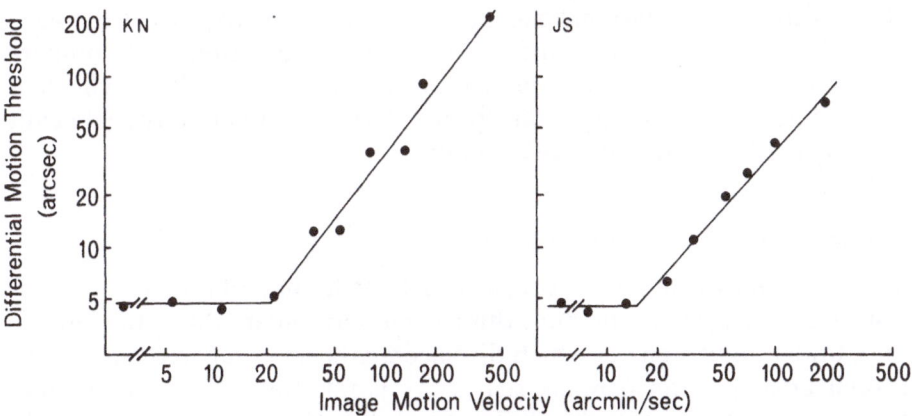

Fig. 13-11. Effect of common image motion amplitude on differential motion thresholds in a random dot pattern for two observers (refer to Figure 13-10 to see the addition of the two motion components). Duration of motion is 100 msec. Pixel size is 2 min arc. (From Nakayama, K. Differential motion hyperacuity under conditions of common image motion. *Vision Research*, 1981, *21*, 1475–1482.)

when the common image motion is along the same axis as the differential motion (Nakayama, 1981). This angular selectivity indicates that the interference is not just a general effect accompanying stimulus change—one that has been hypothesized, for example, to account for saccadic suppression (MacKay, 1970).

In summary, the psychophysical findings using differentially moving random dots show three essential differences between relative motion and relative position sensitivity. First, motion parallax is mediated by a system requiring a minimum differential velocity, whereas relative position sensitivity requires only a minimum recognizable spatial offset. Second, the spatial characteristics are entirely different. Motion parallax is best when comparing widely separated areas of movement relative to others, whereas relative position sensitivity is best when comparing spatial offsets over much shorter distances. Third, relative motion sensitivity cannot withstand very small amounts of common image motion. These three differences, taken alone and together, indicate a profound and qualitative difference in the neural processing of relative position and relative motion. They contradict, for example, any supposition that the coding of relative motion is a second serial stage that follows measured forms of position sensitivity. They support the view that relative motion sensitivity is wired in parallel to relative position sensitivity.

Implications for Space Perception

Earlier in this chapter we discussed the concept of pure visual decomposition—the hypothesized ability of the visual system to fractionate the composite optical velocity field into separate components, the rotational and the translational field, and, in particular, to extract the edges of real three-dimensional surfaces by a local operation that compares velocities in one region relative to its surround (as in Figure 13-2). At this point, we examine this concept in relation to psychophysical results obtained with the random dot stimuli.

Spatial Properties of Motion Sensitivity

The psychophysical results confirm that the human observer possesses a pure velocity system and that this system has remarkable sensitivity. In fact, under conditions in which displacements are compared over a wide retinal area, motion parallax sensitivity is far superior to relative position sensitivity, having thresholds an order of magnitude lower. This finding suggests that the second spatial differentiation process implied by the convexity model is approximate and is best considered in physiological rather than in mathematical terms. Rather than take a strict Laplacian of the field, for example, it is more useful to think of the wiring diagram shown in Figure 13-2, emphasizing the conclusion that the receptive field centers of such cells are very large.

This would indicate that the depth sensation mediated by relative motion is best between nonadjacent features in the visual field. Closely spaced points in the visual field at different depths, for example, would tend to be "smeared" or averaged in their depth value, a point made earlier by Rogers and Graham (1979) for motion parallax sensitivity and originally by Tyler (1975) for stereopsis.

Interfering Effects of Common Image Motion and a New Hypothesis

When there is no overall image motion, the sensitivity to differential motion is as good as, or better than, the sensitivity to differential position. With added image motion above about 2 min arc, motion parallax sensitivity decreases, showing a threshold rise proportional to common image velocity (see also Nakayama, 1981).

This result indicates that there is a distinct limitation in the processing of motion parallax information that would not be predicted had one viewed motion parallax as being derived from a series of successive vernier acuity-like judgments. In particular, vernier acuity is not impaired by common image velocities exceeding 3°/sec. Under comparable conditions motion parallax sensitivity is decreased at least 10-fold.

How does this limitation affect the visual system's ability to extract depth and figure–ground information from the optical velocity field? There are two important sources of common image motion if we consider the analysis to occur within a relatively local region of the visual field, as implied by the model shown in Figure 13-2. First, there is a common image motion imposed by the rotational velocity field, occurring every time the animal rotates its head or eyes, and this can be at relatively high velocities. Second, a common image motion is also produced by the translational velocity field. Points adjacent to each other in the visual field move in the same direction and with the same velocity to the extent that they are in the same or in an adjacent depth plane.

Either or both of these sources of common image motion could obliterate information regarding small differences in distance between environmental points (see Figure 13-11). Two well-known oculomotor mechanisms, however, may help to cancel such common image motion. First are the vestibulo-ocular and the optokinetic reflex systems, well designed to reduce image blur induced by observer rotation over a wide range of angular velocities and frequencies. To the extent that these reflexes perform their functions with the appropriate compensatory gain of −1, the rotational velocity field, as described by Equation (13-1), is removed in its entirety.

A second source of retinal image motion, that coming from the translational velocity field, can also be effectively canceled by the oculomotor pursuit system, but only in a relatively local retinal area (i.e., the fovea). By stabilizing a particular portion of the translational velocity field by eye movements, one destabilizes or increases common image motion at some other points. If, for example, one travels in a straight line while fixating a point to the right of the direction of motion, objects to the left have greater optical velocities than would be the case if the eyes remained immobile. Thus, one can never null a translational velocity field in its entirety with eye movements, one can only add a rotational component and thereby stabilize a local region of the visual field, perhaps for the sake of coding an object of focal interest (Nakayama & Loomis, 1974, Mathematical Appendix). Such stabilization might be of great value for the foveal or parafoveal discrimination of small differences in depth, however.

The above discussion does not establish the fact that the oculomotor system is essential for the extraction of depth information from the optical velocity field. It only establishes that it has this capability and that it could prevent a great loss in motion parallax information. Westheimer and McKee (1975) showed that image smear does not impair visual acuity and hyperacuity. If image stabilization is of such little importance for vernier acuity, why have an oculomotor system perform this task with any degree of accuracy? Walls (1961) suggested that the original biological purpose of the oculomotor system was not to move the eyes but to keep them from moving—maintaining visual acuity by reducing blur during head movements. In this context, Westheimer and McKee's (1975) finding is puzzling. Coupled with the present results, it prompts a modification of Walls' original hypothesis, emphasizing the importance of motion parallax sensitivity rather than visual acuity.

We propose that motion parallax sensitivity provides an important raison d'etre for the oculomotor system. Motion parallax may have provided a principal evolutionary selective pressure on this control system, ensuring that common image velocity remains sufficiently low to protect the pick-up of relative motion information. Both systems—the visual holding reflexes (Walls, 1961) of the oculomotor system and the motor parallax system—are phylogenetically very old. The optomotor response appears in phylogeny before directed saccades and vergence. Motion parallax also shows phylogenetic and ontogenetic priority. Flies, for example, show figure–ground recognition on the basis of motion parallax (Reichardt & Poggio, 1979), and it has been reported that neonatal rats avoid the deep side of a visual cliff using only monocular motion parallax (Walk, 1965). In addition, it is clear that frontal binocular vision is a relatively recent development.

Thus, it is not unreasonable to suppose that both the visual holding reflexes and the motion parallax systems were among the major functional components in the visual system of our phyletic ancestors, appearing well before the emergence of more elaborated types of visual processing and motor response. It is conceivable that the two systems evolved together and that the oculomotor system was shaped by the visual system's need to process motion parallax information for the important task of encoding three-dimensional space.

Acknowledgments

This work was supported in part by National Institutes of Health Grants 5 R01 EY-01582, 5 P30 EY-01186, and 2S07RR 05566, and the Smith-Kettlewell Eye Research Foundation. I wish to thank Dr. Jack Loomis for his helpful comments on the manuscript.

References

Attneave, F. Informational aspects of visual perception. *Psychological Review*, 1954, *61*, 183–193.

Barlow, H. B., Blakemore, C., & Pettigrew, J. D. The neural mechanism of binocular depth discrimination. *Journal of Physiology* (London), 1967, *193*, 327–342.

Brown, J. F. The thresholds for visual movement. *Psychologische Forschung*, 1931, *14*, 232–249.

Frost, B. J., & Nakayama, K. Single visual neurons code opposing motion independent of direction. *Science*, 1983, (in press).

Gibson, J. J. *Perception of the visual world*. Boston: Houghton Mifflin, 1950.

Gibson, J. J. *The senses considered as perceptual systems*. Boston: Houghton Mifflin, 1966.

Goldstein, H. *Classical mechanics*. Cambridge, Mass.: Addison-Wesley, 1950.

Graham, C. H., Baker, K. E., Hecht, M., & Lloyd, V. V. Factors influencing thresholds for monocular movement parallax. *Journal of Experimental Psychology*, 1948, *38*, 205–223.

Hubel, D. H., & Wiesel, T. N. Receptive fields and functional architecture in two non-striate visual areas (18 and 19) of the cat. *Journal of Neurophysiology*, 1965, *28*, 229–289.

Julesz, B. *Foundations of cyclopean perception*. Chicago: University of Chicago Press, 1971.

Koenderink, J. J., & Van Doorn, A. J. Local structure of movement parallax of the plane. *Journal of the Optical Society of America*, 1976, *66*, 717–721.

Kowler, E., & Steinman, R. The effect of expectations on slow oculomotor control. II. Single target displacements. *Vision Research*, 1979, *19*, 633–646.

Leibowitz, H. The relation between rate threshold for the perception of movement and luminance for various durations of exposure. *Journal of Experimental Psychology*, 1955, *49*, 209–214.

Longuet-Higgins, H. C., & Prazdny, K. The interpretation of moving retinal images. *Proceedings of the Royal Society of London; B.*, 1980, *208*, 385–397.

MacKay, D. M. Elevation of visual threshold by displacement of retinal image. *Nature*, 1970, *225*, 90–92.

Nakayama, K. Differential motion hyperacuity under conditions of common image motion. *Vision Research*, 1981, *21*, 1475–1482.

Nakayama, K., & Loomis, J. Optical velocity patterns, velocity sensitive neurons, and space perception: A hypothesis. *Perception*, 1974, *3*, 63-80.

Nakayama, K., & Tyler, C. W. Psychophysical isolation of movement sensitivity by removal of familiar position cues. *Vision Research*, 1981, *21*, 427-433.

Poggio, G. F., & Fischer, B. Binocular interaction and depth sensitivity in striate and pre-striate cortex of behaving rhesus monkey. *Journal of Neurophysiology*, 1977, *40*, 1392-1405.

Prazdny, K. Egomotion and relative depth map from optical flow. *Biological Cybernetics*, 1980, *36*, 87-102.

Reichardt, W., & Poggio, T. Figure-ground discrimination by relative movement in the visual system of the fly. *Biological Cybernetics*, 1979, *35*, 81-100.

Rogers, B. J., & Graham, M. Motion parallax as an independent cue for depth perception. *Perception*, 1979, *8*, 125-134.

Tyler, C. W. Periodic vernier acuity. *Journal of Physiology* (London), 1973, *228*, 637-647.

Tyler, C. W. Spatial organization of binocular disparity sensitivity. *Vision Research*, 1975, *15*, 585-590.

Walk, R. D. The study of visual depth and distance perception in animals. In D. S. Lehrman, R. A. Hinde, & G. Shaw (Eds.), *Advances in animal behavior* (Vol. 1). New York: Academic Press, 1965.

Walls, G. The evolutionary history of eye movements. *Vision Research*, 1961, *1*, 111-123.

Westheimer, G., & McKee, S. P. Visual acuity in the presence of retinal image motion. *Journal of the Optical Society of America*, 1975, *65*, 847-850.

Chapter 14
Perceptual Consequences of Experimental Extraocular Muscle Paralysis

Leonard Matin, John K. Stevens, and Evan Picoult

When we change our direction of gaze from one point to another in the visual field, the positions of images on our retinas are changed correspondingly. Nevertheless, we do not normally see any movement or displacement of the visual field or of objects in the field, as we do when the direction of gaze is held steady and either the entire field or objects within it are moved. Several important aspects of the question regarding how stability of our perception of space is maintained when we turn our eyes have recently been clarified by our observations and measurements of the "oculoparalytic illusion." The main findings leading to the clarification have been reported elsewhere (Matin, Picoult, Stevens, Edwards, Young, & MacArthur, 1980, 1982). In this chapter we show how these results yield an important aspect of the solution.

Since Helmholtz (1866) originally proposed that the "effort of will" employed in changing the direction of gaze is taken into account in judging whether the changed retinal image location of objects is due to the eye movement produced by the will's effort or by movement of the objects themselves, several alternative sources of extraretinal eye position information (EEPI) have been proposed as the basis for perceptual stability. However, the algebra implied by the cancellation mechanism suggested by Helmholtz (Figure 14-1) has been common to all of the proposals. The four conclusions drawn from our recent work do not decide on the source of EEPI but do demonstrate the existence of cancellation and of important constraints placed on the operation of cancellation by the presence of illuminated and structured visual fields.

Conclusion 1: For an observer viewing in a normally illuminated and structured visual field, EEPI-driven cancellation mechanisms (Figure 14-1) are not normally involved in the determination of visual localization of objects either relative to each other or relative to visual norms such as the perceived eye-level horizontal or perceived median plane.

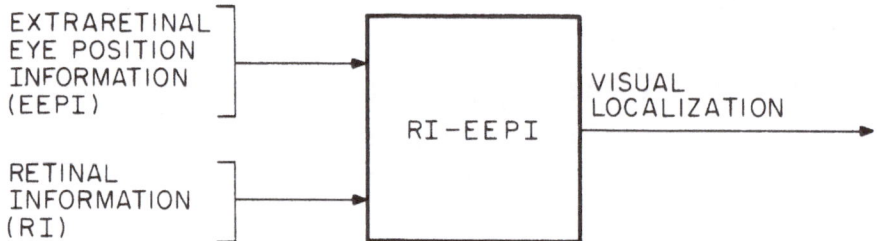

Fig. 14-1. Cancellation theory for visual localization in the presence of eye movements.

Conclusion 2: In an otherwise dark surround, EEPI-driven cancellation mechanisms play a central role in determining visual localization of objects relative to visual norms.

Conclusion 3: EEPI-driven cancellation mechanisms are critical for intersensory localization (e.g., matching locations of sound and light) in either darkness or light if visual capture of the sensory information regarding location via the other (nonvisual) modality is prevented.

Conclusion 4: In normally illuminated visual fields EEPI is not supprssed, nor is the output of the cancellation mechanism. However, what is suppressed in the normally illuminated field is the contribution of EEPI to visual localization. Thus, under some conditions the EEPI-driven cancellation mechanism may contribute to other aspects of perception (e.g., to the comparison of localization of auditory and visual targets) although it may have no influence on visual localization.

The experiments that have led to these conclusions employed subjects who were made paretic (partially paralyzed) by systemic injections of curare (*d*-tubocurarine). The partial paralysis thus produced at the neuromuscular junctions of peripheral cholinergic musculature (Goodman & Gilman, 1977), including the extraocular muscles, did not alter cognitive abilities. The influence on the extraocular musculature produced a dosage-dependent limitation on the range of ocular positions. Within this reduced range, however, maintenance of fixation in any assigned direction was not accompanied by any greater "sense of effort" than maintenance of fixation in any other direction. Thus, the experience of maintaining fixation in any given visual direction between the extremes of gaze was no different than it is for the normal observer who maintains fixation in some given visual direction within his more extended range of possible gaze directions: Although a normal observer who fixates a point 20° to the left of primary position readily reports that he is looking further to the left than when he fixates a point 10° to the left of primary position, he does not report any increase in effort, as he would if he lifted a 20-lb weight relative to when he lifted a 10-lb weight. The act of looking in different directions is not accompanied by sensations of different *magnitudes* of effort—an observation that holds for the partially paralyzed observer.

The Kernel Observation under Partial Paralysis:
Perceived Change in Elevation

The concomitant influences of the paresis on visual localization were substantial; however, with two exceptions,[1] they only occurred in darkness. The observation of greatest significance—the one that provided the key to the subsequent measurements—was the following: In normal room illumination, with the eyes in any given direction of gaze, everything appeared perfectly normal and indistinguishable from its appearance when curare was not administered. However, as soon as all lights were extinguished, except for the single fixation target at eye level, that target appeared to move slowly in the direction of the (invisible) floor. When it reached a position that appeared near or at the floor, movement essentially ceased and the target appeared to remain stationary. Surprisingly, when the room was reilluminated the fixation target immediately appeared as it had originally—at eye level—and the room appeared perfectly normal. This sequence of observations could be repeated as often as desired with successive reilluminations and extinctions of illumination.

In order to test the possiblity that the visually perceived drop of the fixation target was produced because extinction of illumination resulted in the observer's feeling that his body was tilted backward, we had him extend his arm and point his finger in the direction of the horizontal.

---

[1] The first sign of drug action was an uncontrollable diplopia. This appeared before the subject felt any other sign of weakening. Since the diplopia persisted throughout curarization, all viewing was carried out monocularly (eye patch over the other eye) during paralysis.

Furthermore, when voluntary saccades were performed during partial paralysis a brief transient "jumping" of the visual field was typically observed. However, after this the visual field appeared indistinguishable from its appearance before the saccade; it also appeared indistinguishable from its appearance in the unparalyzed state. We have not yet determined which of the following two explanations for "jumping" is the correct one: (a) It is due to a transient failure of an EEPI-driven cancellation mechanism in conjunction with a transient failure of Type B suppression (Matin, 1981, 1982; also see below). (b) It is related to the eye's failure to reach the saccade's goal at the end of the initial saccade: immediately subsequent saccades carry it to the goal, and the visual field then appears normal; the "jump" itself corresponds to the discrepancy between the saccade's goal and the eye's position at the end of the initial paralysis-shortened saccade. The latter explanation is based on related observations made during parametric adjustment of normal eyes: During a saccade from an original fixation point A to a goal at point B, both A and B are extinguished and point C—a target between A and B—is illuminated. On the first such trial, "jumping" of the target goal is observed; by means of immediately subsequent saccades the observer "corrects" eye position to point C, and everything looks normal. After several such trials the observer's attempt to reach B from A leads to reaching C in a single saccade; when this occurs no "jumping" is seen, and if the procedure is carried out in darkness the observer is not aware that his saccade has not carried his eye to B.

Although he could barely raise his arm, the finger always pointed as accurately in darkness toward the horizontal as it did in full illumination.

A second possible explanation of the illusory drop was also readily eliminated: it was possible that the perceived drop was due to a loss of fixation in darkness, with an accompanying eccentric retinal placement of the fixation target. Since perception of visual direction during involuntary eye movements is not compensated by an EEPI-driven cancellation mechanism (Matin, Pearce, Matin, & Kibler, 1966; Matin, Pola, Matin, & Picoult, 1981), the localization errors might be consequent on such involuntary eye movements. However, identical observations resulted when we substituted a 24 X 24 minute (visual angle) transilluminated E for the original 9-minute circular transilluminated fixation target. Since the E always remained clear and sharp even when it appeared near the floor in darkness, and since clarity of the E would drop off rapidly with retinal eccentricity, so that at eccentricities between 2° and 5° the E would become unresolvable, it was possible to conclude that fixation of the single target remained at the central fovea in darkness and loss of fixation was not involved in producing the dramatic drop.

We began identifying the basis for the illusion as soon as we changed the direction of tilt of the observer's head and body relative to horizontal. The original observations described above were made with the head tilted back and the angle between head orientation and horizontal, β (Figure 14-2), set at about 20°. When β was set at a smaller angle, the perceived drop of the fixation target did not proceed as far. When the

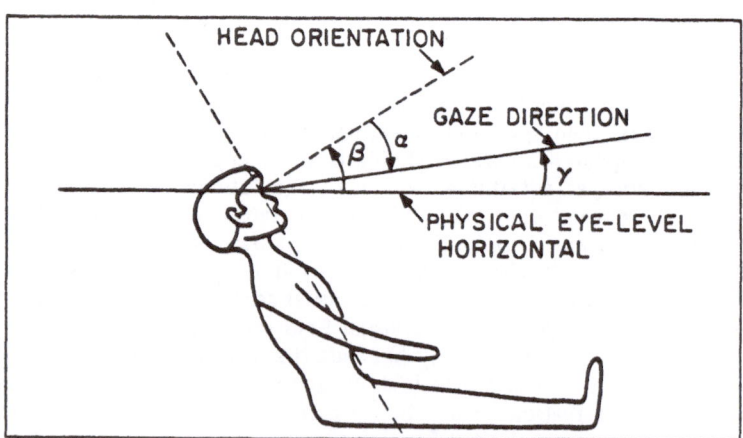

Fig. 14-2. Parameters of experimental observations: relation between the vertical fixation direction (*gaze direction*) relative to the head (angle α), the orientation of the head with respect to the physical horizontal (angle β), and the direction of gaze with respect to the physical horizontal (angle γ).

head was tilted forward, instead of appearing to drop, the light appeared to rise to a position near the (invisible) ceiling when room illumination was extinguished.

Although this result was absolutely clear, it did not yet yield an unequivocal interpretation. Changing the tilt of the subject's head and body also required that he change the position of his eye relative to his head in order to maintain fixation of the target (itself always at physical eye level). Thus, we could not yet decide whether what was important was the variation in the angle of the head and body relative to gravity (angle β) or the variation of the angle of the eye in the head (angle α). We separated these two possibilities by the next set of observations.

With the head and body fixed at a given value of β, the single target visible in darkness was placed at different heights above and below the physical eye level of the observer. For each such setting the room was alternately illuminated and darkened while fixation was maintained on the single target. We now discovered that the larger the deviation of α from some particular value, the larger was the distance of the apparent ascent or descent of the fixated target. The particular setting of α at which neither ascent nor descent occurred was called the no-illusion direction (NID). The NID remained fixed although the tilt of the subject's head and body (angle β) was itself changed, implying that the magnitude of angle α was responsible for the apparent change in position of the fixation target between normal illumination and darkness. At this point more extensive quantitative measurements of the phenomenon seemed to be required in order to further isolate the controlling variables.

Quantitative Measurements of Perceived Eye-Level Horizontal

Two sets of parametric experiments proved that the angle α alone influenced the perceived change in elevation, and that neither angle β nor the physical height of the target itself was involved. In both of these experiments angle α was experimentally set at various values, and at each the subject set a light to the elevation that he perceived to be horizontal eye level.

One-Light Experiment

In the first experiment the subject's head and body were set at a particular value of β. Fixation was maintained on the single target visible in darkness, and the subject directed the experimenter to adjust the target's elevation until it appeared to lie at the eye-level horizontal. This was repeated with β set at each of a number of different values. With β set so that head orientation was pointed in a direction considerably above the physical horizontal, the visual target also had to be set considerably above the physical eye-level horizontal in order to appear to lie at eye level. Lowering the head and body (lower β values) led to a monotonic lowering of the physical elevation of the setting. From this observed relation we extracted the relation between α and the error in the observer's setting from true eye level. The latter relation was linear with a substantial negative slope; with α directed further upward, the observer's setting of the perceived eye-level horizontal (and therefore the error) was further downward. Although the direction of this relation between α and the subject's setting was predicted from the previous measurements of the NID, the simple linear relation could not have been predicted from the NID measurements.

Although these measurements are extremely informative, by themselves they do not resolve the question regarding whether β or α is the critical variable determining the illusory change in perceived eye-level horizontal. They do mesh well with the previous measurements of the NID, but in order to resolve the issue regarding the critical variable it was necessary to carry out an additional experiment—one employing two lights.

Two-Light Experiment

The one-light experiment did not permit unambiguous identification of the critical variable that controlled the illusory change of visual localization. Although the experimenter could vary β systematically he could not simultaneously hold α constant, since for any given value of β, α was varied with and was determined by the elevation at which the observer set the fixated target. The two-light experiment provided independent and simultaneous control of β and α, and completely resolved the issue: α alone determines the magnitude of the illusory localization change.

After a particular value of β was experimentally set by adjusting the position of the head and body, a particular value of α was fixed experimentally by setting a visual fixation target to a particular elevation. With α and β thus fixed, a second visual target was introduced along the same vertical line as the fixation target. The elevation of this second target was adjusted according to the subject's instructions so as to appear at eye level. With β remaining fixed, the value of α was changed by the experimenter who adjusted the elevation of the fixation target, and the second target was set to appear at eye level according to the subject's instructions. Thus we were able to determine directly the relation between the angle of gaze relative to the head (α) and the physical elevation of the perceived eye-level horizontal that held with head and body position fixed. This relation is shown in Figure 14-3a and b. The fact that this relation was unchanged at several different values of β

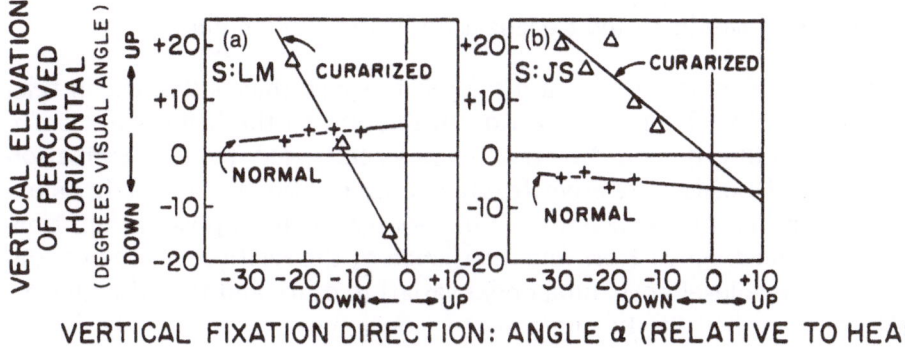

Fig. 14-3. Psychophysical localization measurements by observers LM and JS. *Lines through data:* least square fits. Each point is an average of two or three settings. (a, b) Perceived eye-level horizontal. In complete darkness the observer fixated a small visible target whose angular elevation with respect to the transverse plane through the head (vertical angle of the eye in the head) is plotted on the abscissa. The transverse plane through the head was itself above the physical horizontal by about $25°$ for LM and $30°$ for JS. While maintaining fixation on this first light the observer set a second peripherally viewed light (which was movable in the same vertical meridian as the first target) to a height that appeared to be at his eyelevel horizontal; the latter height is plotted on the ordinate as vertical elevation of perceived horizontal. (c, d) Perceived median plane. In complete darkness the observer fixated a small visible target whose horizontal deviation from the physical median plane through his body (horizontal angle of gaze) is plotted on the abscissa. While maintaining fixation on this first light, the observer instructed the experimenter to set a second visual target (which was movable in the same horizontal plane as the first target) to the perceived median, plotted on the ordinate.

demonstrates that β is uninvolved in the illusion of localization, and supports the conclusion that the variation of the illusion is determined by variation of the position of the eye in the head—angle α.

Illusory Changes in Visual Localization of the Median Plane

The eye-level horizontal is a direction in space that is defined in relation to gravity. Thus eye position in relation to the head is not sufficient to define it as a direction in physical space; some information about head position relative to gravity must also be involved. As expected, the partial paralysis did not modify the head position information (deduced above from the invariance of the relation between α and perceived eye-level horizontal under variation of β), and thus the illusory effects described in the previous section appear to be entirely due to influences of the paralysis on EEPI. It was nevertheless desirable to measure visual localization in a case in which head position relative to physical space (angle β) was not involved in the judgment at all but EEPI was. We chose to measure the visually perceived location of the erect subject's own median plane for this purpose.

Perceived median plane measurements were made in an otherwise dark room. The observer directed the experimenter to move the visual target to the position at physical eye level perceived by the subject as the median plane. The horizontal eccentricity of gaze direction was determined with the aid of a fixation target at eye level. A result analogous to the one obtained for the eye-level measurements was obtained: Increasing gaze eccentricity leftward from a "zero point" produced perceived median plane settings that increased to the right of the true median plane; increasing gaze eccentricity rightward yielded settings that increased leftward; the relation is linear.

This experiment was carried out in two variations. For both, β was set approximately as displayed in Figure 14-2. In one, the settings to the median plane were made with a light at true eye level that appeared near the floor. The results are shown in Figure 14-3c and d. The second set of measurements was made with the fixation target that determined horizontal gaze direction set at an elevation above the true eye level for which the target appeared to lie at eye level. The results of the two variations were indistinguishable.

Influence of Level of Paralysis

Increasing curare dosage level produced an increase in the slope of the relation of error of localization to eccentricity of eye position both for eye-level horizontal and for median plane settings. One set of such data

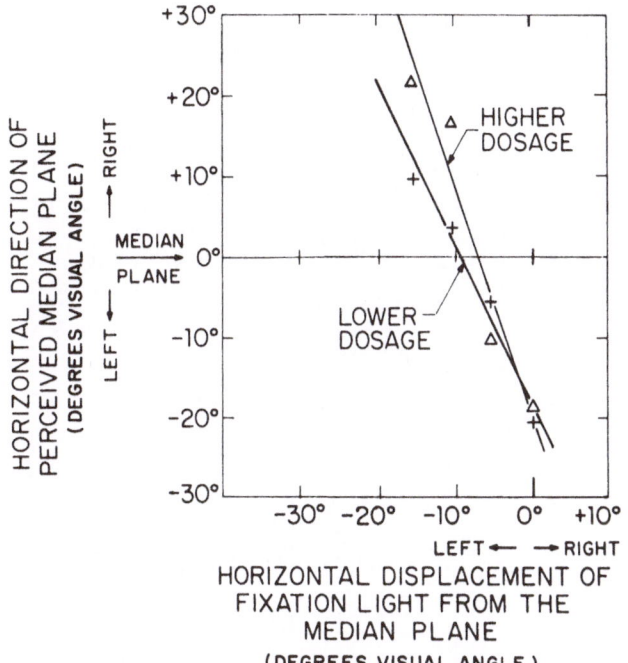

Fig. 14-4. Mislocalizations of the median plane in darkness at two different levels of partial paralysis.

is shown in Figure 14-4. Of considerable interest is the fact that the slopes for both eye-level horizontal and median plane settings for a given individual (not shown) were extremely similar; the change of slope with dose level was also extremely similar for both settings.

Three-Parameter Model of the Influence of Paralysis on Visual Localization and Oculomotor Control

The four main features of the consequences of experimental partial paralysis of the extraocular muscles are depicted in Figure 14-5. Our present thinking also leads us to believe that this model will apply to naturally occurring pathological paretic states, such as those that occur in myasthenia gravis, and we are currently examining this extension of the model.

The four main features encompassed by Figure 14-5 are the following:

1. The center of coordinates for the relation between direction of the perceived eye-level horizontal and vertical gaze direction (Figure 14-5a) has been set at the NID, as has the center of coordinates for the relation between perceived median plane and horizontal gaze direction (Figure 14-5b). For gaze directions above the NID, the perceived eye-level hori-

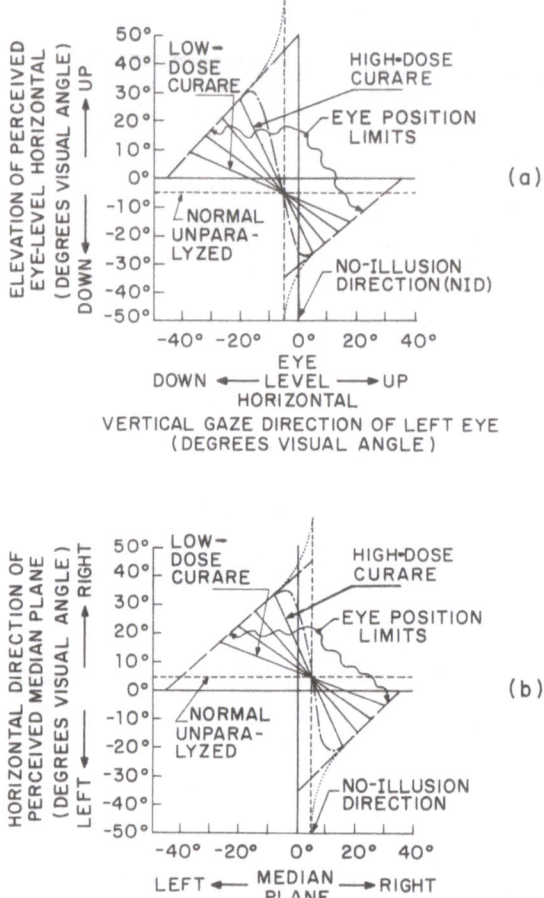

Fig. 14-5. Theoretical functions relating error of visual localization to direction of gaze under various levels of paralysis. (a) Visually perceived direction of eye-level horizontal as a function of vertical gaze direction. (b) Visually perceived direction of horizontal gaze direction. Illusion magnitude I (used in text) is equal to the negative of the ordinate value. Our uncertainty regarding the outcome around total paralysis ($0°$ range of possible eye movement) is indicated by the *dotted lines* (which suggests that the errors increase asymptotically), the *dot-dash lines* (which suggests a drop to zero of the localization errors), and the *tilted dashed lines*, showing a simple intersection of the limit functions with the NID at the $0°$ abscissa value.

zontal deviates downward relative to its direction when gaze direction is at the NID (Figure 14-5a). A similar relation holds between the perceived median plane values and the horizontal deviations of gaze direction from the NID (Figure 14-5b).[2]

2. The results of all of the experiments we have done strongly suggest that the relation of visual localization to gaze direction of either perceived median plane or eye-level horizontal is linear at any given dosage level up to near total paralysis and that the slope of the linear relation is the same on both sides of the NID. Our measurements have not yet been sufficiently precise to deal with the possibility of small deviations from linearity, although large deviations are excluded.

3. The experimental results are clear in establishing that an increase in dosage level produces an increase in illusion magnitude (see Figure 14-4), and this is represented in Figure 14-5 by the increase in slope. It is also clear that the increased dosage has similar influences on the relations involving both directions of change of gaze direction; this is implied in Figures 14-5a and b. The separation between the functions for different dosages in Figure 14-5 depends on the specific quantitative relation of dosage to level of paralysis. We have not yet explored this sufficiently to be able to provide any further information on it beyond data such as that shown in Figure 14-4.

4. Most importantly, the model in Figure 14-5 includes the observation that the paretic state involves a reduction in the range of eye positions that can be attained and that the reduction is greater with an increase in dose level (dashed tilted lines mark the end points of gaze). This decrease in the range of possible eye positions with increase in dose is correlated with the increase in the slope of the relation of visual localization to eye position.

These results suggest the following simple expression relating eye position and illusion magnitude:

$$I = (K - 1)\, \alpha$$

where I is illusion magnitude (specified as the negative of the ordinate value in Figure 14-5), α is direction of gaze (specified as an angle of the

---

[2] What relations the NID has to other measures of the "position of rest of the eyes" or primary position of gaze have not yet been dealt with experimentally. It is clear, however, that some simple relations should be expected. We have not yet dealt with the questions of whether or not the NID for perceived eye level is invariant with changes in horizontal gaze direction (we have some indications that it is not), the perceived median plane is invariant with vertical gaze direction, or the NID is invariant with level of paralysis. In Figure 14-5 we do assume invariance; this assumption does not influence the main line of the present treatment, but the question has yet to be dealt with experimentally.

eye relative to the head), and $K - 1$ is the slope of a straight line in Figure 14-5.[3]

We do not yet know the quantitative function that describes the way in which the limits of gaze change with dosage. This is the function that determines the shape of the dashed curves in Figure 14-5. Two aspects of the observations suggest that the functions will be approximately straight lines over the large region of α as shown: First, at any given level of paralysis, the illusion magnitude was larger if the most extreme eye position possible was less deviant from NID. Given the increase of the absolute value of the slope of the main straight line relations between visual direction and gaze direction in Figure 14-5, this fact tells us that the dashed lines must at least all increase monotonically from left to right in Figure 14-5. Second, the magnitude of the illusory change in visual localization in any given curarized state never exceeded the difference between the end point of the range of eye positions in the normal state and the end point in the curarized state. Indeed, it appeared that this reduction in the eye position limit was approximately equal to the illusion magnitude when the eye was at this limit.

Considerations regarding the gaze limitations in Figure 14-5 lead to the following equations:

The direction of the perceived median plane is equal to $-I$, and

$$I = (K - 1)\,\alpha. \tag{14-1}$$

At the limiting direction of gaze, α_m, I is at its maximum I_m, and

$$I_m = (K - 1)\,\alpha_m. \tag{14-2}$$

We have found K and α_m to vary inversely; if we assume that this relation is a power function, then

$$\alpha_m = \alpha_{m_0}/K^n \tag{14-3}$$

where α_{m_0} is the limit of gaze for the normal, unparalyzed eye. Substituting Equation (14-3) in Equation (14-2), we have

$$I_m = [\alpha_{m_0} K^{(1-n)}] - \alpha_m \tag{14-4}$$

[and, noting from Equation (14-3) that for any n, when $K = 1$, $\alpha_m = \alpha_{m_0}$ for $K \to \infty$, $\alpha_m \to 0$], from Equation (14-4),

---

[3] In order to avoid complicating issues and notation unnecessarily, for the remainder of this chapter α will be employed to refer to either a vertical or horizontal deviation of gaze direction from the NID. (Up to now α referred only to vertical deviations of gaze direction from the eye level horizontal.) The finding that the NID is not generally coincident with the primary viewing position is represented in Figure 14-5 by displacing the NID from the center of coordinates.

$$\text{Lim } I_m = \begin{cases} \infty; n < 1 \\ \alpha_{m_0}; n = 1 \\ 0; n > 1 \end{cases} \qquad (14\text{-}5)$$

$$K \to \infty$$

Thus, as n varies, the three different limiting functions near-zero values of α_m can be approximated: the dotted line with $n < 1$, the dashed line with $n = 1$, the dot-dashed line with $n > 1$ (Figure 14-5). [Although the relation of α_m to K may not be a power function, most useful functions can be approximated by a power series for which Equation (14-3) is the first term.] We cannot yet tell which limiting function is correct. By fitting the relation of α_m and K in the partially paralyzed state with the power function, we should be able to predict how it will behave under total paralysis ($\alpha_m = 0$); however, we are also beginning work with total paralysis directly.

This description suggests a simple interpretation: The linear relation between eye position and error for a given level of paralysis implies that EEPI is increased by a constant scale factor. In the above equations, K can be interpreted as this scale factor. The curare-induced reduction in neuromuscular efficiency required that in order for the eye to reach a given position the pattern of motor signals to the muscles must be equivalent to the pattern employed to turn the unparalyzed eye to a more eccentric location. The EEPI associated with this (which could be either outflowing or hybrid) incorrectly corresponds to the more eccentric eye position.

Illumination Versus Darkness

It is of major interest that our explorations have so far found that the entire illusion of visual localization under paralysis is present only in darkness. In normal illumination, visual localization is entirely accurate. It appears likely that accuracy in visually localizing the median plane is largely based on the curarized observer's ability to see his own body and visually locate the target relative to this view. However, simply seeing his body could not be sufficient to correctly localize the eye-level horizontal. Nor would seeing the entire room per se necessarily produce correct localization of eye-level horizontal; it is possible for the room to have been differently tilted relative to gravity than it was in fact, and this would then require a setting that was differently related to the main lines of organization of the room as available to vision. The visible presence of the room must have influenced and determined setting of visually perceived eye-level horizontal in a marked way.

We have not yet explored these views regarding the influence of room illumination and of sight of the body on visual localization, although experiments to do so are in progress. We have dealt with prior issues. Under partial paralysis visual localization based on the view of the illuminated room (including sight of the body) was entirely different than visual localization based on an EEPI-driven cancellation mechanism as described above. In order to be accurate in room illumination, the localization information from the cancellation mechanism had to be either suppressed or modified. This raised an important question: Was EEPI itself suppressed or was the information from the cancellation mechanism suppressed? Further experiments were conducted to answer this question.

Auditory-Visual Matches

Two facts gave us the tool with which to deal with the question at the end of the immediately preceding section.

First, audition was not influenced by the curare. Curare passes the blood–brain barrier in only miniscule amounts (Matteo, Pua, Khambatta, & Spector, 1977), and since the inner ear lies within this barrier it remains unaffected. Although the auditory muscles are affected, the effect of this on audition is very likely to be small and symmetrical, and hence uninvolved in auditory localization. In any case, we directly determined auditory localization in our curarized subjects by requiring them to select an auditory stimulus that appeared to lie in the median plane from a horizontal array of 25 loudspeakers (Figure 14-6). The choice was entirely unaffected by curare. In addition, we employed a "name the speaker" technique in which the subject reported the number assigned to the speaker ("1"–"25"). The pattern of accurate reports and errors made by the partially paralyzed observer was indistinguishable from the pattern made when he was in the normal, unparalyzed state, although the pattern was different for different individuals.

Second, the normal, uncurarized individual was able to match the location of a sound and light with a reliability of about 2°, giving us a reliable baseline against which to compare the curarized observer.

The errors in matching the light to the sound as a function of horizontal gaze eccentricity in darkness are shown in Figure 14-7. It is clear that the partially paralyzed observer makes very substantial errors in matching, and that these errors are linearly related to gaze eccentricity: For any given departure of gaze from the NID (determined by the location of the fixated light), the observer matched the light to a sound source that was even more eccentrically placed. This error increased with departure in each direction from the NID. In subsequent experiments we determined that, for any given level of curarization, the magnitudes of the errors in auditory–visual matching were indistinguishable from the magnitudes of the errors in visual localization of the median plane alone described earlier.

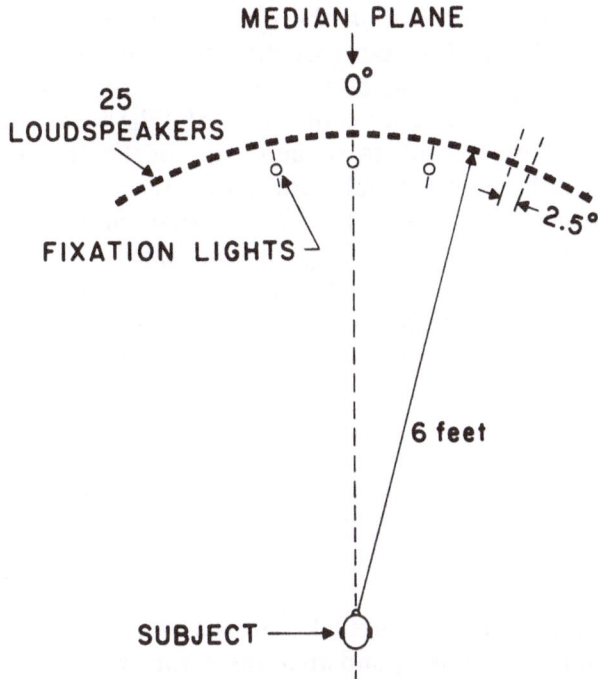

Fig. 14-6. Spatial relations between observer and stimuli for auditory-visual matches. The observer and fixation lights could be rotated in a horizontal plane around the center.

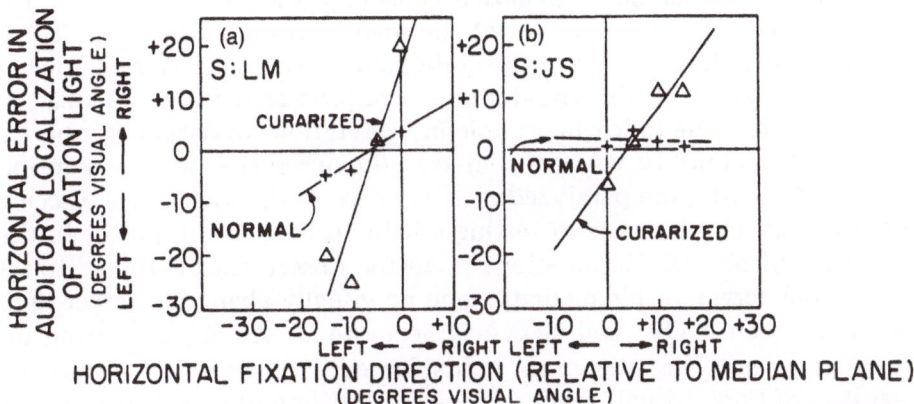

Fig. 14-7. Auditory-visual matches by observers LM (a) and JS (b). The observer matched the perceived horizontal location of a sound to that of a fixated light. The physical location of the fixated light with respect to the median plane is plotted on the abscissa. The error in the auditory localization of the fixated light—the difference between the point of subjective equality of the auditory localization of the fixated light and the physical location of that light—is plotted on the ordinate. Room illumination was left on. The results were indistinguishable when the experiment was carried out in total darkness. *Lines through data:* least square fits. Each point is an average of two or three responses.

Most important in answering the question presently at issue, however, was the fact that the errors of the curarized observer in matching a sound to a light were identical whether normal room illumination was present or whether the match occurred in total darkness. Our procedure eliminated the normal capture of auditory localization by visual context: the observer had no visual information regarding which of the 25 loudspeakers produced the sound. When visual capture was present, it controlled auditory localization by the curarized and uncurarized observer equally well. For example, regardless of gaze eccentricity, the paralyzed observer localized the source of speech by the experimenter as emanating from the experimenter's mouth when the experimenter was in view.

EEPI and Cancellation Are Not Suppressed in Illumination; Visual Localization and Intersensory Localization Are Guided by Cancellation in Darkness

The facts that the auditory–visual matches were identical in room illumination and in darkness, and that the errors were substantial, indicate that neither EEPI nor the output of the EEPI-driven cancellation mechanism was suppressed per se; both EEPI and the output of the EEPI-driven cancellation mechanism were unchanged by the presence or absence of illumination. Furthermore, the function relating errors in the auditory–visual match to gaze eccentricity was the same as the error function for visual localization of the median plane in darkness. These results imply that both errors have the same basis and that the basis lies in the way in which the visual stimuli are processed for visual localization. We offer the following simple interpretation: In darkness, where a single light cannot be visually related to the observer's visual perception of his own body, the paralyzed observer's overestimation of gaze eccentricity leads to his errors in setting a light to the median plane. In the presence of normal illumination, when the observer is instructed to set the visual target to his median plane he visually aligns the target light with his seen body, and hence is accurate in visually localizing his median plane. This alignment holds for each eccentricity of gaze regardless of how "visual localization is itself differently displaced relative to veridicality" for different gaze directions. The phrase in quotes, however, refers to a relation that cannot be directly measured by a comparison of locations simultaneously observed by vision alone, since, if the "translation of visual localization" is uniform across the visual field, all purely visual relations are unchanged. The change in the relation of visual localization to veridicality can be indirectly measured by comparing visual localization to localization by a sense modality whose relation to veridicality is unchanged both by curare and by variations in gaze eccentricity. The auditory–visual matches have provided these measurements.

Thus, although visual localization in normal illumination appears to be uninfluenced by the EEPI-driven cancellation mechanism, we find that the cancellation mechanism is itself intact and available for intersensory localization. The suppression of the output of EEPI-driven cancellation by visual context is thus specific to its use for visual localization.

A Paradox and Its Resolution

Although the interpretation of our results so far follows in a reasonably straightforward way from the measurements, an extremely interesting problem regarding the observations in normal illumination remains for our further exploration. While we favor the interpretation presented below it is not the only possible one.

For the case in which the partially paralyzed observer fixates a visual target to the left of his median plane (Figure 14-8, point F), the results may be summarized as follows: As described above, when setting a light to his visually perceived median plane, the observer does so accurately in normal illumination (setting to point A), but with a systematic error in darkness (setting to point D). When choosing the loudspeaker whose perceived horizontal location matches the perceived horizontal location of the fixation target, he makes the same systematic error in darkness

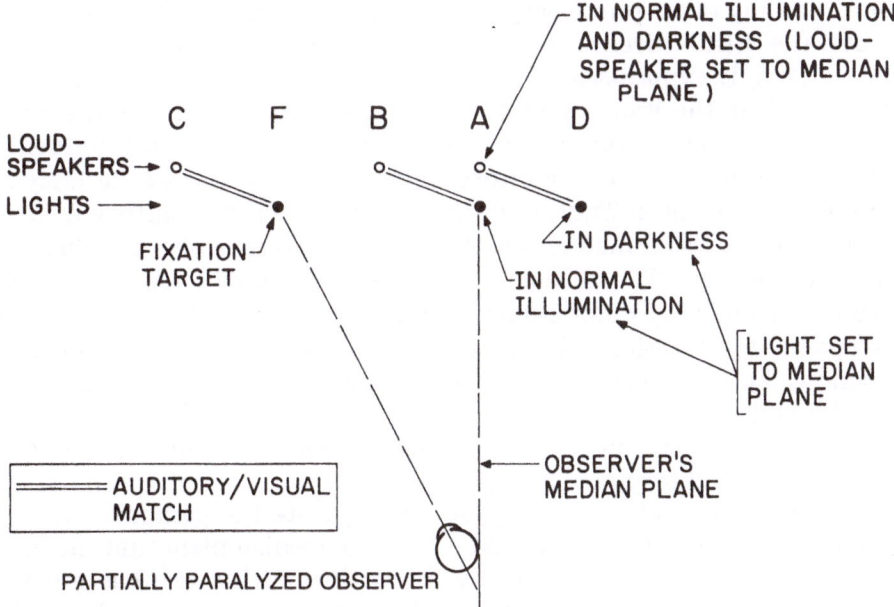

Fig. 14-8. Spatial relations involved in the paradox. See text for description.

and in normal illumination (loudspeaker at C chosen to match fixation light at F with distance AD approximately equal to distance CF). In addition, when the observer chooses the loudspeaker that sounds as if it is in his median plane, he accurately chooses the speaker at A in both darkness and in normal illumination.

Thus, a paradox exists, and may be stated in either of two essentially equivalent ways: (a) In normal illumination, the partially paralyzed observer says that a sound whose location appears to match the location of a light that is visually localized in his median plane (e.g., sound at B matched to light at A) is not itself auditorily localized in his median plane. (b) In normal illumination the partially paralyzed observer says that a sound that is auditorily localized in his median plane is matched in location to a light that does not visually appear to lie in his median plane (e.g., sound at A matched to light at D).

The resolution of the paradox follows directly from the treatment given earlier: The gaze-dependent errors in EEPI are unmodified by the presence or absence of illumination, but they do not lead to errors in visual median plane settings. Although under paralysis the cancellation mechanism driven by the abnormal EEPI has shifted the entire visual coordinate structure relative to the auditory coordinates when a change of gaze direction occurs, the purely visual spatial relations within the visual field are unmodified. Since the visual median plane settings in normal illumination are unaffected by gaze direction they appear to be an outcome of a purely visual comparison with the observer setting the target to visual alignment with the view of his own body. (Neither EEPI nor any other proprioceptive information is involved; as described above the EEPI-driven cancellation mechanism's influence on visual localization is suppressed.) However, the auditory judgments are not influenced by the visual field of view since visual capture for the loudspeakers has been eliminated. Since curarization did not influence auditory localization, the auditory median plane is set as accurately as without curarization. Thus, in normal illumination the auditory median plane and the visual median plane are each set accurately, although for different reasons. The gaze-dependent shift of visual coordinates relative to the auditory coordinates is unnoticeable to the observer visually except for the transient "jumping" around the time of the saccade (see footnote 1). The shift leads to the gaze-dependent errors in auditory-visual matches.

It will be desirable in subsequent observations to require the observer to compare auditory and visual median planes themselves in normal illumination. The observer may be able to note the paradox directly. The deviation between the sound set to the median plane and the light set to the median plane is clearly discriminable. But, he has not yet been confronted with the task of setting them simultaneously to the median plane. Under this constraint it is possible that he will employ the two different criteria simultaneously and recognize that he is doing so.

Source of EEPI

There has been considerable controversy regarding whether EEPI derives from outflow, inflow, or hybrid sources (for reviews see Matin, 1972, 1976, 1982; Stevens, Emerson, Gerstein, Kallos, Neufeld, Nichols, & Rosenquist, 1976). The present experiments do not resolve the issue. They do, however, suggest that EEPI-driven cancellation mechanisms have different access to visual localization in darkness and in normal illumination; the failure of the previous work to arrive at a simple conclusion may be a result of not dealing with this issue. Thus, following Helmholtz, it used to be argued that if attempts were made to turn a totally paralyzed eye, outflow theory (EEPI is assumed to be derived from feedforward signals) required that an apparent movement be observed in the direction of the attempted eye turn; no apparent movement implied inflow theory. Although Kornmuller's (1930) work with partially immobilized eyes (in which observers saw movement when they made attempts to turn their eyes) was taken to support outflow theory, observers in experiments by Brindley, Goodwin, Kulikowski, and Leighton (1976), Seibeck (1954), Siebeck and Frey (1954), and Stevens et al. (1976) failed to observe any movement when attempts were made to turn totally paralyzed eyes. Although this appears to argue for an inflow or hybrid theory, no observations were made in total darkness. The results of the experiments described above suggest that if an outflowing or hybrid source of EEPI were involved, observations in normal illumination would be different from those in darkness. In fact, in one report (Stevens et al. 1976), the totally paralyzed observers noted that when they attempted to turn their eyes there was a feeling that if they were to attempt to touch a given point they would have to reach in a different direction under eccentric gaze than they would if no attempt at eye turn was made. This may be the precursor to the mislocalizations we have measured and described above.

We are beginning experiments employing total paralysis; these will be carried out in both normally illuminated environments and in darkness and should resolve the question of whether EEPI is from an outflowing or hybrid source. Although the present evidence remains somewhat in favor of an outflow source, as it has since Helmholtz first argued for it, this result is not assured by any means.

Acknowledgments

This work was supported in part by Contract N 62269 80C 0296 with the Naval Air Development Center and in part by U.S. Public Health Service Research Grant EY 03198 from the National Eye Institute, NIH.

References

Brindley, G. S., Goodwin, G. M., Kulikowski, J. J. & Leighton, D. Stability of vision with a paralyzed eye. *Journal of Physiology*, 1976, *258*, 65-66.

Goodman, L. S., & Gilman, A. *The pharmacological basis of therapeutics* (5th ed.). New York: Macmillan, 1977.

Helmholtz, H. von. [*A treatise on physiological optics*, (Vol. 3).] (J. P. C. Southall, Ed. and trans.). New York: Dover, 1962, (Originally published, 1866).

Kornmuller, A. E. Eine experimentelle Anesthesie der auberen Augenmuskein am Menschen und ihre Auswirkungen. *Journal für Psychologie und Neurologie*, 1930, *41*, 354-366.

Matin, L. Eye movements and perceived visual direction. In D. Jameson & L. Hurvich (Eds.), *Handbook of sensory physiology* (Vol. VII, Part 4). Heidelberg: Springer-Verlag, 1972, pp. 331-380.

Matin, L. A possible hybrid mechanism for modification of visual direction associated with eye movements—The paralyzed eye experiment reconsidered. *Perception*, 1976, *5*, 233-239.

Matin, L. Suppression of the use of extraretinal eye position information (EEPI) for visual localization is normal in normally illuminated visual fields. April Supp. *Investigative Ophthalmology and Visual Science* (Suppl.), April 1981, *20*, 55.

Matin, L. Visual localization and eye movements. In A. H. Wertheim, W. A. Wagenaar, & H. W. Leibowitz (Eds.), *Tutorials on motion perception*. New York: Plenum Press, 1982.

Matin, L., Pearce, D. G., Matin, E., & Kibler, G. Visual perception of direction: Roles of local sign, eye movements and ocular proprioception. *Vision Research*, 1966, *6*, 453-469.

Matin, L., Picoult, E., Stevens, J. K., Edwards, M. W., Jr., Young, D., & MacArthur, R. Visual context dependent mislocalizations under curare-induced partial paralysis of the extraocular muscles. *Investigative Ophthalmology and Visual Science* (Suppl.), April 1980, *19*, 81.

Matin, L., Picoult, E., Stevens, J. K., Edwards, M. W., Jr., Young, D., & MacArthur, R. Oculoparalytic illusion: Visual-field dependent spatial mislocalizations by humans partially paralyzed with curare. *Science*, 1982, *216*, 198-201.

Matin, L., Pola, J., Matin, E., & Picoult, E. Vernier discrimination with sequentially-flashed lines: Roles of eye movements, retinal offsets and short-term memory. *Vision Research*, 1981, *21*, 647-656.

Matteo, R. S., Pua, E. K., Khambatta, H. J., & Spector, S. Cerebrospinal fluid levels of *d*-tubocurarinine in man. *Anesthesiology*, 1977, *46*, 396-399.

Siebeck, R. Wahrehmungsstorung und Storungswahrnehmung bei Augunmuskellahmungen. *von Graufes Archiv für Opthalmologie*, 1954, *155*, 26-34.

Siebeck, R., & Frey, R. Die Wirkungen muskeleschlaffender Mittel auf die Augenmuskein. *Anaesthesist*, 1953, *2*, 138-141.

Stevens, J. K., Emerson, R. C., Gerstein, R. L., Kallos, T., Neufeld, G. R., Nichols, C. W., & Rosenquist, A. C. Paralysis of the awake human: Visual perceptions. *Vision Research*, 1976, *16*, 93-98.

Chapter 15
Mechanisms of Space Constancy

Bruce Bridgeman

The problem of space constancy despite movements of the eye is one of information selection. The perceptual system is concerned with the identification of objects, faces, letters of the alphabet, etc., and must be independent of changes in the retinal image that accompany saccadic eye movements. At the same time, the perceptual system must know with great sensitivity about displacements of objects in the world. A space constancy system, therefore, must null retinal information about eye movements while preserving information about movement in the world. Properties of the image that specify events in the world must be processed, while those that specify properties of retinal motion itself must be eliminated.

This analysis applies only to the cognitive or "focal" branch of the visual system (Trevarthen, 1968); the information necessary to maintain coordination of vision with motor activity is assumed to be handled by a lower motor or "ambient" system. These two systems have been associated physiologically with geniculostriate structures, and with the superior colliculus and associated brain stem structures, respectively (Schneider, 1969). Their psychophysical definition (Bridgman, Lewis, Heit, & Nagle, 1979) is based on response mode: symbolic responses such as verbal reports of object identities or motions test the focal system, while open-loop isomorphic responses, in which there is a one-to-one relationship between stimulus position and eye or hand position, isolate the ambient system.

The problem of space constancy in this view reduces to a cognitive problem of stimulus selection. The traditional solution to the problem, however, has been in terms of a neurological subtraction process: during an eye movement, the oculomotor control signal is also sent to higher brain centers to be subtracted from the resulting change in the retinal signal, so that movements of the eye can be canceled while movements of objects are preserved. Such a process was originally proposed by Helmholtz (1866), who suggested that the "effort of will" (*Willensanstrengung*) required to move the eyes was compared with the retinal signal. If the eyes move while the environment remains stationary, the resultant will be zero, and no motion should be apparent. A movement in the world, however, should be perceived because it creates a mismatch between the two signals.

Helmholtz did not differentiate between focal and ambient systems. Some of his evidence we would now ascribe to the focal system, such as his observation that the world appears to move when the eye is pressed with a finger. This observation originated with Descartes (1664), who showed that the apparent movement of the world was due to a displacement of retinal images. Other evidence implicated both the focal and the ambient systems, such as Helmholtz's observation that patients with eye muscle paresis or paralysis experience apparent motion of the visual world during eye movements (focal) and also display overshoot in their pointing to a target in the paretic field when their hand is not seen (ambient).

The effort of will principle was rediscovered by Sperry (1950), who explicitly postulated a neural subtraction mechanism, and used the idea in a more mechanistic way to show why fish with inverted eyes display forced circling. He called the internal signal a *corollary discharge*, the term which will be used in this paper. Sperry conceived of the corollary discharge as a signal flowing out from oculomotor centers to coordinate eye position with retinal image position. By the above definitions he limited the corollary discharge idea to the ambient system. Holst and Mittelstaedt (1950) independently proposed a similar signal, called the *Efferenzkopie* or efference copy, and supplied a control theoretical diagram to translate the idea into a neurological model.

These quantifications of the corollary discharge theory made it more precise and subject to experimental verification, for the theory makes some nonintuitive predictions. One of these asserts that displacements of targets in the world during eye movements should be detectable with nearly the same precision as during fixations, even if no relative-movement cues are available. When the corollary discharge and the retinal signal are equal in magnitude and opposite in sign, the resultant is zero. This is the case in normal perception. A displacement in the world, however, changes the retinal signal without changing the corollary discharge. The resulting composite signal is nonzero and the displacement should be detected. Of course, the neurological process of resetting the visual frame of reference due to the saccade might itself result in a small threshold increase, but the model requires that this increase not be greater than the precision of the stabilization process. Because the cognitive stabilization of the visual world is very effective, the world not seeming to jump even slightly during saccades, this error should be very small.

Beyond Corollary Discharge

This prediction of the corollary discharge theory has now been disproved by several investigators, beginning with Ditchburn (1955), who noted that target movements on an oscilloscope correlated with the subject's eye movements were not perceived. Wallach and Lewis (1965) used an optical, in contrast to Ditchburn's electronic, method and independently obtained qualitatively similar results. Brune and Lücking (1969) rediscovered the phenomenon from a more ecological perspective. The phenomenon has been repeatedly confirmed using modern electronic and psychophysical techniques (Beeler, 1967; Bridgeman, Hendry, & Stark, 1975; Mack, 1970; Mack, Fendrich, & Pleune, 1978; Stark, Kong, Schwartz, Hendry, & Bridgeman, 1976). There is also a distortion of perception of rapid motion during saccades (Orban, Duysens, & Callens, 1973). This literature provides quantitative estimates of the minimum detectable ratio of target movement to eye movement. This ratio (the *displacement ratio* of Wallach) can be as high as 30% (Bridgeman et al., 1975), although its exact value depends upon experimental conditions.

In general, two methods of determining the displacement ratio have been used: In one, an eye movement record is fed back to change a target position with a variable gain ("dynamic feedback"). In the second, a target is abruptly displaced during a saccade. It is now clear why the first method generally yields somewhat lower displacement ratios than the second. Bridgeman et al. (1975) measured displacement ratios throughout the time course of the saccade using the abrupt displacement method, and expressed their threshold as the minimum of the resulting saccadic suppression function. The dynamic feedback method, however, entails stimulus motion throughout the saccade, even in periods which were shown, by the displacement method, to have less effective saccadic suppression. If target motion is available to the system throughout the saccade, it will be detected with greatest sensitivity at a time when the saccadic suppression function is weakest. Thus, the dynamic feedback method gives the system an opportunity to detect the motion signal at times when saccadic suppression is relatively ineffective and lower thresholds are expected.

These studies have shown that the corollary discharge theory is inadequate by itself to explain motion perception during saccades, but the experiments apply only the focal system. However, information about displacement during saccades continues to be available to the ambient system, as measured with an open-loop pointing procedure, even when saccadic suppression prevents target displacement from being reported (Bridgeman et al., 1979). Subjects could point accurately to a target whether its displacement had been detected or had been masked by saccadic suppression of displacement. Thus the corollary discharge idea might still be effective for the ambient system, which requires great stability but less accuracy than the focal system.

MacKay (1972, 1973) suggested a modified form of the corollary discharge theory which is not contradicted by the above evidence. His theory applies equally well to inflow (from proprioception in the orbits and eye muscles) and outflow (from brain stem oculomotor centers) as sources of extraretinal information about eye movement, maintaining only that an extraretinal signal informs the system of self-produced image movement. A subtraction process is not necessary because the extraretinal signal changes the rules of movement perception to warn the system that a displacement is occurring. The accompanying change in retinal signal then serves as a confirmation of the motion, and any movement that is roughly consistent with the extraretinal signal can be interpreted as an eye movement rather than a displacement of the visual world. In this way sensitivity to displacements can be maintained during fixation or pursuit tracking.

Implicit in MacKay's theory is the requirement that it be applied only to the focal system, which does not need an absolute calibration between the image of an object on the retina and its position in space. Eye–hand coordination cannot be maintained by this mechanism because the confirming signal is qualitative rather than quantitative.

Like the corollary discharge, MacKay's theory makes nonintuitive predictions that can be tested experimentally. The most important of these is that the rules of perception change during saccades. While the corollary discharge theory maintains that the rules of perception remain constant, with only a subtraction process preventing perception of displacement, the MacKay theory requires that the visual system be less sensitive to retinal displacements during saccades than it is to the same retinal displacements during fixation.

Rules of Motion Perception During Saccades
Versus During Fixation

Do the Rules Change?

This prediction of the MacKay theory has now been tested in my laboratory. We started by analyzing the retinal conditions in a previous two-alternative, forced-choice study of saccadic suppression of displacement (Bridgeman & Stark, 1979). In this experiment subjects made two successive 18° saccades, during one of which a visual target was displaced 2°. The subjects' task was to guess which of the two saccades was accompanied by displacement. In retinal terms, the subjects saw one displacement of 18° and another of either 16° or 20°, depending on whether the displacement was in the same direction as the eye movement or in the opposite direction. The retinal problem reduces to a determination of difference thresholds, asking whether an 18° saccade-induced displacement can be distinguished from a larger or smaller displacement.

To test the null hypothesis that difference thresholds during saccades and during fixation are identical, we simulated Bridgeman and Stark's (1979) retinal conditions during fixation by displacing a target 18°, extinguishing it, re-presenting it in the original position, and displacing it again by a different amount. All of this occurred during fixation, so that any differences in results between Bridgeman and Stark (1979) and the current experiment would be due to the effects of an extraretinal signal (the retinal signals being the same in both cases).

Under computer control, a projected slide 18° wide could be displaced by flipping a mirror in the projection system. The experiment began with the subject looking into a homogeneous field 180° wide by 60° high. When a shutter opened, the subject fixated on the right edge of the projected image, a random dot texture with sharp edges. After .75 seconds the target jumped to a new position 16°, 18°, or 20° to the right, and was extinguished after 1.25 seconds. The subject was instructed to maintain fixation when the target jumped. After a 1-second dark interval, the same sequence was repeated with a target displacement of a different size, and the subject then pressed one button if the first of the two displacements had been larger, or another if the second had been larger. The pair of displacements was given in one of four randomly presented sequences (16° vs. 18°, 18° vs. 16°, 18° vs. 20°, or 20° vs. 18°) so that the subjects could use only the relative sizes of the two displacements as a basis for judgment. Eye movements were continuously monitored with a photoelectric system (Bahill, Clark, & Stark, 1975).

The results of the experiment were clear: subjects could detect differences in displacement as small as 15 min arc at above random rates, even though the Bridgeman and Stark (1979) study showed that displacements as large as 2° were undetectable under the same retinal conditions (Figure 15-1). We can conclude that the rules of perception are indeed different during saccades than they are during fixation, because retinal displacements that are easily detectable during fixation are not detectable during saccades. A decrease in sensitivity to retinal displacement seems to accompany saccades themselves and must be caused by an extraretinal signal.

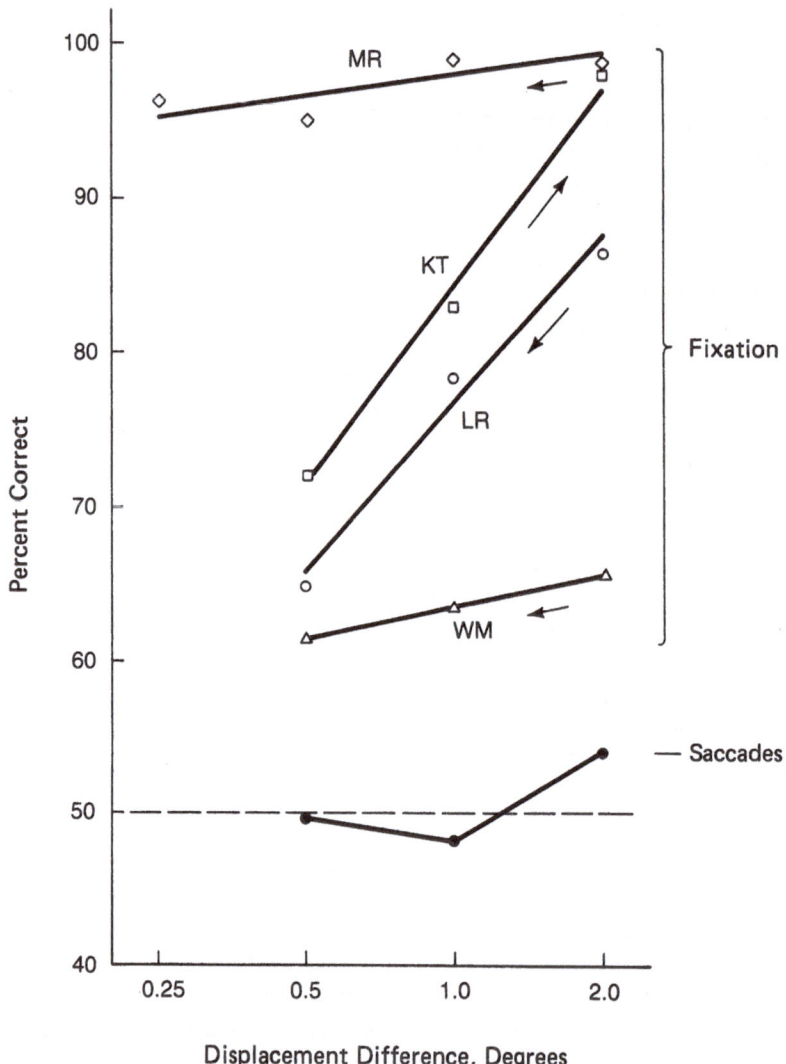

Figure 15-1. Detectability of differences in target displacements during fixation in successive exposures, using a two-alternative forced-choice technique. *Upper four lines:* detection during fixation. Each line represents one subject, and each point summarizes data from a block of 200 trials. *Arrows:* temporal order in which the data were collected for each subject; practice effects were not important. Subject WM had an unsteady fixation pattern with saccadic intrusions. *Bottom line:* detection during a saccade. The retinal conditions are similar to those in the fixation condition. Average data for two subjects, replotted from Bridgeman and Stark (1979).

Why Do the Rules Change?

The next step in understanding space constancy is to seek the source of this difference in sensitivity during saccades. Hints about the source of the difference might come from a closer examination of the nature of sensory events surrounding displacements during saccades, using the psychophysical strategy of pushing the limits of a perceptual system until it begins to break down. It is in the location and the nature of the breakdown that the system reveals its characteristics. Because space constancy is a property of the focal system, it makes sense to probe the limits of the space constancy system using complex and meaningful patterns that are particularly suited to the focal system.

In a little-known paper by Brune and Lücking (1969), the breakdown in space constancy was examined with naturalistic landscape scenes as well as the more common but perceptually degenerate dots, lines, and edges. By feeding an EOG signal back into a mirror which controlled the horizontal position of a slide, these authors replicated the saccadic suppression of displacement found by others. Then they went further:

> If one presents to the subject a landscape and lets him observe it continuously, it will move back and forth with optical exploration because of the coupling between eye and object movement. This movement will not be perceived as long as the excursion of the object does not exceed twenty percent of the eye excursion. It is not important whether the movement occurs in the same direction as the eye or in the opposite direction. If one exceeds this value, prominent objects in a complex landscape release themselves from the background and begin to move. They wander synchronously back and forth with the eye movement, though with subjectively small amplitude. The subject can direct attention sometimes to one and sometimes to another object, and the attended ones then belong to the figure, while the others remain attached to the background and take no part in the movement. (Brune & Lücking, 1969; translation by B.B.)

This important result implies that space constancy in the focal system is achieved at least partially through the mechanism of selective attention. The implication is that constancy breaks down first for the attended objects, since retinal signals of equal magnitude are somehow ignored for the background as a whole. Might this process of ignoring background motion account for space constancy itself?

Our first step in pursuing this idea was to replicate Brune and Lücking's findings with more precise methods. Signals from the paired photocell system for recording eye movements, which offers high temporal and spatial resolution, were fed back with variable gain into a mirror that could displace a projected image using the apparatus described above. When the gain of this system was set just below the threshold for seeing the entire image jump with the eye, three of four practiced psychophysical observers saw parts of a complex landscape jump while the scene as a whole remained stable. When relative movement was seen, it was always in the true direction of slide motion, whether the feedback was in the same phase as the eye movement or in the opposite phase; the phase was reversed every few minutes to avoid adaptation effects (Mack et al., 1978). The fourth subject always saw either motion of the whole slide or stability. The identity of the moving objects was somewhat more variable than experienced by Brune and Lücking's subjects: one subject (the author) consistently saw relative motion of the object located at the goal of a saccade, while the other observers saw jumps of unpredictable objects in the scene.

Exploring Space Constancy with Figure–Ground Reversible Images

The source of these selective movements remained unknown. Although the jumps may have been controlled by attentional factors, they might have been mediated by such image-related variables as the brightness of the objects, their spatial frequency distribution, their contrast, etc. It is easy to develop a hypothesis wherein greater stimulus brightness, for instance, leads to a different retinal latency and therefore to a difference in space constancy during a saccade. In order to control all of these variables simultaneously, we introduced patterns with reversible figure–ground relationships. If attention or figure–ground organization rather than image variables were controlling the appearance of movement, it should be possible to see movement in either one part or another by intentionally reversing figure and ground in the image.

The stimuli we chose were a series of tesselations of planes by the Dutch artist M. C. Escher, each consisting of a series of two kinds of interlocking figures that completely filled the plane (Figure 15-2). Subjects could concentrate on one set of forms as the "figure" so that the other became the "ground," and reverse the two at will.

Using these reversible images we asked the same subjects to form a stable percept of one set of patterns as the figure and introduced feedback as they looked repeatedly across the image. After recording their observations we asked the subjects to reverse the figure–ground relationship and to repeat the same procedure. All subjects were able to form stable figure–ground organizations and to reverse them for one or another of the prints we used, and the same three subjects who saw relative motion in the landscapes saw reversible relative motion in at least one of the Escher prints. Subjects usually maintained a stable figure–ground percept by fixating on the parts of the image perceived as the figure, although the patterns of eye movements were not distinguishable in the two cases.

We replicated the results in a more formal study on 12 naive under-

Figure 15-2. One of the stimuli with reversible figure–ground segregation, by M. C. Escher. By reversing figure–ground relationships voluntarily, subjects could perceive selective motion of either the lighter or the darker figures with no change in stimulus parameters. From The Graphic Work of M. C. Escher (J. E. Brigham, trans.). New York: Meredith Press, 1967. Reprinted by permission.

graduates. Following calibration, each subject was shown two landscape slides, one a complex scene of Roman traffic and the other a simpler alpine landscape. After practicing with a double-staircase threshold procedure on these slides, each subject saw a series of four Escher prints (four prints were used because the symmetry of figure–ground segregation in the prints is idiosyncratic and reversal is easier in some images than in others). The experimental design is described in more detail elsewhere (Bridgeman, 1981).

One subject saw movement in all three of the catch trials and was discarded. Of the remaining 11 subjects, 8 saw movement of parts of the landscape slides, with 7 of them reporting movement of discrete objects rather than regions of the slide. Among the latter 7 subjects, 4 were able to reverse the figure–ground relationship on one or more of the Escher prints and see movement of only the figure in both conditions. Thus, even among naive, unpracticed observers, space constancy tends to break down according to cognitive variables rather than image-related or oculomotor factors.

There are four possible combinations of observable motion in this experiment (motion of figure, ground, both, or neither). The most striking finding was that one of the combinations, movement of ground without movement of figure, was never reported in any condition by any observer. If it is seen at all, motion of part of the array is always in the figure.

These results are subject to one rather uninteresting interpretation. It is possible that displacement thresholds in general are lower for the figure than for the ground regardless of eye movements, so that the space constancy mechanisms involved at the time of a saccade would not be uniquely related to the figure–ground threshold phenomenon. We recently tested this possibility directly, displacing an extended target by a threshold amount during visual fixation. Using the same Escher stimuli employed in the saccadic study, we found that displacement threshold is not related to figure–ground segregation (Delgado & Bridgeman, 1982).

The results of these experiments are consistent with the idea that space constancy is an adaptation; but that idea itself implies potential complications for a mechanism that must maintain constancy over a wide variety of visual conditions and leads to another experimentally verifiable prediction. Specifically, the system should show the same degree of constancy regardless of the size of the visual image and the amount of texture it contains. Thresholds for displacement of targets of varying sizes should be constant because the system could not maintain a separate adaptation state for every possible size of target.

Although the prediction is intuitively appealing from a cognitive standpoint, two other theoretical analyses predict different effects of target size on displacement ratio. The standard psychophysical analysis would invoke probability summation, because in a threshold task of this sort the threshold for detecting displacement during a saccade should equal the threshold for detecting displacement of any element in a texture. A larger number of elements in a texture, that is, a larger pattern of the same texture density, should be more easily detectable because there is a higher probability that one or another of the texture element displacements would be detected as the number of elements increases. Thus probability summation predicts that displacement ratio should decrease as target size increases with a constant texture density.

The theory of Gibson (1950, 1966), in contrast, makes the opposite prediction. According to Gibson, the visual system assumes the world to be stable because of its large retinal extent. Therefore, a large target should be assumed to be stable even though a small target may undergo motion with respect to a stable world. This theory allowed Gibson to explain the notorious failure of space constancy for very small objects, such as lights in darkened rooms, where the autokinetic phenomenon and apparent movements during saccades are common. The more closely the visual world assumes the characteristics that normally specify stability in ecological objects, that is, the larger the size of the texture, the more stable the texture ought to appear and the larger the displacement necessary to inform the system of a violation of that assumption. Thus a Gibsonian theory predicts larger thresholds for larger textures.

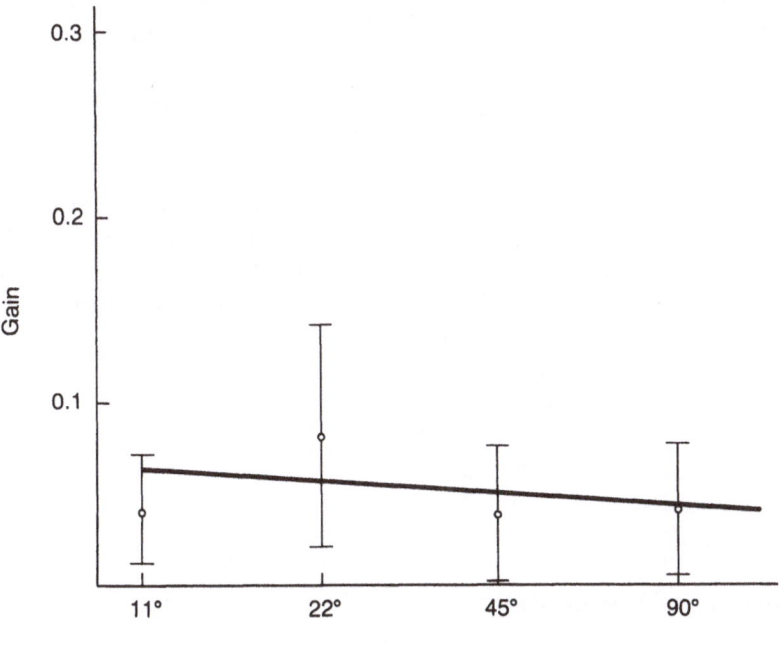

Figure 15-3. Relationship of target size to gain of the feedback of eye position on target position. Because the targets had a constant height, target width indicated on the abscissa also indicates the ratios of target areas. Gain expresses the ratio of eye movement magnitude to magnitude of displacement; that is, at a gain of .1, a $10°$ eye movement will result in a $1°$ image movement with similar dynamics. Gain is functionally equivalent to Wallach's displacement ratio, although the stimulus movements are dynamically similar to saccades and are not step displacements. The relatively low gains obtained here are probably due to the relatively slow motion of the targets, which are driven by amplified saccadic eye movements rather than by voltage steps. Target motion is further slowed by the mirror system used to displace the projected targets, which had a half-amplitude band width of 120 Hz. Thus the target could still have been moving slightly at the end of a saccade. Slope of the best fit regression line is $-.0043$.

In summary, probability summation predicts that the threshold displacement ratio should decrease as target size increases, a Gibsonian analysis predicts that the threshold should increase, and the adaptation theory proposed here predicts that the threshold should remain constant over a wide range.

In order to measure this phenomenon we displaced targets varying in width from 11° to 90° and measured the threshold displacement ratio for each size of displacement by the dynamic feedback method. The experiment used the same apparatus and 11 naive subjects as the figure-ground experiment described above. The four targets were random dot patterns with a rather coarse texture density of one element per square degree, 59° high and 11°, 22°, 45°, or 90° in width. Following threshold determinations with a double-staircase method on at least five practice trials, each target size was presented five times in a random order. The direction of feedback (positive or negative) was reversed every 2-3 minutes to minimize space constancy adaptation effects (Mack et al., 1978), and two catch trials were included in the sequence.

The results (Figure 15-3) support the adaptation hypothesis: although there was considerable variability in displacement ratios, probably due in part to short-term adaptation effects, there was no relationship between displacement ratio and target size. The-best-fit-least-squares regression line through the average threshold displacement ratios for the 11 pooled subjects had a slope of $-.0043$, a relationship which accounts for 7% of the variance in the data.

This result extends the recent result of Scobey and Johnson (1980), who showed that threshold displacements during fixations are constant over a wide range of sizes smaller than those used here, as long as the fovea is included in the stimulated area.

Mechanisms of Stabilization

These experiments show that the basis of space constancy is cognitive. It is an adaptation to retinal displacement during saccades—a selective inattention to a range of image displacements—rather than a cancellation or subtraction of the incoming signal. Several older observations, previously unexplained, can now be seen to fit into the same framework. This family of observations about the phenomenology of space constancy has been displaced because it has never fit very well into the control theoretical mold. The central observation is that stabilization is too perfect: the world doesn't seem to jump even slightly during saccades under normal conditions, even in activities such as reading, in which saccades occur with great frequency for long periods. If extraretinal signals alone were responsible for stabilization, one would expect a slight error because the biological system is not quantitatively perfect.

Feedforward schemes such as the corollary discharge would be expected to drift, and feedback should oscillate slightly. Only some additional, higher level factor not related to the correcting of position vectors can explain such hyperstability.

Despite great stability, space constancy is paradoxically very fragile, being upset by the slightest abnormality in sensory or motor conditions. Saccading from one point at the edge of the eye's fixation range to another results in an apparent displacement of the world in the direction opposite the movement (Helmholtz, 1866), seeming to imply that the "compensation" mechanism is so simple that it assumes the same gain in the extraocular muscles throughout their range despite the mechanical disadvantages encountered at the extremes of the range. Even in the "smaller fixation space" of Hering (1868), where oculomotor gains are nearly linear and eye movements normally take place, a rapidly repeated alternation from one fixation point to another results in a breakdown of stability after a few cycles. Might this effect occur because the experiments draw attention away from the objects themselves and toward the eye movements? Even making a single large, deliberate saccade can result in apparent displacement if attention is directed to the image rather than the world.

In order to demonstrate the role of different parts of the image in stabilization much more simply than in the above feedback experiments, we can return to the old trick of pushing gently with a finger on the outer canthus of one eye. The world slides around, but carefully observed it slides in a peculiar way, with the center moving more than the periphery (I thank O. -J. Grüsser for this observation). The object of attention, at the fovea, again seems less subject to stabilization.

We recently used the eye-press technique to clarify the role of corollary discharge and found that it has no influence on position perception in a structured visual field except for a transient impression of motion. Corollary discharge governs position perception in structureless fields, and it is the only source of visual position information available for interaction of vision with motor and auditory systems (Bridgeman & Stark, 1981). This work extends paralysis experiments by Matin et al. (Chapter 14, this volume).

On the sensory side, stabilization is equally ephemeral. According to existing theories, including MacKay's, a simplified visual world such as a single luminous point or a group of points in darkness should be stabilized in the same way as more complex images, yet apparent motion during saccades is the rule under these conditions, even within the smaller fixation space. How are image properties taken into account? Attention again seems to be the only answer.

Sensory and motor properties of the system were linked in a recent series of experiments by Mack et al. (1978), who found that prolonged exposure to the sort of feedback provided in the Escher experiments described above results in both sensory and motor adaptation. Moving an image vertically in response to horizontal eye movements, they showed that the threshold range for detection of image displacement shifts toward the exposed conditions. At the same time, motor adaptation shifts saccadic patterns so that saccades are made to the predicted vertical position of the saccadic target at the conclusion of the movement, not to the actual position before the movement. There is no room for such flexibilities in the classical theories, and any system that drives adaptations as subtle as these could perform the stabilization itself.

Such adaptations are necessary in the natural world because the gain of the oculomotor system varies with growth and with aging and can change radically with paresis of the eye muscles. Patients with paresis of the extraocular muscles generally complain that the world seems to jump every time they attempt to look into the affected field, an observation which was one of the bases for proposing the *Willensanstrengung* theory. More puzzling, however, is the disappearance of this complaint after a few weeks without any improvement in oculomotor abilities. The patients adapt to the reduced gain or even the nonlinearity of the system. The corollary discharge theory, in postulating a feedforward mechanism to compensate for retinal image displacements during saccades, cannot account for adaptations that logically require feedback. MacKay's modification of the corollary discharge theory can handle adaptation only because the postulated mechanism is itself imprecise, using a qualitative rather than a quantitative corollary discharge signal, but otherwise being logically similar. If space constancy is itself an adaptation effect, these alterations in the adaptation state can follow immediately.

The adaptation effects for image motion during saccadic eye movements can be quite extensive, as shown in a long series of experiments by Wallach (reviewed by Wallach, 1981). By altering oculomotor feedback gain with magnifying and minifying lenses, Wallach showed stable adaptations with the expected negative after-adaptation effect. The adaptation to space constancy thus may belong to the same family of observations as adaptation to displacing prisms, inversion of the visual world, displacement of part of the visual world, and other perceptual compensations (Kohler, 1951). The present experiment and analysis merely add space constancy to this list.

Recent results by Heywood (1981; Heywood & Churcher, 1981) clarified the process of stabilization during saccades. In these experiments subjects were presented with an array of two or more dots, one of which was displaced during a saccade from one dot to another. As expected, the threshold for dot displacement increased—an example of saccadic suppression of displacement. When asked which dot had been displaced, however, the subjects made frequent mistakes or misattributions of displacement. If the dot constituting the saccadic goal had been displaced, the subjects frequently attributed the displacement to the starting point of the saccade or to an intermediate dot, but the reverse misattribution was rare. It seems that the frame of reference had shifted during or just before the saccade to the new fixation point, and target positions were now judged with respect to this point, which was assumed to be stable. This phenomenon may also have an attentional component, for attention is directed more strongly to the saccadic goal. In experiments that require the subject to make two successive saccades to two different points, this stabilization feature is shared by both of the saccadic goals, even if the second goal is displaced during the first saccade.

A figure-ground interpretation of this experiment might interpret the saccadic goal points as the figure and the remaining point as the relatively less attended ground. Paradoxically, however, stabilization in Heywood's experiment seems more effective for the figure, the saccadic goal points, that it is for the ground, the remaining points. There are some subtle differences between Heywood's conditions and the present ones, however, which may lead to an explanation of the apparent contradiction. In Heywood's experiment only one dot is moved on a given trial, while in the experiments reported above an entire complex scene is displaced as a unit. Heywood's simple array of dots might be considered as a figure in itself against an ambiguous homogeneous background, so that his results may reflect processing within a unit perceived as the figure rather than a figure-ground segregation. The contrast between these results reveals some significant gaps in our understanding of space constancy. We are on the threshold of understanding the events mediating perception and space constancy during saccades, but the goal has not yet been achieved.

Acknowledgments

This research was supported by NSF Grant BNS-7906858.

References

Bahill, A. T., Clark, M. E., & Stark, L. Dynamic overshoot in saccadic eye movements is caused by neurological control signal reversals. *Experimental Neurology*, 1975, *48*, 95-112.

Beeler, G. Visual threshold changes resulting from spontaneous saccadic eye movements. *Vision Research*, 1967, 7, 769-775.

Bridgeman, B. Cognitive factors in subjective stabilization of the visual world. *Acta Psychologica*, 1981, *48*, 111-121.

Bridgeman, B., Hendry, D., & Stark, L. Failure to detect displacement of the visual world during saccadic eye movements. *Vision Research*, 1975, *15*, 719-722.

Bridgeman, B., Lewis, S., Heit, G., & Nagle, M. The relationship between cognitive and motor-oriented systems of visual position perception. *Journal of Experimental Psychology: Human Perception and Performance*, 1979, *5*, 692-700.

Bridgeman, B., & Stark, L. Omnidirectional increase in threshold for image shifts during saccadic eye movements. *Perception and Psychophysics*, 1979, *25*, 241-243.

Bridgeman, B., & Stark, L. Efferent copy and visual direction. *Investigative Ophthalmology and Visual Science* (Supplement), 1981, *20*, 55.

Brune, F., & Lücking, C. H. Oculomotorik, Bewegungswahrnehmung und Raumkonstanz der Sehdinge. *Der Nervenarzt*, 1969, *40*, 413-421.

Delgado, D., & Bridgeman, B. Foreground/background constancy during visual fixation. *Investigative Ophthalmology and Visual Science Supplement*, 1982, *22*, 86.

Descartes, R. [*Treatise of man*] (T. H. Hall, trans.). Cambridge, Mass.: Harvard University Press, 1972. (Originally published, 1664.)

Ditchburn, R. Eye-movements in relation to retinal action. *Optica Acta*, 1955, *1*, 171-176.

Gibson, J. J. *The perception of the visual world*. Boston: Houghton Mifflin, 1950.

Gibson, J. J. *The senses considered as perceptual systems*. Boston: Houghton Mifflin, 1966.

Helmholtz, H. von. [*A treatise on physiological optics*] (J. P. C. Southall, Ed. and trans.). New York: Dover, 1962. (Originally published, 1866.)

Hering, E. [*The theory of binocular vision*] (B. Bridgeman, trans., and B. Bridgeman & L. Stark, Eds.). New York: Plenum Press, 1977. (Originally published, 1868.)

Heywood, S. Detection of displacement during saccades: Spatial and functional differences allied to preprogramming. *Acta Psychologica*, 1981, *48*, 141-149.

Heywood, S., & Churcher, J. Direction-specific and position-specific effects upon detection of displacements during saccadic eye movements. *Vision Research*, 1981, *21*, 255-261.

Holst, E. von, & Mittelstaedt, H. Das Reafferenzprinzip: Wechselwirkungen zwischen Zentralnervensystem und Peripherie. *Naturwissenschaften*, 1950, *37*, 464-476.

Johnson, C. A., & Scobey, R. P. Foveal and peripheral displacement thresholds as a function of stimulus luminance, line length and duration of movement. *Vision Research*, 1980, 709-715.

Kohler, I. [*The formation and transformation of the perceptual world*] (H. Fiss, trans.). New York: International Universities Press, 1964. (Originally published, 1951.)

Mack, A An investigation of the relationship between eye and retinal image movement in the perception of movement. *Perception and Psychophysics*, 1970, *8*, 291-298.

Mack, A., Fendrich, R., & Pleune, J. Adaptation to an altered relation between retinal image displacements and saccadic eye movements. *Vision Research*, 1978, *18*, 1321-1327.

MacKay, D. Visual stability. *Investigative Ophthalmology*, 1972, *11*, 518-524.

MacKay, D. M. Visual stability and voluntary eye movements. In R. Jung (Ed.), *Handbook of sensory physiology* (Vol. VII, Part 3): *Central visual information A*. Berlin: Springer-Verlag, 1973.

Orban, G., Duysens, J., & Callens, M. Movement perception during voluntary saccadic eye movements. *Vision Research*, 1973, *13*, 1343-1353.

Schneider, G. E. Two visual systems. *Science*, 1969, *163*, 895-902.

Sperry, R. Neural basis of spontaneous optokinetic response produced by visual inversion. *Journal of Comparative and Physiological Psychology*, 1950, *43*, 482–489.

Stark, L., Kong, R., Schwartz, S., Hendry, D., & Bridgeman, B. Saccadic suppression of image displacement. *Vision Research*, 1976, *16*, 1185–1187.

Trevarthen, C. Two mechanisms of vision in primates. *Psychologische Forschung*, 1968, *31*, 299–337.

Wallach, H. Eye movement and motion perception. In A. Wertheim, W. Wagenaar, & H. Leibowitz (Eds.), *Tutorials on motion perception*. New York: Plenum Press, 1982.

Wallach, H., & Lewis, C. The effect of abnormal displacement of the retinal image during eye movements. *Perception and Psychophysics*, 1965, *1*, 25–29.

Chapter 16

Visual Information Processing for Saccadic Eye Movements

John M. Findlay

The appearance of a target in the human peripheral visual field frequently leads to the response of a saccadic eye movement to that target. In many cases this appears to have a reflex nature, whereas in others the movement may best be described as voluntary target following. It is suggested later that a dichotomous classification into reflex and voluntary responses is oversimplified, but for the most part the analysis in this chapter ignores this question and concentrates on the details of the process whereby the visual information about the target position leads to an accurate saccade at a particular time. Although this analysis makes use almost entirely of behavioral observations on human subjects, the explanatory concepts used are, in many cases, derived from neurophysiological knowledge of the visual and oculomotor systems.

A visual target conveys information about its position in space in terms of the *spatial* pattern of activation on the retina. In order to move the eye to a particular position, in a saccade, a particular pattern of activation must be applied to the appropriate muscles. The amplitude of the saccade is coded largely in terms of the *duration* of the pulse of activation applied (Robinson, 1973). Thus it is clear that at some stage a conversion from the spatial to the temporal pattern of coding is required. However, this is only part of the complexity of the saccade production mechanism. It is possible to ask three questions about the occurrence of any particular saccade: When is it initiated? What direction will it take? How large will the movement be? The theme of this chapter is that, corresponding to these questions, three different routes for visual processing are involved and the visual information is processed in different ways in each case.

Processes Involved in Saccade Generation

Dissociation of the Spatial and Temporal Characteristics of Saccadic Eye Movements

Saccadic eye movements can readily be elicited by asking a subject to follow a target that moves in a series of steps. In the following experiment, the target moved in a sequence on the horizontal axis from a central position to a peripheral position, then back to the center and out again to the periphery, and so on. Horizontal eye movements were recorded using an infrared differential reflection technique based on that described previously (Findlay, 1974). The interval between target jumps varied unpredictably, taking values from 350 to 750 msec. At these rates tracking took on an automatic character and anticipatory saccades were often made when the target moved back to the center. Full details are reported elsewhere (Findlay, 1981a), but it may be noted that despite the rapid saccade rate, the accuracy of the saccades was comparable with that reported for saccades of similar size by Timberlake, Wyman, Skavenski, and Steinman (1972), whose subjects made saccades to single target movements.

Results were obtained for a variant of this task, in which the central target was always a single spot but the peripheral target could be (a) a single spot either on the right or on the left at 2°, (b) two spots at 2°, one right and one left, or (c) two spots on the same side, either right or left, one at 2° and one at 3°. The subjects were instructed to follow the spot and, if two spots occurred, to move to one of them. The latency and amplitude of the first saccade following each target jump was measured, the former with a resolution of 10 msec and the latter with an accuracy better than 0.1°. The results are shown in Figure 16-1. When the two spots appeared on the *same* side of the fixation point, the saccade *latencies* were virtually identical with those to a single spot. However, the saccade *amplitudes* were increased significantly. This did not come about because the saccades were sometimes directed at one spot and sometimes at the other, as can be seen from examination of the amplitude standard deviations. When the two spots were presented on opposite sides of the fixation point, essentially the opposite pattern of results emerged: in this case the second spot affected the latency, but the amplitude was unaffected by the presence of the opposite spot.

This pattern of results is consistent with previous reports. Coren and Hoenig (1972) showed that when a pair of targets are presented, saccades are directed at an intermediate position, and Lévy-Schoen (1969) showed that when two targets are presented on opposite sides of a fixation point, then the resultant saccades are slowed by about 40 msec. The latter finding is clearly inconsistent with the idea that the two target stimuli on opposite sides of the fixation point act independently and the saccade results from a type of horse race between the two centers of activation.

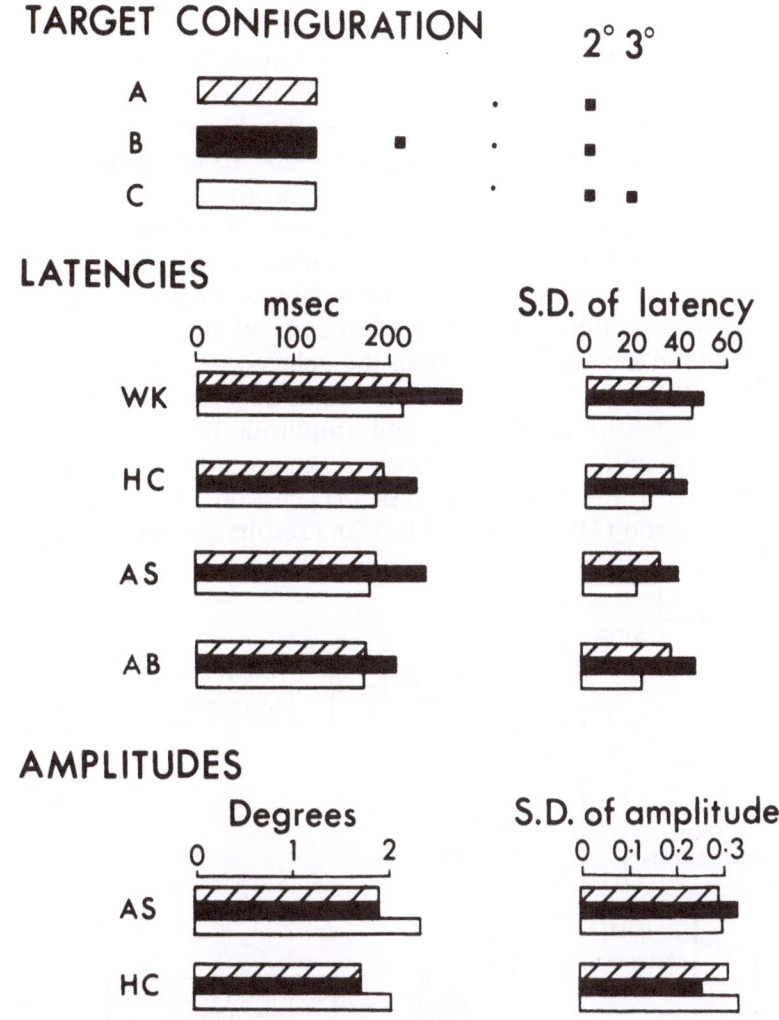

Fig. 16-1. Dissociation of spatial and temporal aspects of saccades. Results from a repetitive rapid tracking experiment where subjects (*WK*, *HC*, *AS*, and *AB*) alternated gaze between a central position and a peripheral position on either right or left. Three types of peripheral targets could appear (*A*, *B*, or *C*). Saccade latencies are slowed when two stimuli appear on opposite sides of the point of fixation (condition *B*), but saccade amplitudes are identical to those for a single target (condition *A*). When two stimuli occur on the same side of the fixation point (condition *C*), saccade latencies are the same as those for a single stimulus, but saccade amplitudes are increased. Calibration records showed that accurate amplitude estimations were not possible for subjects *WK* and *AB*.

Working Model for Saccade Generation

The results discussed above may be accommodated by a model of the type shown in Figure 16-2, which is very similar to that proposed by Becker and Jürgens (1979). The characteristic feature of this type of model is that a decision is made to produce a saccade in a particular direction *separately* from the decision concerning how large the saccade will be. In Figure 16-2 the cross-coupled "initiate" boxes are concerned with the first decision. When one or the other becomes activated, there follows a reference to the appropriate "map" to calculate the saccade amplitude. The "map" is activated by visual input, and such input also activates the timing ("initiation") mechanism. It will be shown that the "initiate" mechanisms are subject to a variety of sources of excitation and inhibition. In particular, if it is assumed that they are subject to mutual reciprocal inhibition that for a brief period after visual stimulation is stronger than any excitation, then the delayed saccades resulting from dual stimulation on the left and on the right may be accounted for.

A similar separation of timing and amplitude has been suggested explicitly or implicitly in many models of the saccade generation process, originating with the "sampled data" suggestion of Young and Stark (1963). Robinson (1973) showed how the results on double-step experi-

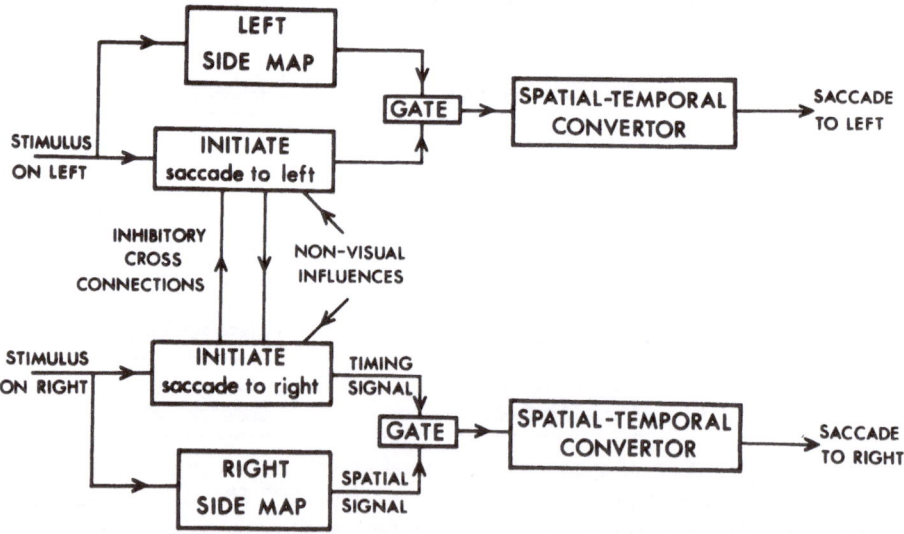

Fig. 16-2. Proposed model for the process of saccade production to visual stimuli on the horizontal axis. Visual stimulation, on the left or on the right, activates either of two parallel processes; the "initiation" process, when complete, determines that a saccade will occur in the specified direction. This acts as a gate to the spatial information simultaneously being registered in the spatial "map." The value obtained by this "look-up" process determines the size of the pulse transmitted to the oculomotor system.

ments could be explained with a model, similar to that described in Figure 16-2, having separate stages concerned with the initiation and the amplitude calculation. He exhorted model builders to take account of spatially distributed processing within the visual system. Ironically, this also led him to suggest that "models of the saccadic system should abandon the artificial constraints inherent [in models without spatially distributed processing] such as the separation of timing and amplitude into different circuits." Nonetheless, in his subsequent sketch for a model that does take spatially distributed processing into account (Robinson 1973, Fig. 7), a stage is included that "processes nonvisual data" concerned with the initiation of a saccade, for example, allowing its inhibition by nonvisual factors. It may be argued that such a mechanism does not need much elaboration to perform the functions of the "initiation" stage of the model in Figure 16-2. In later, more physiologically oriented, work, some form of trigger signal is also postulated to initiate the spatial processing for saccade generation (van Gisbergen & Robinson, 1977; Keller, 1977).

Peripheral Vision and the Magnification Factor

The following discussion of the way in which peripheral visual information is processed makes considerable use of the variable of stimulus eccentricity (i.e., distance from the visual axis expressed in terms of visual angle). Many properties of the human visual system show approximate radial symmetry, and thus eccentricity occupies the role of a major variable. Linked with this is the concept of magnification factor. It is well established that the visual projection to the cortex selectively magnifies the central region of the retina. The quantitative expression of this function (in terms of area of cortical surface representing unit visual angle on the retina) has been evaluated in a variety of subhuman species (Hughes, 1977). Work on phosphenes elicited by cortical stimulation in a blind patient has enabled a reasonably precise estimate to be made of this function in man (Cowey & Rolls, 1974).

The significance of the magnification factor in studies of peripheral vision cannot be overestimated. It is possible on the basis of this function alone to account for the variation found with retinal eccentricity in many psychophysical results; examples are visual acuity (Cowey & Rolls, 1974), and detection of lines, disks, and gratings (Koenderink, Bouman, Bueno de Mesquita, & Slappendel, 1978; Virsu & Rovamo, 1979). These workers linked their results to the *cortical* magnification factor. Because of the strong involvement of the superior colliculus in saccadic eye movements, it would be very desirable to know the corresponding function for the mapping onto this visual center. The data that are available (for the cat, Feldon, Feldon, & Krueger, 1970; for the owl monkey, Lane, Allman, Kaas, & Miezin, 1973; and for the cat, McIlwain, 1975) suggest that the selective magnification of the central regions of space may well be identical to that in the visual cortex.

In the following sections we discuss the effects of retinal eccentricity on the various parameters associated with saccadic eye movements in human observers with a view to establishing the effects of the differential magnification of the central visual area. The absence of any effect of retinal eccentricity on saccade latencies is used to argue that the initiation stage of the saccade merely requires an adequate trigger signal, whose nature and amplitude are unimportant. On the other hand, effects of stimulus eccentricity may be demonstrated on the choice of saccade direction and saccade amplitude.

Retinal Eccentricity and Saccade Direction

Surprisingly, it is far from easy to observe any effects of the magnification factor on the initiation of human saccadic eye movements. It is possible to argue that saccades can be made with equal facility to any point in the visual field, at least within the central 20° of visual space. Nevertheless, it is possible under some circumstances to demonstrate effects. As first shown by Lévy-Schoen (1969, 1974), when two targets are presented simultaneously in the visual field, there is a very strong tendency to saccade to the closer of the two. A recent experiment (Findlay, 1980) quantified this tendency. A subject was presented with the task shown in Figure 16-3. When the subject was fixating, two target squares were presented on opposite sides of the fixation point (the axis was randomly chosen from left–right, up–down, or the intermediate diagonals). Shortly following this, two small digits were briefly presented at the same locations as the targets. The subject's task was to read off one digit. The spatial and temporal parameters were such that he could only do this if he made a saccade to one of the targets and thus viewed the digit in central vision. No other constraints were placed on the subject except that he was asked to try not to anticipate any particular target position. The identity of the digit showed the target to which the subject made a saccade. Over a repeated series of trials the visual characteristics of the targets could be varied to determine the way in which these characteristics affected the choice of direction.

It turns out that the tendency to fixate the closer stimulus can be overcome if the more distant one is made more "intense." The stimuli used consisted of computer-generated point elements, and this variation was accomplished by varying the number of points in the target square (and concomitantly the size and overall luminous flux) while maintaining stimulus luminance constant. A trade-off function between salience, expressed as the relative size of the stimuli, and retinal eccentricity

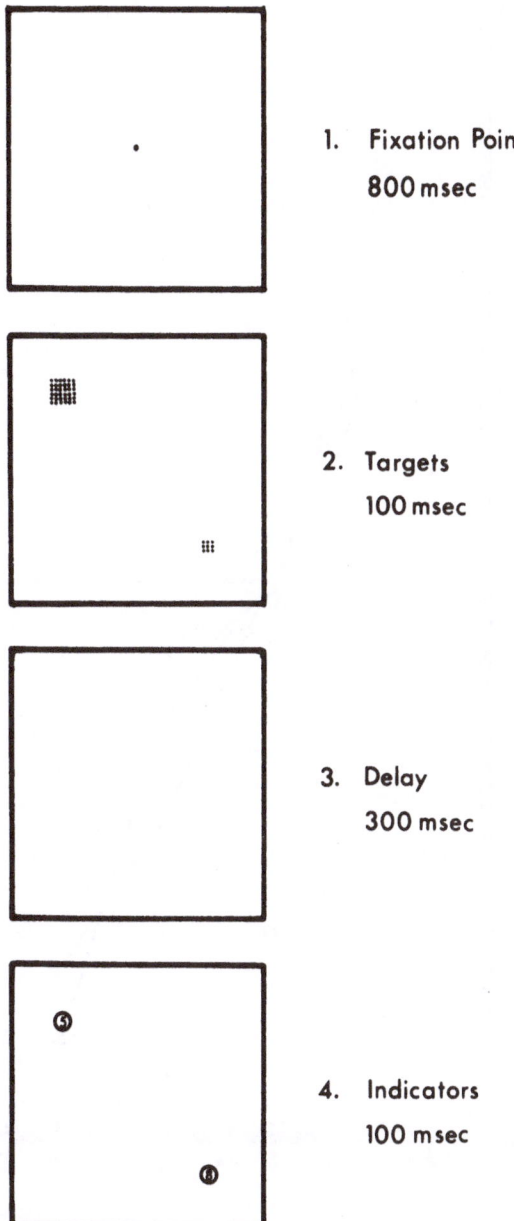

1. Fixation Point
   800 msec

2. Targets
   100 msec

3. Delay
   300 msec

4. Indicators
   100 msec

Fig. 16-3. Stimulus sequence used to demonstrate the effects of retinal eccentricity upon the choice of saccade direction. Following an 800-msec presentation of a fixation point (a), two square clusters of dots (the target stimuli) were presented for 100 msec (b). The interdot spacing was 6 min arc, and the maximum square side used was 54 min arc (10 X 10). After a delay during which the subject's saccade occurred (c), two indicator digits were presented briefly (d). The subject's task was to read off one of these. On different trials the target stimuli appeared on different axes (horizontal, vertical, and the principal diagonals). The relative sizes of the squares were varied using a staircase procedure to obtain a balance point at which saccades were equally likely to go to either stimulus. (From Findlay, J. M. The visual stimulus for saccadic eye movements in human observers. *Perception*, 1980, 9, 7–21.)

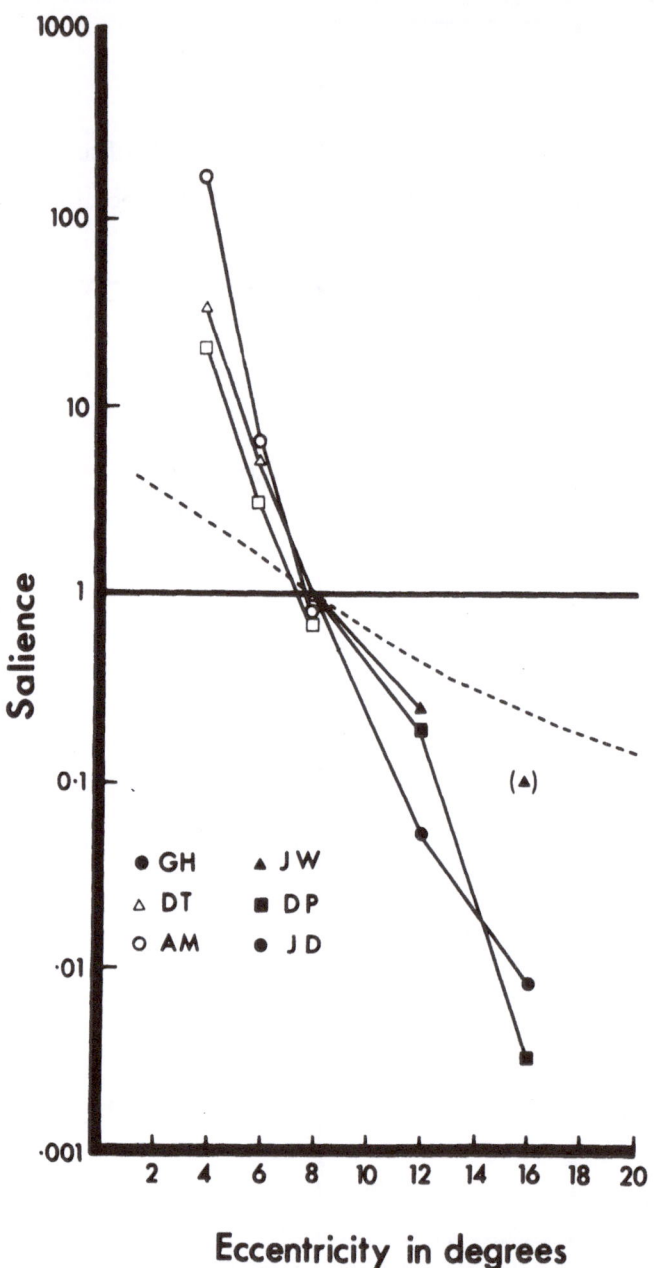

Fig. 16-4. Effect of retinal eccentricity upon the choice of saccade direction: results from experimental procedure shown in Figure 16-3. Each point corresponds to the outcome of an experimental block in which two target eccentricities were presented on each trial, one of these being 8°. The "salience" measure was derived from the ratio of the two square sizes when each was equally effective at eliciting saccades (thus the more distant target was larger). Results are shown for three subjects at each eccentricity. *Broken line:* cortical magnification factor (for area), normalized to 8°, derived from Cowey and Rolls (1974). Note that the axis is logarithmic; thus the two functions differ in a way that cannot be subsumed by a multiplicative scaling transformation.

could then be established. This trade-off was established using a psycho-physical staircase procedure over a series of trials in which the size of one stimulus was varied until saccades were directed to either one with 50% probability. The results are presented in Figure 16-4; a striking effect of retinal eccentricity may be seen. Over the range of eccentricities studied (4°-16°) doubling the eccentricity required an increase by a factor of about 60 in the number of elements in order to maintain equally effective saccade elicitation.

The cortical magnification factor established by Cowey and Rolls (1974) shows that doubling the eccentricity results in a decrease of about one-quarter in the area of cortex activated by a particular stimulus area. Consequently, although the result found suggests an explanation in terms of magnification factor, the fall off with increasing eccentricity is even greater than can be accounted for by increased area of cortex stimulated. It is tempting to postulate an explanation in terms of increased transmission time for more peripheral stimulation; however, this receives no support from studies of saccade latencies to single targets.

Saccade Latency

Retinal Eccentricity and Saccade Latency

A number of studies have examined the way in which saccade latency varies as a function of target eccentricity, and a selection of results is presented in Figure 16-5. Since, as discussed below, nonvisual effects such as temporal uncertainty can exert a profound effect on saccade latency, comparisons can only legitimately be made within studies rather than across studies. This may be done by comparing linked points in Figure 16-5. Several points of interest emerge. Above about 20° eccentricity, latency is generally found to increase with retinal eccentricity. Below this value, however, no consistent effect is found except for the very small saccades studied by Wyman and Steinman (1973) and Young (1971). Thus, in the range of naturally occurring saccades (Bahill, Adler, & Stark, 1975) it appears that any movement can be produced equally rapidly. Exceptions to this have been reported and seem to depend on some additional detection requirement, such as when the target is dim (Cohen & Ross, 1977, Experiment 2) or if sudden target onset is absent (Todd & van Gelder, 1980).

Some other qualifications must also be made. First, the results all apply to saccades along the horizontal axis, with the exception of those of Heywood and Churcher (1980). Second, it is known that anticipation of a particular target position can decrease latencies by a small amount (see later discussion). Thus when a range of target latencies is studied within a block of trials, targets at the middle of the range may receive preferential attention and thus show decreased latencies (Bartz, 1967). Finally, a possibility that cannot be ruled out in most of the studies is that the trigger to the timing mechanism comes from the fixation point offset, rather than from the target onset. Saslow (1967)

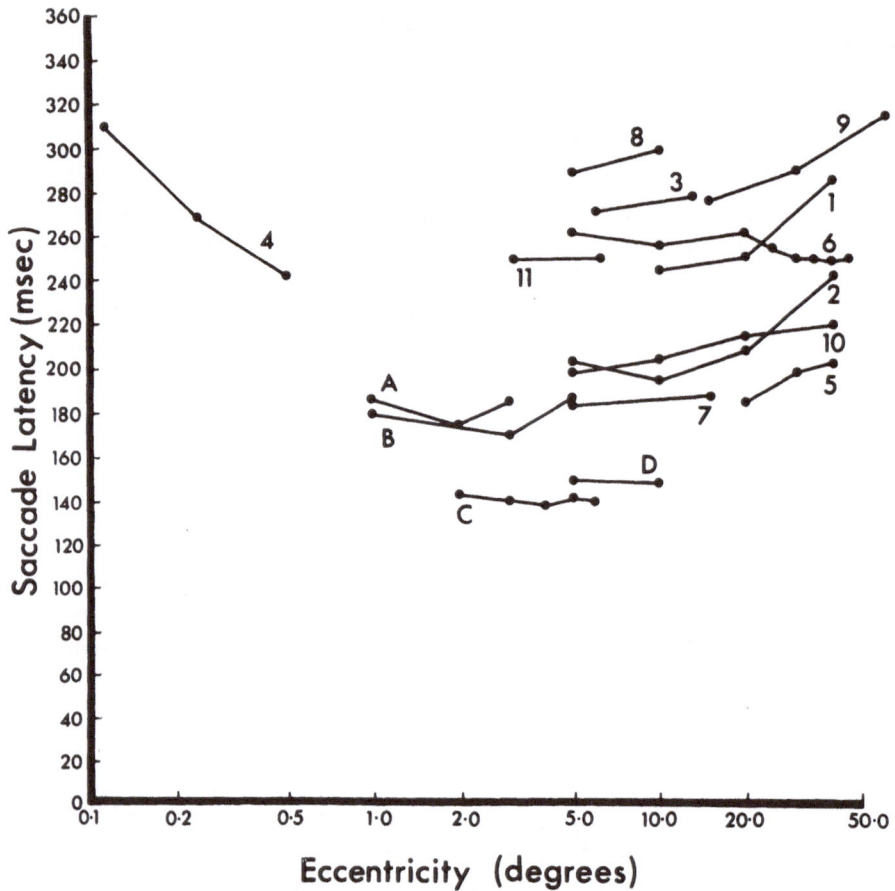

Fig. 16-5. Studies relating saccade latency to retinal eccentricity. Each set of points represents latencies derived from a particular study (in some cases averages have been made over different conditions at the same eccentricity). The studies are the following: 1, White, Eason, and Bartlett (1962); 2, Bartz (1962); 3, Miller (1969); 4, Wyman and Steinman (1973); 5, Megaw (1975); 6, Frost and Pöppel (1976); 7, Cohen and Ross (1977); 8, Prablanc and Jeannerod (1974); 9, Becker and Jürgens (1979); 10, Sharpe, Lo, and Rabinovitch (1979); 11, Heywood and Churcher, (1980); A,B,C, and D, unpublished measurements of the author. A and B were obtained in a rapid tracking experiment similar to that shown in Figure 16-1 (Findlay, 1981a); C and D were obtained using the experiment described in Figure 16-8 (Findlay 1981b). The unusually low latencies occurring in studies C and D are attributable to the fact that in these studies the target initiation was under the immediate control of the subject. The aggregated data suggest that retinal eccentricity does not affect saccade latency for saccades in the range 1°–15°.

showed that a substantial (100-msec) increase in saccade latency results from delaying the fixation point offset relative to the target onset, and conversely. However, if this were the case, the increased latency above 20° would be difficult to account for.

Thus the central 20° of retina appears to be equipotential when saccades elicited by a single target are considered. This suggests that the "initiation" stage of the model in Figure 16-2 requires merely an adequate trigger signal to set the process in motion and operates in an all-or-nothing rather than any quantitative fashion. Other considerations probably apply above 20°, where a double saccade is generally programed (Frost & Pöppel, 1976). The strong effects of retinal eccentricity found when competing stimuli are present cannot then be related to differences in processing speed of the stimuli. To retain the model proposed in Figure 16-2 the ad hoc assumption must be made that the effects result from differential activation of the cross-inhibitory channels. It might be conjectured to be advantageous in terms of speed of response for a system to have evolved in which any peripheral signal could act as a trigger for saccadic movements, decision time only being required if some competitive inhibitory influence appears. It is of interest here to note two recent reports of patients with damage to the geniculostriate pathways who were able to make saccades into "blind" regions. Latencies for saccades into the blind regions were no longer, and in some cases appeared to be shorter, than those for saccades to corresponding intact regions (Sharpe, Lo, & Rabinovitch, 1979; Zihl, 1980).

Other Influences on Saccade Latencies

The absence of effects of target position on saccade latencies is one instance of a more general finding that the characteristics of the target have very little effect on the latency of a saccade to it. Thus, for example, when the target intensity is varied, substantial increases in latency only occur at extremely low-intensity values (Wheeless, Cohen, & Boynton, 1967). In contrast to this, nonvisual factors can profoundly affect the measured latency of a saccade. Variations in the degree of temporal uncertainty about the instant of target appearance can result in latency changes of 100 msec or more (Cohen & Ross, 1977; Michard, Têtard, & Lévy-Schoen, 1974). If the occurrence of a target is actually initiated by the observer, although its location is unpredictable, the latency may be reduced from the "classical" figure of 200 msec to 150 msec or less (Figure 16-5; Lévy-Schoen, 1981).

It is generally found that foreknowledge of the target position can result in a decrease in latency of some 10–20 msec (Michard et al., 1974; Posner, Nissen, & Ogden, 1978; but see Heywood & Churcher, 1980). Interestingly, this effect shows no interaction with the effects of temporal uncertainty (Findlay, 1981a; Michard et al., 1974). When two factors can each influence a response latency and these are found to operate independently, an interpretation made frequently is that this constitutes evidence for two separate serial stages in the latency determination (Sternberg, 1969). In this case, accepting this interpretation would entail abandoning the idea that the process determining the latency is not spatially dependent. Further work seems to be required here to disentangle the effects of prior information about position and those of prior information about direction.

Finally, studies of saccades made in more complex situations, such as search tasks, suggest that the decision to initiate a saccade depends upon the amount of information to be processed at the fixation location. Gould (1973) measured fixation durations in a task in which a subject scanned a set of characters and had to respond if the character was an exemplar of a previously learned target set. The duration of fixations in this task increased with the size of the target set in an approximately linear fashion. Likewise, Lévy-Schoen (1981) showed that fixation durations can be shortened if a subject can carry out useful perceptual processing in peripheral vision at the preceding fixation location.

Saccade Amplitude

Calculation of Saccade Amplitude

Acquisition of a peripheral target is frequently be carried out in stages, a primary saccade being followed by a secondary, corrective saccade. Although the properties of these secondaries are of considerable interest (Becker, 1976; Prablanc & Jeannerod, 1975), attention is focused here on computation of the amplitude of primary saccades.

Important information is provided by the variability of saccade amplitude. The problem of measurement errors becomes particularly acute when the variation in saccade amplitudes is under consideration, and it is possible that some reported figures may be overestimates. A measurement of saccade variability may be made by recording a number of saccades to the same target and then expressing the standard deviation as a proportion of the saccade amplitude. Since saccades are, on the average, slightly hypometric (i.e., undershot), this measure differs from a measure of deviation from the target distance. Variability is reported to decrease from about 20% for very small saccades (Wyman & Steinman, 1973) to a figure between 5% and 10% for saccades of 5° and over (Frost & Pöppel, 1976; Prablanc & Jeannerod, 1974). Leushina (1965) reported that a trade-off can be made between latency and accuracy; however, the latencies in that study were unusually long (320–450 msec), and such a trade-off may not be characteristic of saccades made in more normal circumstances. No such effect was found in the experiment reported below.

How is visual information processed to achieve the necessary accuracy? In order to direct $10°$ saccades with an accuracy of $1°$, it might appear necessary for visual space to be analyzed with a "window" or "receptive field" $1°$ in width. This would evidently be true if the array of receptive fields were nonoverlapping. However, as has been pointed out on various occasions (e.g., Erickson, 1968), an equivalent accuracy could be be achieved in principle by locating the peak of activation in an ordered array with windows of a much larger size. Evidence to support the second alternative comes from the study of saccades made to double targets. The essential finding is that when two targets are presented simultaneously in the visual field, saccades land at a position intermediate between the two. It appears that the saccade system does not completely resolve the individual targets, and even voluntary effort cannot entirely overcome this tendency (Findlay, 1981a).

One example of this result has already been given. It was examined in more detail in the following experiment. The subject viewed an oscilloscope screen, initially showing a small fixation cross. When he released a hand-held button, a target appeared, either on the right or on the left, which consisted of either one or two squares. On each trial the stimulus was chosen from one of the eight possibilities shown in Figure 16-6. On one-half of the trials, the target contained a small gap (3 min arc) in the side of one of the squares. The subject's task was to detect the gaps and the task was designed to require subjects to fixate the individual squares.[1] In an attempt to find out the extent to which the subject's eye movements were under the control of factors specific to the task, blocks were also run in which the task was changed so that only a gross discrimination of square size was required.

The amplitudes of the first saccades made to each different target configuration are shown in Figure 16-7. It is clear that, as in the tracking experiments discussed earlier, for double-square configurations the saccades were directed to an intermediate location relative to the two squares. Furthermore, the relative sizes of the squares systematically influenced saccade amplitudes. The difference between saccade amplitudes in the two tasks was statistically significant, F $(7, 21) = 5.32$; $p = .016$. Surprisingly, the task of gap detection, designed to elicit accurate fixation, resulted in saccades landing, on the average, further from the individual squares. In this task it was invariably the case that further saccades occurred before the decision was made. Related to this

---

[1] Partly for technical reasons, smaller gap sizes were not used. Although if a subject was given unlimited time, some degree of gap detection could occur at $5°$ in peripheral vision (a test on one subject gave a detection rate of 37/50 and a false positive rate of 2/50 in a forced choice procedure), it seems unlikely that different results would have been obtained if smaller gaps were used.

0°   5°   10°

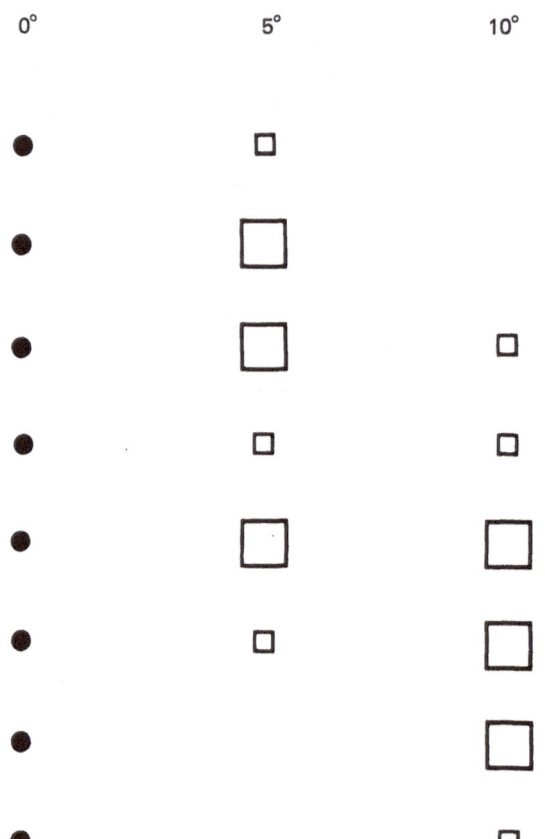

Fig. 16-6. Experiment to investigate the determination of saccade amplitude. The subject fixated the central point (0°), then on each trial was presented with one of the eight possible stimulus configurations shown in the figure, which could either be on the right or on the left. On one-half the trials a small gap was present in the side of one square and the subject's task was to detect this gap. Results were also obtained from a second task in which the same stimuli were used but the subject was required to ignore the gaps and indicate whether the stimulus configuration was the same as, or different from, that occurring on the previous trial. (From Fisher, D. F., Monty, R. A., & Senders, J. W. (Eds.). *Eye Movements, Cognition and Visual Perception.* Hillsdale, N.J.: Erlbaum, 1981.)

difference in amplitude, there was also a difference in saccade latency between the tasks. The latency averaged 133 msec for the gap detection task and 165 msec for the other task.

This shows that the amplitude computation process arrives at a saccade of intermediate amplitude when two discrete peripheral targets are presented. Because of the ballistic nature of saccades it seems impossible for any "averaging" to occur within the oculomotor system, and thus attention is directed to the sensory stages of visual processing. We had originally anticipated that the amplitude of saccades directed to

Mean and S.D. of First Saccade Amplitude

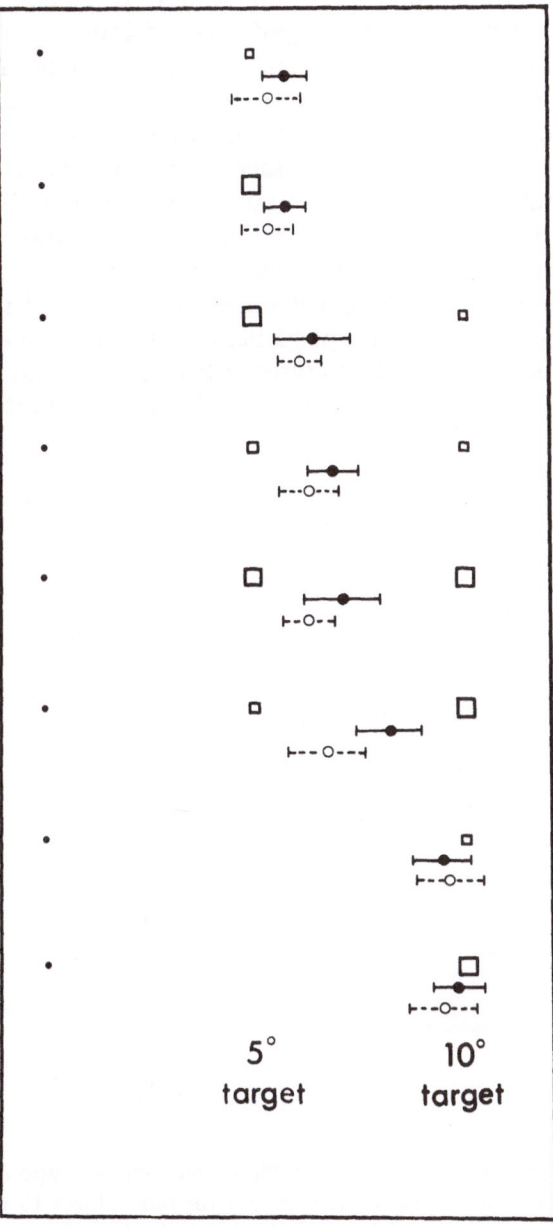

Fig. 16-7. Saccade amplitudes (mean ± SD) obtained in the experiments described in Figure 16-6. *Filled circles*, amplitudes in the gap detection task, plotted relative to the actual positions of the targets; *open circles*, amplitudes obtained in the comparison task. The standard deviation measures are the means of within-block standard deviations (this produces a more realistic measure because of variation of the calibration scaling factor across blocks).

intermediate positions might be predictable from an averaging of the stimulation at each eccentricity, perhaps using the idea of a "center of gravity" of the target configuration (Coren & Hoenig, 1972) and a weighting factor related to the magnification factor.

Further analysis showed that this attempt to elaborate a computational model for saccade amplitudes is misleading unless dynamic factors are also taken into consideration. There is no unique location to which saccades are directed. The saccade amplitude to double-square targets turns out to depend on the latency with which the saccade happens to be elicited. The more delayed the saccade, the closer it comes to the near target, and it seems a reasonable assumption that if the saccade is delayed for long enough it might be directed at the near square with no influence of the far square being apparent. An example of the evidence for this is shown in Figure 16-8 (see also Findlay, 1981b). In this figure the amplitude of an individual saccade is plotted against the

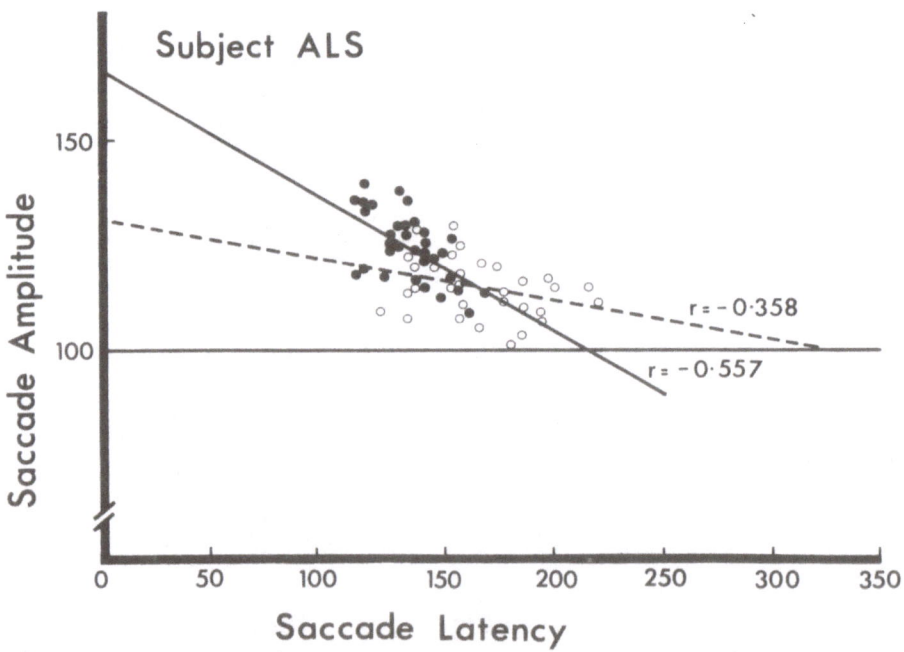

Fig. 16-8. Amplitude of saccades to double targets depends upon their latency of initiation. The plot shows individual saccades for one subject to all double-square targets with both squares equal in size. Amplitude is plotted as a percentage of the amplitude recorded to single squares at 5° in each block (to obviate calibration scaling factors). *Filled circles*, results from gap detection task; *solid line*, fitted least squares regression; *open circles* and *broken line*, results from comparison task. Similar results were obtained from other subjects (Findlay, 1980a). (From Fisher, D. F., Monty, R. A., & Senders, J. W. (Eds.). *Eye Movements, Cognition and Visual Perception*. Hillsdale, N.J.: Erlbaum, 1981.)

latency of that saccade. The plot shows, for one subject, all saccades recorded to double-square targets with the components identical (two large or two small squares). It demonstrates that short-latency saccades had larger amplitudes than those of longer latency. This effect occurred in both tasks, and to a large extent the variation in mean amplitude between the tasks appears to be related to the corresponding latency difference. The fitted regression lines, however, show that the dependence of amplitude upon latency was more marked in the detection task (for three subjects; the fourth subject showed a nonsignificant regression in the detection task, attributable to a very narrow range of saccade latencies.)

Retinal Eccentricity and Saccade Amplitude

These results give a dynamic view of the visual processing stages involved in the processing of saccade amplitude. The earliest information to arrive is not capable of resolving the individual targets in the two-target configuration. With succeeding milliseconds gradually more detailed information becomes available. The fitted regression line suggests that for the earliest information no differential weighting with retinal eccentricity occurs (the saccades are directed at the midpoint of the targets). This suggests that this early information might be identical with that initiating the timing process described above.

The principle that the more coarse-grained information (in visuospatial terms) is processed more rapidly is supported by results of several recent studies of visual system functioning. Interest has, in general, been concentrated on the difference in propagation rates between sustained and transient channels, and the studies have used sine wave gratings of varying spatial frequency. Breitmeyer (1975) measured the manual reaction time to gratings of equal subjective contrast and showed that this was a monotonically decreasing function of spatial frequency. Parker and Salzen (1977) showed a similar trend in the early components of the human visual evoked potential.

If any attempt is made to relate such stimuli to single-cell responses, geometrical considerations require that a close inverse relationship exists between receptive field size and optimum spatial frequency. It is well established that in the superior colliculus receptive fields of visually driven cells increase in size with increased depth below the surface (Goldberg & Wurtz, 1972). McIlwain (1975) suggested that "it is difficult to avoid the conclusion that many premotor elements of the colliculus have large receptive fields." It is thus tempting to speculate that the present findings may be related to the results of Mohler and Wurtz (1976), who showed that intermediate-layer cells in the superior colliculus with large receptive fields actually fire prior to more superficial cells with smaller fields. If it proves to be a general result that cells with large receptive fields, receiving converging input from many retinal receptors, are actually the earliest to be activated, the implications for visual functions are considerable.

Receptive fields such as the ones shown in the Mohler and Wurtz (1976) study, with sizes measured in tens of degrees, would be of the order of magnitude necessary to account for the lack of resolution shown in the double-square experiment. The limit of the integration process proposed has been explored in an extension of the double-square gap detection experiment in which the separation between the squares was varied. The near square was always at $2°$ eccentricity. With the distant square at $4°$, the results followed the same patterns as those described for the $5° + 10°$ combination. However, when the distant square was further away ($5°$ or $6°$), the first saccade amplitude was less than that in the $2° + 4°$ condition, although some effect of the distant target still appeared. In addition, the variances of the amplitude increased considerably in the $2° + 5°$ and the $2° + 6°$ conditions.

Two-Step Tracking

The experiments reported here in which two targets were presented simultaneously have some similarities to a limiting case of the task of tracking a spot that moves in two steps successively, when the interval between the steps is reduced. Saccadic responses in this situation have been studied by several workers. A comprehensive investigation was reported by Becker and Jürgens (1979), and their interpretation has many similarities to that given here. Certain differences in substance and in emphasis may be noted. Becker and Jürgens postulated a separate "decision mechanism," which is concerned with direction and timing of the saccade, together with an "amplitude computation," which occurs in parallel. They suggested that the amplitude computation is performed by an integration of the stimulation over a "temporal window" in order to account for saccades landing at intermediate positions between the targets. This has formal similarities with the integration over a "spatial window" suggested here, and it is probable that both forms of integration occur. Becker and Jürgens found the number of saccades to intermediate positions to be much greater when the two steps were on the same side of the original fixation point than when the second step was on the opposite side. However, in the latter case, a small number of saccades occurred in the direction of the first target but with reduced amplitude. Similar saccades have been seen during rapid predictive tracking (Findlay, 1981a). Becker and Jürgens suggested that the second step can be taken into account in the calculation of the saccade amplitude. Perhaps a more parsimonious possibility is that these hypometric saccades represent cases in which the inhibitory block from the opposite hemifield arrived while the saccade was being generated and this produced only partial suppression.

Two further points made by Becker and Jürgens may be noted. First, they suggested that efference copy information can be used in the computation of amplitude to account for short-latency corrective saccades. The involvement of an efference copy signal in the process of saccade generation has received much attention recently (see Sparks and Mays, Chapter 4, this volume). The issue appears to be logically independent of the ones discussed in this chapter. Second, they pointed out that modifications that shorten a saccade can be effected more quickly than modifications that increase the amplitude. This finding is suggestive but has so far been shown only with saccades in excess of $20°$ (e.g., Lévy-Schoen & Blanc-Garin, 1974), which, as has already been argued, show certain differences from smaller ones. It would be of interest to attempt a replication using smaller saccades.

Conclusions

The following facts emerge from a study of the way saccade generation is affected by the variable of retinal eccentricity. The *latency* to generate a saccade to a peripheral target does not depend upon the eccentricity of that target in the range $1°$–$20°$ of eccentricity. If two targets are presented simultaneously at different peripheral locations, both in the same direction away from the fovea, the *amplitude* of the saccade depends upon the latency. The faster the saccade is produced, the more closely the saccade takes the eye to the midpoint of the two locations. Saccades that are more delayed have amplitudes that show greater weighting for the central regions. However, when two stimuli are simultaneously presented in locations that compete for the choice of saccade direction, the *direction* of the resultant saccade is very strongly dependent upon the relative eccentricity of the two stimuli.

The interpretation offered is in terms of the model shown in Figure 16-2. This envisages parallel pathways concerned with timing and direction, on the one hand, and with saccade amplitude on the other. The earliest visual information to arrive acts as a trigger to initiate the timing and direction mechanism, which at some later instant is ready to look up a value for the amplitude in the pathway associated with this. The earliest information to arrive on both pathways shows no differential weighting for the central retina. If processing is delayed, however—for example, by competing stimulation in different directions—then effects due to differential weighting occur.

A huge simplification of the actual system has been achieved by restricting attention to the horizontal dimension. Here the results are consistent with the dissociation of direction and amplitude such that, once the decision is taken to move the eyes either left or right, the amplitude is determined solely by stimulation in the direction the eyes are to be moved. An obvious aim for future experimental studies is to discover the conditions governing the production of vertical and oblique eye movements, extending the preliminary results of Hou and Fender (1979).

Acknowledgments

This work was in part supported by Medical Research Council (Great Britain) Grant G977/865/N. Some of the experiments reported here were carried out while the author was a Sabbatical Visitor at the Laboratoire de Psychologie Expérimentale, Université René Descartes, Paris. The advice and assistance of Dr. Ariane Lévy-Schoen are gratefully acknowledged, as are the valuable comments of Laurence Harris on an earlier draft of this chapter.

References

Bahill, A. T., Adler, D., & Stark, L. Most naturally occurring human saccades have magnitudes of 15 degrees or less. *Investigative Ophthalmology*, 1975, *14*, 468–469.

Bartz, A. E. Eye movement latency, duration and response time as a function of angular displacement. *Journal of Experimental Psychology*, 1962, *64*, 318–324.

Bartz, A. E. Fixation errors in eye movements to peripheral stimuli. *Journal of Experimental Psychology*, 1967, *75*, 444–446.

Becker, W. Do correction saccades depend exclusively on retinal feedback? A note on the possible role of non-retinal feedback. *Vision Research*, 1976, *16*, 425–427.

Becker, W., & Jürgens, R. An analysis of the saccadic system by means of double step stimuli. *Vision Research*, 1979, *19*, 967–983.

Breitmeyer, B. G. Simple reaction time as a measure of the temporal response properties of sustained and transient channels. *Vision Research*, 1975, *15*, 1411–1412.

Cohen, M. E., & Ross, L. E. Saccade latency in children and adults: Effects of warning interval and target eccentricity. *Journal of Experimental Child Psychology*, 1977, *23*, 539–549.

Coren, S., & Hoenig, P. Effect of non target stimuli upon length of voluntary saccades. *Perceptual Motor Skills*, 1972, *34*, 499–508.

Cowey, A., & Rolls, E. T. Human cortical magnification factor and its relationship to visual acuity. *Experimental Brain Research*, 1974, *21*, 447–454.

Erickson, R. P. Stimulus coding in topographic and non-topographic afferent modalities; On the significance of the activity of individual sensory neurons. *Psychological Review*, 1968, *75*, 447–465.

Feldon, S., Feldon, P., & Kruger, L. Topography of the retinal projection on the superior colliculus of the cat. *Vision Research*, 1970, *10*, 135–143.

Findlay, J. M. A simple apparatus for recording microsaccades during visual fixation. *Quarterly Journal of Experimental Psychology*, 1974, *26*, 167-170.

Findlay, J. M. Local and global influences on saccadic eye movements. In D. F. Fisher, R. A. Monty & J. W. Senders (Eds.), *Eye Movements, Cognition and Visual Perception*. Hillsdale, N.J.: Erlbaum, 1981. (b)

Findlay, J. M. The visual stimulus for saccadic eye movements in human observers. *Perception*, 1980, *9*, 7-21.

Findlay, J. M. Spatial and temporal factors in the anticipatory generation of saccadic eye movements. *Vision Research*, 1981, *21*, 347-354. (a)

Frost, D., & Pöppel, E. Different programming modes of human saccadic eye movements as a function of stimulus eccentricity; Indications of a functional subdivision of the visual field. *Biological Cybernetics*, 1976, *23*, 39-48.

Goldberg, M. E., & Wurtz, R. H. Activity of superior colliculus in behaving monkey. I. Visual receptive fields of single neurons. *Journal of Neurophysiology*, 1972, *35*, 542-559.

Gould, J. D. Eye movements during visual search and memory search. *Journal of Experimental Psychology*, 1973, *98*, 184-195.

Heywood, S., & Churcher, J. Structure of the visual array and saccade latency; Implications for oculomotor control. *Quarterly Journal of Experimental Psychology*, 1980, *32*, 335-341.

Hou, R. L., & Fender, D. H. Processing of direction and magnitude by the saccadic eye movement system. *Vision Research*, 1979, *19*, 1421-1426.

Hughes, A. The topography of vision in mammals of contrasting life style: Comparative optics and retinal organization. In F. Crescitelli (Ed.), *Handbook of sensory physiology* (Vol. 7-5): *The Visual System in Vertebrates*. Berlin: Springer, 1978.

Keller, E. L. Control of saccadic eye movements by midline brainstem neurons. In R. Baker & A. Berthoz (Eds.), *Control of gaze by brain stem neurons*. Amsterdam: Elsevier/North-Holland, 1977.

Koenderink, J. J., Bouman, M. A., Bueno de Mesquita, A. E., Sleppendel, S. Perimetry of contrast detection thresholds of moving spatial sine wave patterns. III. The target extent as a controlling parameter. *Journal of the Optical Society of America*, 1978, *68*, 854-860.

Lane, R. H., Allman, J. M., Kaas, J. H., & Miezin, F. M. The visuotopic organization of the superior colliculus of the owl monkey. *Brain Research*, 1973, *60*, 335-349.

Leushina, L. I. On the estimation of position of photostimulus and eye movements. *Biofizika*, 1965, *10*, 130-136.

Lévy-Schoen, A. Détermination et latence de la réponse oculomotrice a deux stimulus. *L'Année Psychologique*, 1969, *69*, 373-392.

Lévy-Schoen, A. Le champ d'activité du regard: Données expérimentales. *L'Année Psychologique*, 1974, *74*, 43-66.

Lévy-Schoen, A. Flexible and/or rigid control of oculomotor scanning behaviour. In D. F. Fisher, R. A. Monty, & J. W. Senders (Eds.), *Eye Movement, Cognition and Visual Perception*. Hillsdale, N.J.: Erlbaum, 1981.

Lévy-Schoen, A., & Blanc-Garin, J. On oculomotor programming and perception. *Brain Research*, 1974, *71*, 443-450.

McIlwain, J. T. Visual receptive fields and their images in the superior colliculus of the cat. *Journal of Neurophysiology*, 1975, *38*, 219-230.

Megaw, E. D. Factors underlying distributions of eye fixation times. In A. Lavilla, C. Teiger, & A. Wisner (Eds.), *Age et constraintes de travail*. Jouy-en-Josas, France: NEB Editions Scientifiques, 1975.

Michard, A., Têtard, C., & Lévy-Schoen, A. Attente du signal et temps de réaction oculomoteur. *L'Année Psychologique*, 1974, *74*, 387-402.

Miller, L. K. Eye movement latency as a function of age, stimulus uncertainty and position in the visual field. *Perceptual Motor Skills*, 1969, *28*, 631-636.

Mohler, C. W., & Wurtz, R. H. Organization of monkey superior colliculus: Intermediate layer cells discharging before eye movements. *Journal of Neurophysiology*, 1976, *39*, 722-744.

Parker, D. M., & Salzen, E. A. Latency changes in the human visual evoked response to sinusoidal gratings. *Vision Research*, 1977, *17*, 1201-1204.

Posner, M. I., Nissen, M. J., & Ogden, W. C. Attended and unattended processing modes. In H. L. Pick & I. J. Saltzman (Eds.), *Modes of perceiving and processing information*. Hillsdale, N.J.: Erlbaum, 1978.

Prablanc, C., & Jeannerod, M. Latence et précision des saccades en fonction de l'intensité, de la durée et de la position retinienne d'un stimulus. *Revue d'Electroencephalographie et de Neurophysiologie Clinique*, 1974, *4*, 484-488.

Prablanc, C., & Jeannerod, M. Corrective saccades: Dependence on retinal afferent signals. *Vision Research*, 1975, *15*, 465-470.

Robinson, D. A. Models of the saccadic eye movement control system. *Kybernetik*, 1973, *14*, 71-83.

Saslow, M. G. Effects of components of displacement step stimuli on latency for saccadic eye movement. *Journal of the Optical Society of America*, 1967, *57*, 1024-1029.

Sharpe, J. A., Lo, A. W., & Rabinovitch, H. E. Control of the saccadic and smooth pursuit systems after cerebral hemidecortication. *Brain*, 1979, *102*, 387-403.

Sternberg, S. The discovery of processing stages: Extensions of Donders' method. In W. G. Koster (Ed.), *Attention and performance II. Acta Psychologica*, 1969, *30*, 276-315.

Timberlake, G. T., Wyman, D., Skavenski, A. A., & Steinman, R. M. The oculomotor error signal in the fovea. *Vision Research*, 1972, *12*, 1059-1064.

Todd, J. T., & van Gelder, P. Implications of a transient-sustained dichotomy for the measurement of human performance. *Journal of Experimental Psychology, Human Perception and Performance*, 1980, *5*, 625-638.

van Gisbergen, J. A. M., & Robinson, D. A. Generation of micro- and macrosaccades by burst neurons in the monkey. In R. Baker & A. Berthoz (Eds.), *Control of gaze by brain stem neurons*. Amsterdam: Elsevier/North-Holland, 1977.

Virsu, V., & Rovamo, J. Visual resolution, contrast sensitivity, and the cortical magnification factor. *Experimental Brain Research*, 1979, *37*, 1-16.

Wheeless, L. L., Cohen, G. H., & Boynton, R. M. Luminance as a parameter of the eye-movement control system. *Journal of the Optical Society of America*, 1967, *57*, 394-400.

White, C. T., Eason, R. G., & Bartlett, N. R. Latency and duration of eye movements in the horizontal plane. *Journal of the Optical Society of America*, 1962, *52*, 210-213.

Wyman, D., & Steinman, R. M. Latency characteristics of small saccades. *Vision Research*, 1973, *13*, 2173-2176.

Young, L. R. Pursuit eye tracking movements. In P. Bach-y-Rita, C. C. Collins, & J. E. Hyde (Eds.), *The control of eye movements.* New York: Academic Press, 1971.

Young, L. R., & Stark, L. Variable feedback experiments using a sampled data model for tracking eye movements. *IEEE Transactions in Human Factor in Electronics,* 1963, *HFE-4,* 38–51.

Zihl, J. "Blindsight": Improvement of visually guided eye movements by systematic practice in patients with cerebral blindness. *Neuropsychologia,* 1980, *18,* 71–77.

Chapter 17

Optic Ataxia: A Specific Disorder in Visuomotor Coordination

M. Thérèse Perenin and Alain Vighetto

Since Balint's famous observation in 1909, the term *optic ataxia* has designated a disorder of coordination and accuracy of visually elicited hand movements not related to motor, sensory, visual acuity, or visual field deficits. The term optic ataxia was proposed by Balint as an analogy to tabetic ataxia. His patient could execute body-oriented movements normally, compensating for defective visual control by using somatosensory cues in the same way vision is used by tabetic patients in compensating for a defective muscle position sense. In addition to optic ataxia, Balint's patient exhibited disorders in visual attention, including neglect of the left hemispace and "psychic paralysis of gaze." However, such disorders could not account for optic ataxia, which only affected the right hand of the patient. Balint suggested that visual information arriving in the visual cortex could not be conveyed to the hand motor area. Postmortem examination showed a bilateral softening of the parieto-occipital junction.

Some years later, Holmes (1918) reported cases of "visual disorientation" in subjects with large bilateral parieto-occipital injuries. This syndrome must be regarded as a different entity in that Holmes' patients showed prominent oculomotor disorders. They were unable to orient their gaze toward objects detected in their peripheral visual field, a defect which was not mentioned by Balint. In addition, they could not estimate either the relative or absolute position of objects in space. Such perceptual disabilities were considered by Holmes as directly responsible for the inaccuracies in visuomotor behavior shown by his patients when they reached for objects or avoided obstacles.

Since these two original studies, observations of similar bilateral lesions have been reported, resembling either Balint's case (Hécaen & de Ajuriaguerra, 1954; Luria, 1959) or, more often, Holmes' cases (Godwin-Austen, 1965; Kase, Troncoso, Court, Tapia, & Mohr, 1977; Michel, Jeannerod, & Devic, 1965). Visual disorientation restricted to one hemifield has also been observed following unilateral damage of the posterior parietal cortex (Brain, 1941; Cole, Schutta, & Warrington, 1962; Riddoch, 1935). In these cases, however, it was not always clear whether misreaching reflected a perceptual or a visuomotor deficit.

Only recently has optic ataxia been reported in the absence of perceptual or oculomotor disorders. Such cases have been observed following unilateral posterior parietal lesions (Garcin, Rondot, & de Recondo, 1967; Levine, Kaufman, & Mohr, 1978; Rondot & de Recondo, 1974; Rondot, de Recondo, & Ribadeau Dumas, 1977; Tzavaras & Masure, 1976), and sometimes after bilateral lesions (Damasio & Benton, 1979). Most often, patients are impaired in reaching with either hand in the visual hemifield contralateral to the lesion. However, other patterns of deficit have been encountered. Misreaching may affect only one hand within one hemifield, a symptom that cannot be either a pure motor or a pure perceptual disorder. Such observations suggest that the main impairment in optic ataxia lies in the accessibility of motor centers to visual input. The involvement of the posterior parietal cortex in this syndrome raises important questions concerning the neural organization of reaching behavior.

Case Reports

We have studied six cases of optic ataxia that followed unilateral posterior parietal lesion. The patients showed little or no associated visual, proprioceptive, or motor deficits, and thus these deficits could not account for the incoordination and inaccuracy of visually directed movements. Only the main clinical features are described here; further details are available elsewhere (Perenin, Vighetto, Mauguiere, & Fischer, 1979; Vighetto, 1980).

Lesion Localization

The topography of the lesions was defined by computerized tomography scans in five subjects (PP, PD, AV, JR, JP) and by scintigraphy in one (CB). Lesions appeared to be located on the left side in four cases (PP, PD, AV, JR) and on the right side in two (CB, JP). Except for one lesion of tumoral origin (JP), all the others were vascular. As shown in Figure 17-1, where lesion data from five patients are reconstructed and outlined, lesions were centered on the posterior parietal region, including to varying extents the superior parietal lobule (areas 5 and 7 and part of area 19 of Brodmann) and the inferior parietal lobule (areas 39 and 40 of Brodmann). The medial-dorsal part of the superior parietal lobule (corresponding to area 7a) was extensively damaged in three patients (JR, CB, and JP). The trace left by the surgical removal of

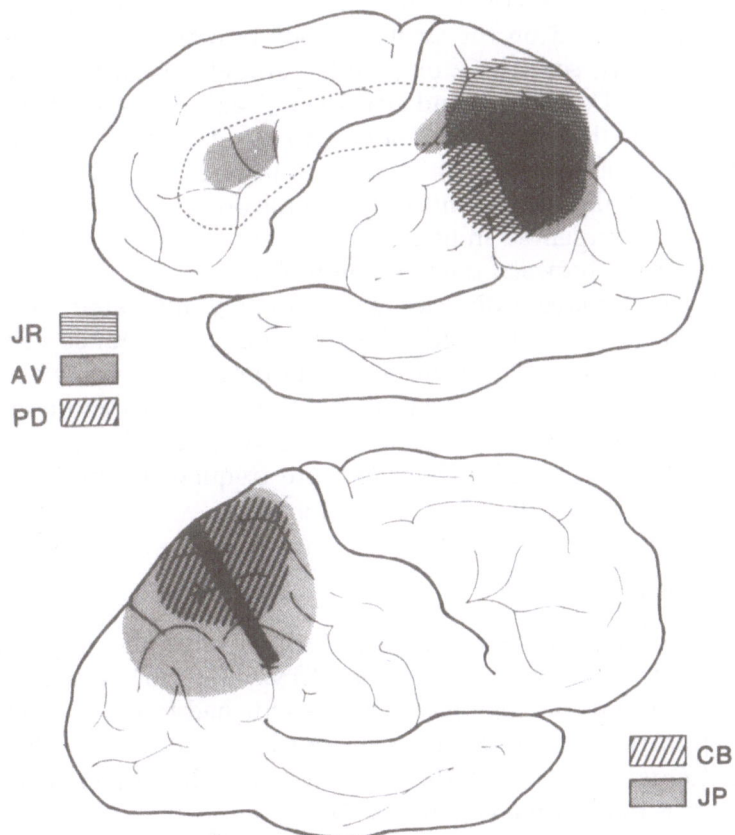

Fig. 17-1. Topography of the lesions in six patients with optic ataxia. Lesions are centered on the posterior parietal region. An associated lesion can be seen in the prefrontal cortex in AV and at the inner face of the left hemisphere in JR (broken line). *Black bar:* trace of the surgical removal of the tumor in JP.

the tumor in patient JP, as observed on postmortem examination, is represented by the black bar in Figure 17-1; the incision was found to encroach upon the posterior parietal lobe, from top to bottom.

Lesions outside the posterior parietal lobe were noted in two patients. In AV, there was a small infarct in the prefrontal region, and in JR there was a larger parasagittal lesion including the anterior third of the corpus callosum, part of the cingulate gyrus, and part of the paracentral lobule.

Vision

Visual acuity was normal or nearly so in all six patients. When tested with either the confrontation technique or the Tubingen perimeter with a large and bright spot the visual field also appeared normal. However, some shrinkage of the visual field contralateral to the lesion could occasionally be observed on the smallest isopters with a corresponding decrease in contrast sensitivity on profile perimetry (PD, CB, and JP). In four patients (PD, JR, CB, and JP) visual attention disorders could also be observed in the field contralateral to the lesion. When these four patients were presented with several objects in that hemifield they often failed to detect the most peripheral ones, although these could be seen when presented individually. No extinction was observed in a rivalry condition between the two hemifields.

Visual space perception was approximately unaltered. All the patients could give a correct verbal estimate of the distance or the relative position of objects within each hemifield. However, when simple stimuli such as single dots or lines were presented tachistoscopically to patients AV and JR, some errors were observed. These errors in location of dots or orientation of lines were more frequent in the field contralateral, than in the field ipsilateral, to the lesion.

Eye Movements

All six patients could fixate objects located anywhere in their visual field. In three of them (JR, AV, and JP) visually elicited saccades were recorded by electrooculography. The patients had to direct their gaze at targets appearing randomly along the horizontal meridian. Although all three patients could direct their gaze toward the targets, oculomotor behavior was not normal (Figure 17-2).

Saccade latency was increased in JR and JP. The maximal increase (up to 450 msec) was observed in saccades directed at the more peripheral targets (30° and 40°) and in JR at those located in the right visual field. In subjects AV and JP, saccades did not differ from normal when directed at targets located in the visual field ipsilateral to the lesion. This was not the case in the opposite field, where target position was reached after a succession of small saccades. Such a staircase pattern was observed when patients were moving their eyes from the midline to the field contralateral to the lesion, or when they were returning to the midline from the field ipsilateral to the lesion.

Somesthesia

Somatosensory sensations were virtually normal. Only a few kinesthetic errors on fingers contralateral to the lesion were observed in subjects PD, PP, and AV. When either thumb was passively moved by the experimenter, the patients had no difficulty in reaching for them with eyes open or closed. In two patients (AV and PP) tactile extinction occurred on the hand contralateral to the lesion when bilateral and symmetrical areas were stimulated. Stereognosis was normal except in AV and PD,

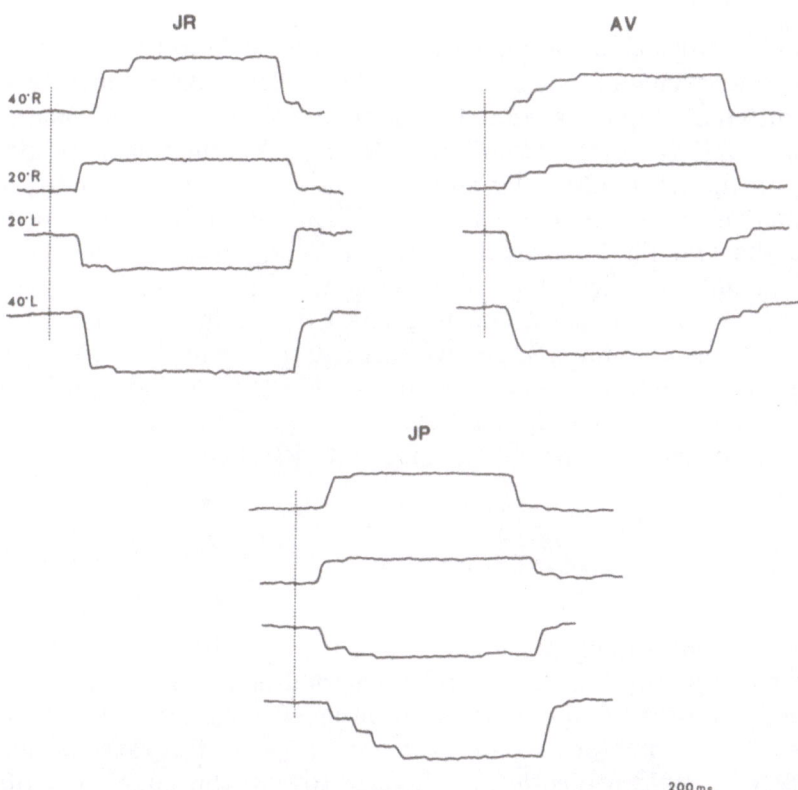

Fig. 17-2. Oculomotor responses (recorded by electrooculography) to step visual stimuli at 20° and 40° eccentricities to the right or left in patients JR. AV, and JP. Upward and downward deflections indicate movements to the right and to the left, respectively. Note the increased latencies in JR, mostly for the 40° right target, and in JP for both 40° targets. Note also the staircase ocular behavior in AV and JP when these patients have to direct the eyes at targets located in the visual field contralateral to the lesion. *Vertical broken line:* the moment at which targets were illuminated.

who made a few errors in identifying three-dimensional shapes with the right hand.

Motor Functions

Elementary motor capacities were largely spared. The right thumb–index grip was slightly weakened in PD and AV. Repetitive finger movements, such as strumming or tapping, were often slowed. Absence of spontaneous use of the arm contralateral to the lesion was frequently observed in all patients with left hemisphere damage and to a lesser degree in CB. In addition, gestures were often altered by various degrees of apraxia in the patients with left-sided lesions (see below).

Neuropsychological Symptoms

Neuropsychological disorders related to parietal lesions could be observed to a varying extent in all six patients. The most frequently seen were different types of apraxia: bilateral ideomotor and ideational apraxia in left-damaged patients (PP, JR. and AV; the former predominating on the right side in PP and AV, and on the left side in JR), and constructive apraxia in two of the left-damaged patients (PP and JR) and in the two right-damaged patients (CB and JP). A marked Gerstmann syndrome (right–left indistinction, finger agnosia, agraphia, acalculia) was noted in JR, and to a lesser degree in AV. Body scheme was altered in JR and CB. In two patients (JR and AV) there was a slight hemispatial neglect on the right side. Finally, CB and JP had some difficulties in route finding on a map. No sign of aphasia was observed in any of the patients. All the patients were right handed.

Proximal and Distal Deficits in Optic Ataxia

Arm movements directed at a visual object, in which subjects with optic ataxia are typically impaired, can be regarded as the superimposition of two major motor components, each designed for a different attribute of the object. The first component, involving mostly proximal joints, consists of a rapid projection of the arm toward the target. The other, distal component is a postural adaptation of the hand to the shape and size of the object in anticipation of the grasp. In our patients with optic ataxia, these two components were analyzed by means of magnetoscopic and cinematographic recordings in two different situations intended to quantify the proximal and distal deficits, respectively.

Proximal Deficit: Reaching for an Object

In the initial examination, the patient had to fixate the camera lens. The patient was asked to reach and grasp as quickly and as accurately as possible a salient object that was presented successively in the visual hemifield ipsilateral to the lesion, in the central field, and then in the contralateral hemifield. First the hand ipsilateral and then the hand contralateral to the lesion was tested. At least 10 trials were performed for each hand–field combination.

Three different types of motor responses could be observed: *normal*, in which the subject reached and grasped the object directly; *corrected*, in which the subject initially missed the target by a variable amount and then corrected; and *noncorrected*, in which the subject missed the target and did not correct the error except when, by chance, his hand hit the object. The distribution of the spatial errors (corrected and noncorrected) in each hand–field combination for each patient is shown in Figure 17-3.

Some findings appeared common to all subjects. Reaching with the hand ipsilateral to the lesion in the ipsilateral hemifield and in the cen-

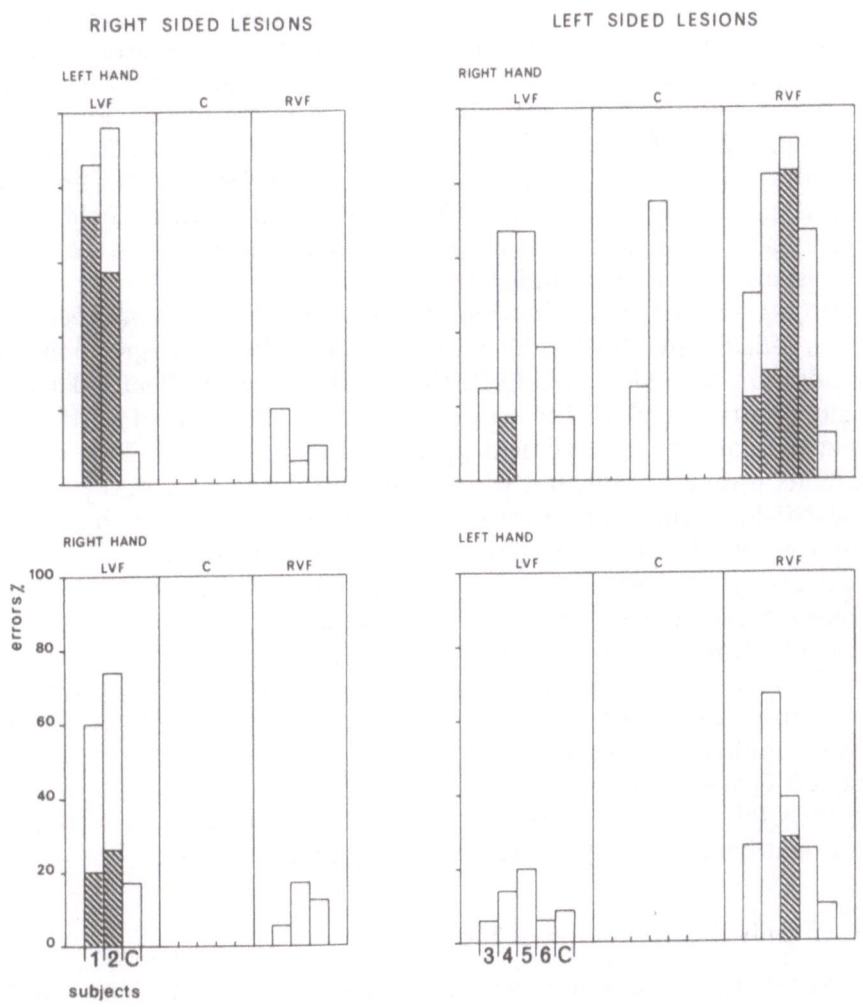

Fig. 17-3. Repartition of the spatial errors across the different hand–field combinations for each subject during reaching at a visual object. The number of errors is shown as a percentage of whole trials in each individual combination and subject. *White bars*, corrected errors; *hatched bars*, noncorrected errors; *LVF*, left visual field; *CF*, central field; *RVF*, right visual field; *1–6*, CB, JP, PD, PP, JR. AV; *C*, control subjects. Note that in both groups of patients there is a "visual field effect" (i.e., errors with both hands in the field contralateral to the lesion) and in the left-damaged group only there is a "hand effect" (i.e., errors with the right hand not only in the right hemifeild but also in the left hemifield, and in two patients in the central field too).

tral field was always similar to that of normal controls. There were only few errors of small amplitude, which always were corrected. In most trials the object could be grasped after a rapid (400–500 msec) projection of the arm. The maximum impairment occurred for reaching with the hand and in the visual field contralateral to the lesion. High percentages of reaches with spatial errors were observed in this combination (from 50% in PD up to 96% in JP) including many noncorrected errors (from 22% in PD up to 83% in JR). Thus, after a rather slow movement of the arm the patients often failed to grasp the object. Finally, the hand ipsilateral to the lesion was also impaired in the hemifield contralateral to the lesion, although to a lesser extent. Most of the errors observed in this combination were corrected. Such a "visual field effect" (both hands affected in the visual field contralateral to the lesion) was a constant finding in all subjects.

Differences in the distribution of errors were observed between right- and left-damaged patients. All four members of the latter group showed, in addition to a visual field effect, a marked "hand effect." The right hand (contralateral to the lesion) was not only affected in the right field, but also in the left field (ipsilateral to the lesion). Two of these patients were also impaired when using their right hand in their central visual field. Errors were mostly of the corrected type. The hand effect observed in the left-damaged subjects was more prominent than the field effect. The sum of errors of the right hand in both hemifields exceeded the sum of errors of both hands in the right field. No such hand effect was seen in the right-damaged patients.

When subjects had to orient eyes and head toward the object, instead of fixating the camera lens, reaching improved dramatically in all cases. Errors could no longer be detected in the right-damaged patients, nor in the left-damaged patients when they used their left hand. In the latter group a few corrected errors occurred with the right hand in both left and right hemispaces (from 11% in AV up to 35% in JR).

Distal Deficit: Orientation of the Hand

Hand orientation was studied with a simplified version of the test used by Haaxma and Kuypers (1974) in the monkey. In our version of the test the target was a white disk with an oval hole through which the extended hand had to pass (Figure 17-4). As in the reaching situation, the different hand–field combinations were tested systematically, but for this test the disk remained at the same place: at arm's length and 30 cm from the midline for each given hand–field session. Only the orientation of the main axis of the hole changed from trial to trial. Four patients were tested with this device, one with a right-sided lesion (JP) and three with a left-sided lesion (PD, JR, and AV).

The task could be accomplished in all cases, although often slowly

(a)   (b)   (c)

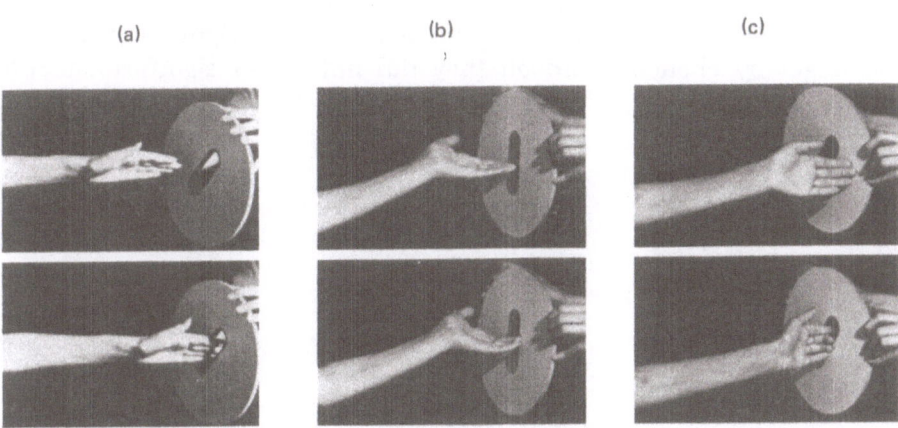

Fig. 17-4. Responses observed in the hand orientation test. (a) Normal response. (b) Orientation error. (c) Spatial error. Left hand of subject JP in the right hemifield (a) and in the left hemifield (b and c).

and awkwardly. As the hand approached the disk, both spatial and hand orientation errors could be observed (Figure 17-4). These two types of errors could occur either separately or, more often, in association. Spatial errors were generally of small amplitude and always of the corrected type. This probably was a reflection of the fact that the target was large and always located at the same place for a given hand–field session. Hand orientation errors were observed in the right-damaged as well as the left-damaged patients. Their distribution in the four patients (Figure 17-5) was very similar to that of the misreaching shown in Figure 17-3. For the right-damaged patients, orientation errors occurred only in the visual field contralateral to the lesion, whichever hand was used. In contrast, the left-damaged patients showed not only a field effect but a hand effect as well. Angular error ranged from $45°$ to $90°$ in most cases, but a total inversion of the normal hand posture (corresponding to a $180°$ error) sometimes occurred.

Frame-by-frame analysis revealed in both testing situations and in all subjects a disturbance of the ability to shape the hand to conform to the target object. This defect had the same distribution across the different hand–field combinations as the spatial errors. In reaching for an object, the hand generally kept a "fanned" aspect, with fingers fully extended and spread out. In addition, the left-damaged patients showed impairments with the right hand not observed in the right-damaged patients. Movements were much more hesitating in groping for the target. Totally unadapted motor behavior was often seen. For example, the hand could remain closed during its whole trajectory, or could be moved into useless alternating prone and supine positions. In the hand orientation task fingers were often flexed and could not be passed together into the hole. Finally, the left-damaged patients often failed to use tactile cues to accomplish the task when the right hand touched the target by chance, although they did not have a significant tactile deficit.

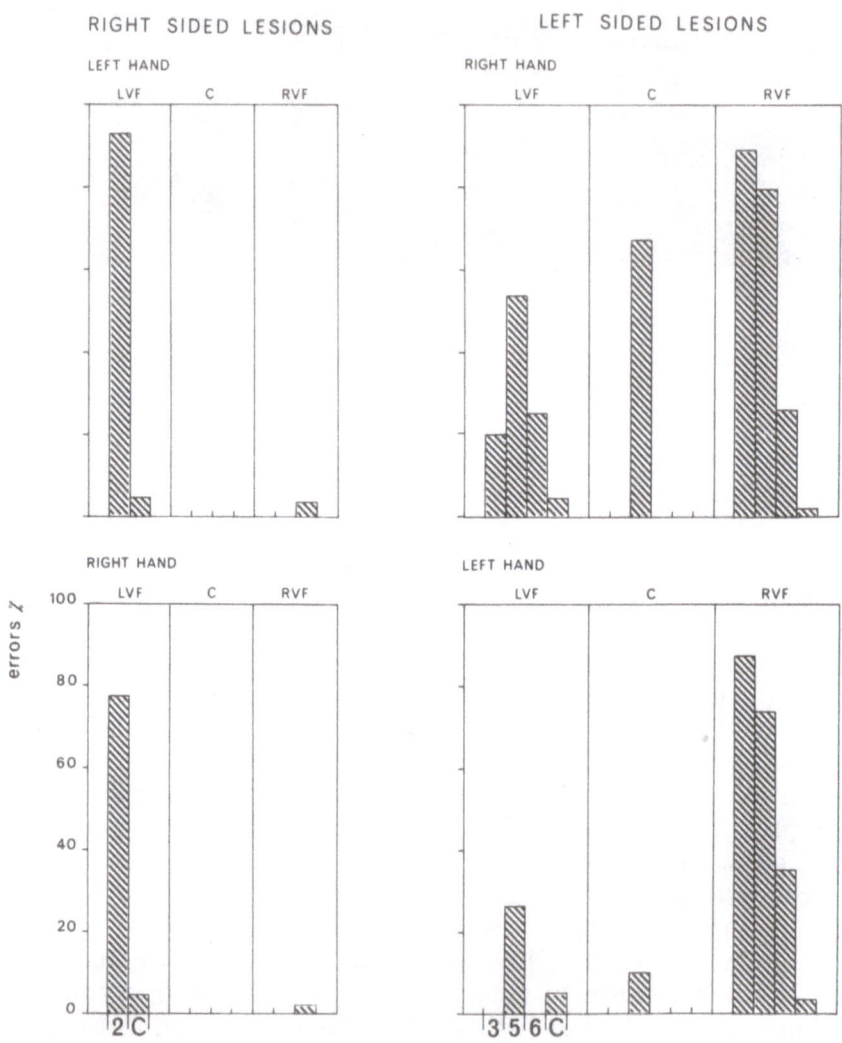

Fig. 17-5. Repartition of the orientation errors of the hand across the different hand–field combinations in four patients required to pass a hand through a hole as shown in Figure 17-4. The number of errors is shown as a percentage of the total number of trials in each individual combination and subject. *LVF*, left visual field; *CF*, central field; *RVF*, right visual field; patients *2, 3, 5,* and *6,* JP, PD, JR, and AV; *C*, control subjects. Visual field and hand effects resemble those observed for spatial errors.

Comparison with Findings from Experiments in Monkeys

Effects of Posterior Parietal Lesions

As observed in our left- and right-damaged patients, optic ataxia consists of both a proximal and a distal deficit of visual goal-directed movements. The former was emphasized in the past, with very little attention paid to the latter. When mentioned, distal deficits mainly concerned independent finger movements during manipulation of the visual target (Damasio & Benton, 1979). Tzavaras and Masure (1976), however, used the same manipulandum as Haaxma and Kuypers (1974) and observed an absence of anticipatory shaping of the finger grip in one left-damaged patient but not in one with a right-sided lesion.

There are now a number of arguments suggesting that the proximal and distal components of reaching movements would result from a parallel central processing of visual input by two different types of visuomotor channels. "Space" and "object" channels, dealing with the spatial location and with the shape of the visual stimulus, respectively, would produce motor outputs in different arm segments (Jeannerod & Biguer, 1982). In fact, according to the two visual system hypothesis, position and shape of visual objects would be analyzed in separate visual pathways (Schneider, 1969; Trevarthen, 1968; however, see Goodale, Chapter 3, this volume). Some segregation also appears on the efferent side. According to studies by Brinkman and Kuypers (1975) in the monkey, distal motricity of the arm is exclusively mediated by the contralateral corticomotoneural, pyramidal pathway, whereas proximal motricity depends on both contra- and ipsilateral corticospinal and subcorticospinal pathways. Thus, interruption of connections between the visual cortex and the motor control area logically should impair the adjustment of finger posture to the target more than other components of the movement. This is what Haaxma and Kuypers (1974) observed after a unilateral leukotomy at the parieto-occipital junction in split-brain monkeys. When using the hand contralateral to the lesion, the animals failed to seize small pieces of food from narrow grooves—a task requiring a precise shaping of the fingers. Larger posterior parietal lesions that include areas 5 and 7 (Lamotte & Acuna, 1978) or area 7 alone (Faugier-Grimaud, Frenois, & Stein, 1978) impair both the directional precision of the arm movement and the postural adaptation of the hand and fingers contralateral to the lesion.

Thus, in the monkey as in man, posterior parietal damage would alter both channels involved in reaching for objects. This finding is concordant with the fact that the posterior parietal lobe receives indirect visual information from the two visual systems. It is connected to the superior colliculus and to the pretectum via the pulvinar (Baleydier & Mauguiere, 1977; Benevento & Fallon, 1975) and to the geniculostriate pathway via the prestriate cortex (Jones & Powell, 1970; Mesulam, Van Hoesen, Pandya, & Geschwind, 1977).

Whereas unilateral posterior parietal lesions in man are generally known to result in misreaching with either hand only in the contralateral hemifield, similar lesions in the monkey produce misreaching only with the contralateral limb, irrespective of the target location in space (Ettlinger & Kalsbeck, 1962; Faugier-Grimaud et al., 1978; Hartje & Ettlinger, 1973; Lamotte & Acuna, 1978). However, this opposition appears to be too schematic for several reasons. In the monkey, the effects of the lesions have always been tested with free-moving eyes, which condition may prevent observation of a visual field effect. In fact, after cooling of area 7, misreaching has been found to predominate in the contralateral part of visual space (Stein, 1978). In man both visual field and hand effects can occur; it should be noted, however, that if the former effect has been more frequently observed, the majority of optic ataxia cases reported previously concerned right-damaged patients. Thus, as confirmed by our findings, an isolated visual field effect appears to characterize optic ataxia resulting from a right-sided lesion. Associated field and hand effects, as seen in four of our patients and in two reported before (Tzavaras & Masure, 1976; Tzavaras, Ozonas, & Chodkiewicz, 1974), appear to characterize optic ataxia resulting from a left-sided lesion. The only exception is a right-damaged patient, reported by Levine et al., (1978) who showed a hand effect—but mostly when prevented from viewing his limb.

Such a difference in the distribution of the deficit according to the side of the lesion in man suggests an asymmetry of functional organization between right and left parietal areas subserving visually elicited movements. This would represent one more example of the well-documented functional asymmetry between the two hemispheres in man. Since the original work of Liepmann (1908) on apraxia, the dominance of the left hemisphere, especially the left inferior posterior parietal lobe, in the control of motor acts has been confirmed by a number of studies (e.g., Heilman, 1979; Kimura & Archibald, 1974; Wyke, 1971). All patients with optic ataxia resulting from a left-sided lesion showed some apraxic disorders, most often predominating in the right arm. We suggest that the additional distal impairments of visually elicited movements observed in these patients, but not in the right-damaged patients, represent a cumulative effect of optic ataxia and apraxia. Further anatomicoclinical correlations should help our understanding of the relationship between these two types of disorders, whose anatomical substrate may either overlap or be contiguous.

Due to the considerable development of the inferior parietal lobule in man (areas 39 and 40 have no equivalent in the monkey) and to the functional asymmetry of the human brain, posterior parietal lesions may result in more complex and varied visuomotor disorders in man than in the monkey.

Properties of Posterior Parietal Cells

In the past few years the properties of cells in areas 5 and 7 of the monkey's parietal lobe have been compared with the behavioral defects following the removal of these cortical areas. Several functional categories of neurons have been identified and related to different aspects of reaching behavior. It has been shown that spatial information about the position of the different body parts, the position of visual objects, and the direction of gaze and arm movements within the immediate surround are represented in parietal areas 5 and 7. According to Mountcastle (1978), these neuronal properties would be compatible with a command function of the parietal lobe "for the selective and directed visual attention into the immediate behavioral surround, for the visual grasping of objects and for skilled, coordinated actions of hand and eye." Alternative or complementary functional correlates have been ascribed to the activity of the parietal cells, which might also subserve high-level sensory association mechanisms (see Lynch, 1980).

Effects of Altering Spatial Cues on Reaching Behavior in Optic Ataxia

In order to demonstrate the effects of selective alteration of the cues normally used for controlling accuracy of visually guided reaching in optic ataxia patients, we have applied, in three patients (JR, AV, JP), an experimental paradigm used by Prablanc, Echallier, Komilis, and Jeannerod (1979) in normal subjects. Patients were tested a few months after the examinations reported in the previous sections. Although some recovery had occurred, optic ataxia was still quite clear on clinical testing in all three patients.

The patients were required to point at small target lights, a task in which distal postural adjustments are minimized and thus "space" channels are principally involved. Targets appeared randomly in the peripheral visual field along the horizontal meridian. During the sessions, when eye and hand positions were recorded, the head was immobilized. Each hand was tested separately. While the hand was pointing at the target, the eyes were either directed at the target (foveal vision condition) or fixated at the midline (peripheral vision condition). These two conditions were used for testing the role of cues related to the direction of gaze in controlling hand movement accuracy.

Other conditions were designed to test the role of visual information hand position, and/or movement, in the accuracy of pointing. The view of the hand could be either allowed or prevented (closed-loop and open-loop conditions, respectively). In a more selective condition, the hand could be seen only before its displacement, not during the movement itself (static closed-loop condition).

Closed Loop, Foveal Vision

When they were allowed to see the hand and to track the target with the eyes, the three patients pointed accurately (Figure 17-6). Means and standard deviations of the hand precision values were in the same range as in normal controls.

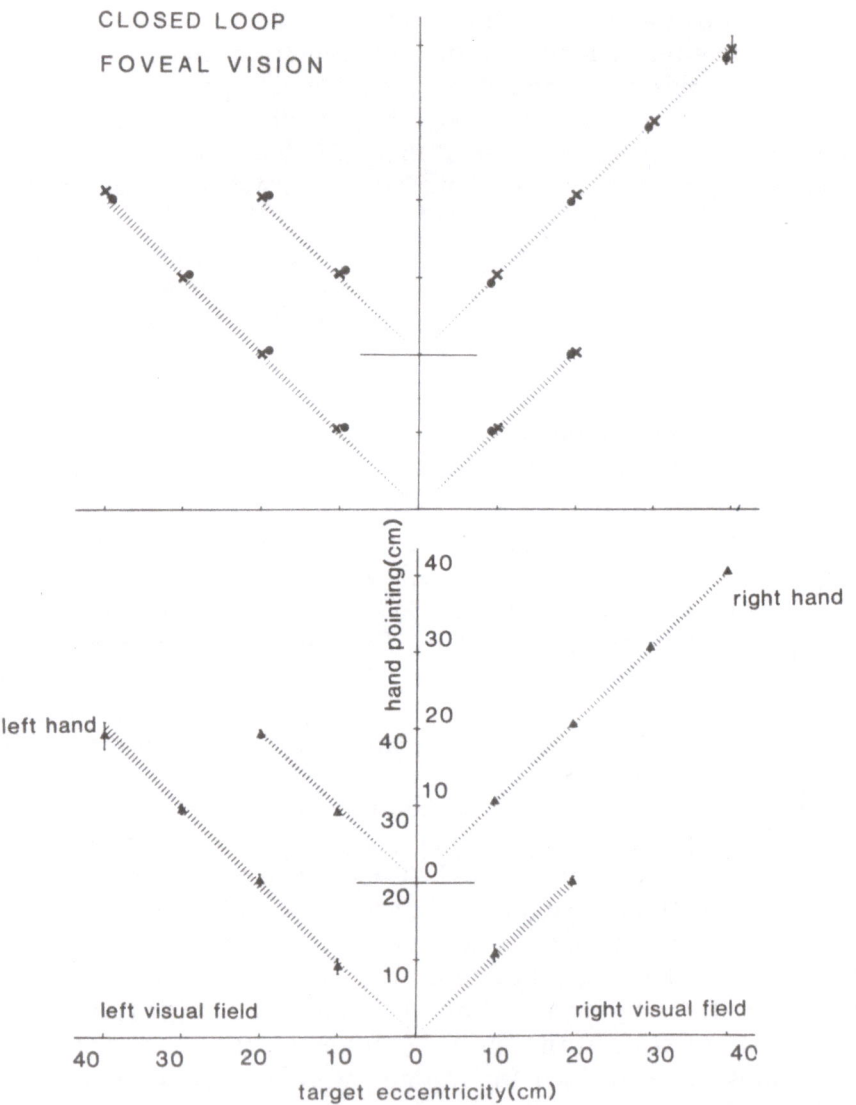

Fig. 17-6. Hand pointing positions (mean±SD) in closed-loop and foveal vision conditions. Patients can see the hand while pointing and have to direct the gaze at the targets that appear randomly in both hemifields. (a) Responses of the two left-damaged patients, JR (*circles*) and AV (*crosses*). (b) Responses of the right-damaged patient, JP (*triangles*). For comparison, the standard deviation of pointing positions of eight normal subjects tested in the same situation are shown by *striped areas*.

Closed Loop, Peripheral Vision

When fixation of the target was prevented and patients were required to reach for it in their visual periphery, pointing was no longer accurate throughout the visual field. Although the level of accuracy remained similar to that of controls in the hand–field combination ipsilateral to

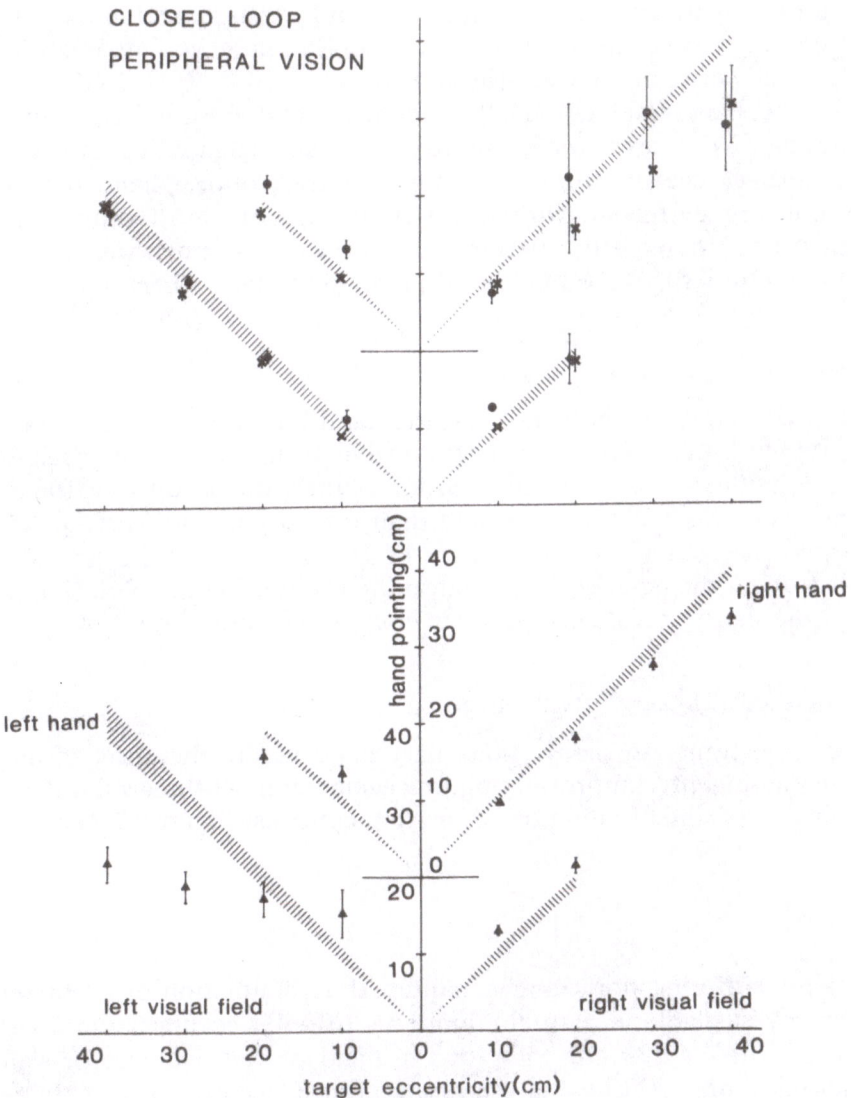

Fig. 17-7. Hand pointing positions (mean±SD) in closed-loop and peripheral vision conditions. Patients can see the hand and have to keep the gaze on a central target while pointing at targets presented in both hemifields. (a), (b), and symbols same as in Figure 17-6. Inaccurate pointing is seen mostly in the visual field contralateral to the lesion with one (AV) or both (JR and JP) hands, although the hand contralateral can also be impaired in the field ipsilateral to the lesion.

the lesion (and for patient AV in the two crossed combinations too), anywhere else precision was impaired. Inaccuracy was maximal in the hand–field combination contralateral to the lesion (Figure 17-7). As mentioned earlier, the pointing inaccuracy could be related either to a visual field effect (affecting the two hands in the field contralateral to the lesion) or to a hand effect (affecting the hand contralateral to the lesion in the two hemifields). In the right-damaged patient (JP), in spite of a slight inaccuracy of the left hand in the right field, the field effect dominated the hand effect, whereas there was no significant difference between the two effects in JR.

In the hand–field combination contralateral to the lesion, undershooting of the target was more frequent than overshooting, at least for the greatest eccentricities. Pointing responses of the hand remained significantly correlated with target position, as in most other experimental conditions. This relationship, however, was much weaker in the patient with a right-sided lesion (JP) than in the two others.

Open Loop, Foveal Vision

When the patients could not see the hand but were free to move the eyes, they became grossly inaccurate when using the hand contralateral to the lesion. In addition, the loss in accuracy due to opening the eye-hand loop was much more severe than for normals. All three patients showed a systematic pointing error of $10°-15°$ toward the side of the lesion when the hand contralateral to the lesion was used. Pointing with the other hand was as accurate as in controls (Figure 17-8).

Static Closed Loop, Foveal Vision

The opportunity to see the stationary hand before the onset of movement significantly improved pointing accuracy in all three subjects; two of them (AV and JP) did not differ from controls (Figure 17-9).

Discussion

Patients suffering optic ataxia require the conjunction of all the cues normally available in natural conditions to point accurately at a visual target. When either the view of the hand or the oculomotor signals resulting from fixating the target are absent, accuracy is lost. In fact, in order to achieve normal precision in the closed-loop, foveal vision situation these patients seem to use different strategies than normal controls. This is suggested by some temporal features of their pointing responses, such as the increased latency and duration of hand movements (Vighetto, 1980). In addition, the severe misreaching produced

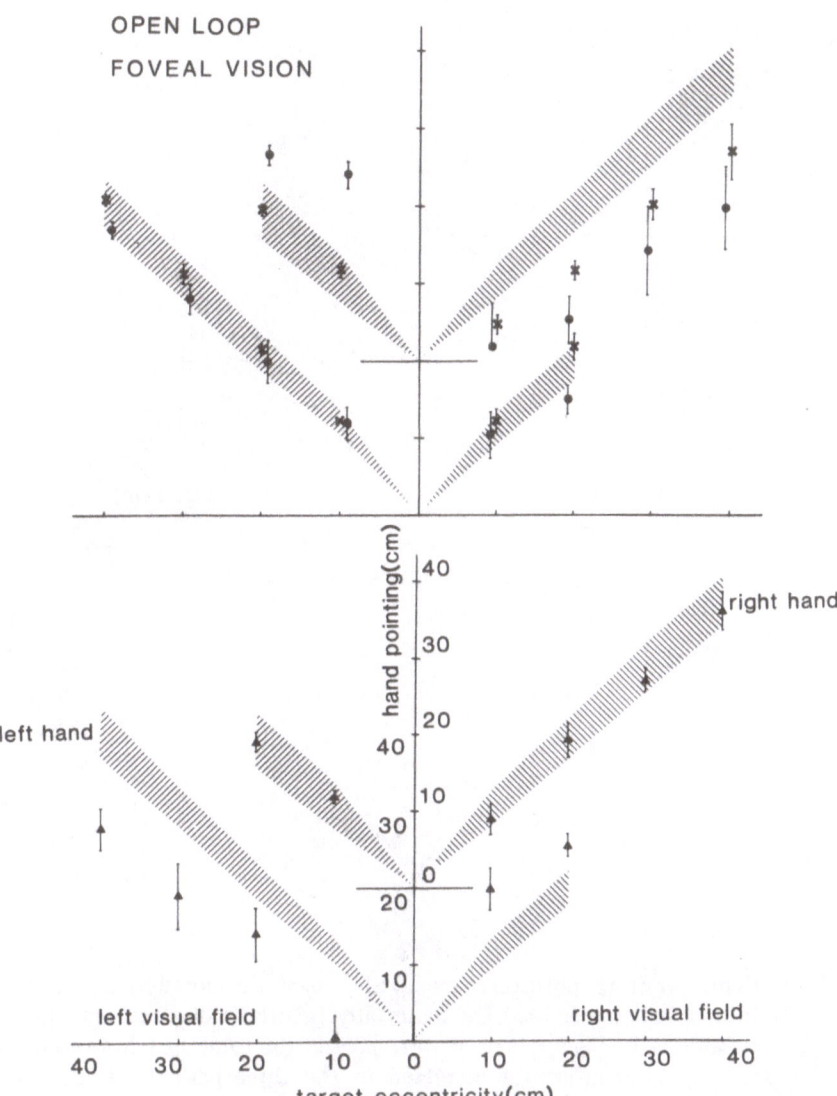

Fig. 17-8. Hand pointing positions (mean ±SD) in open-loop and foveal vision conditions. Patients cannot see the hand while pointing and have to direct the eyes at the targets presented in both hemifields. (a), (b), and symbols same as in Figure 17-6. Note that the inaccuracy is limited to the hand contralateral to the lesion in all three patients, with all pointing responses displaced toward the lesion side.

by opening the visuomotor loop, a condition in which visual feedback is interrupted, indicates that patients with optic ataxia rely mostly on feedback control for directing the hand at the target. This means that such patients lack another mode of control that usually is present even when the visuomotor loop is open. The capacity of normal subjects to

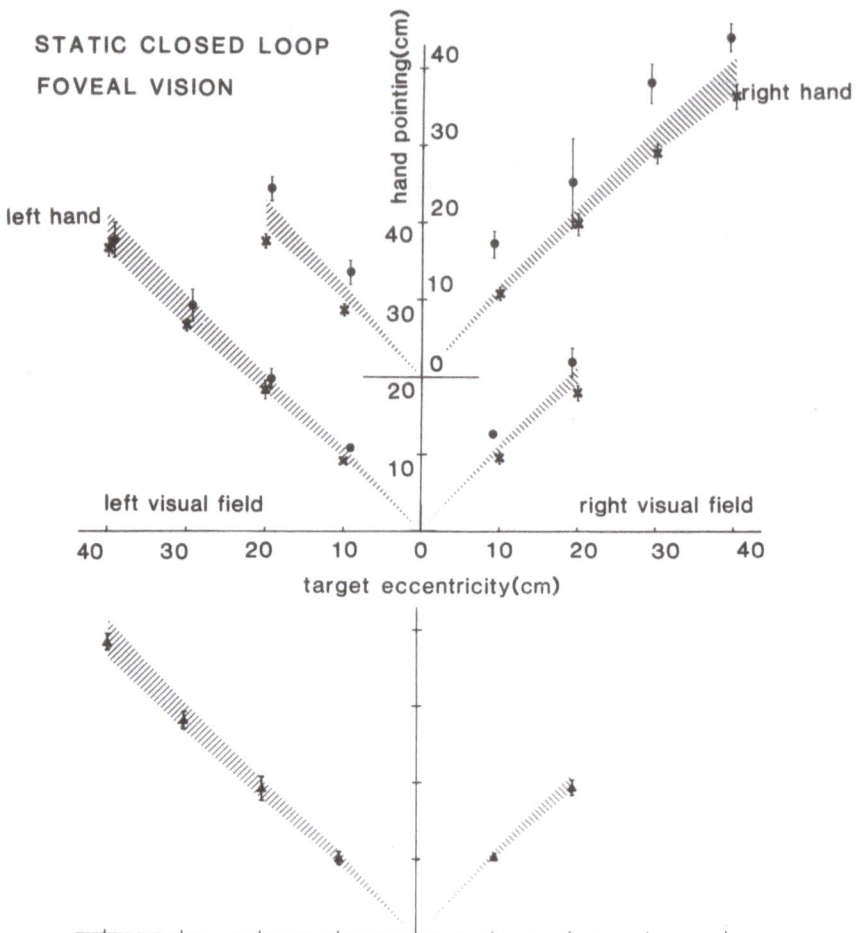

Fig. 17-9. Hand pointing positions (mean±SD) in static closed-loop and foveal vision conditions. Patients can see the hand only before its displacement. (a), (b), and symbols same as in Figure 17-6. With respect to the total open-loop condition, pointing accuracy is significantly increased in the three patients. Two of them behave like the normal controls. For subject JP, only pointing responses of the left hand were recorded in this situation.

direct the hand at a visual target in the open-loop condition is currently ascribed to a central motor program. This central mechanism is supposed to control the initial part of reaching movements, although their terminal precision depends more on peripheral visual feedback. An intermediate level of control has also been hypothesized, by which the ongoing program could be modulated by centrally registered cues more quickly than by sensory feedback. In a study of reaching movements in normal subjects, Prablanc, Echallier, Jeannerod, and Komilis (1979) proposed that an internal visual map of target and hand positions stored at the onset of the movement could serve to exert an open-loop visual control allowing optimization of the program by a feedforward mechanism. Signals related to the position of the gaze axis in the orbit would also represent a cue for an open-loop control of hand movements.

In patients with optic ataxia such mechanisms would still operate, provided they could add to each other. However, their efficacy would be much reduced in conditions in which they must act separately. This is the case when subjects can see the hand but are not allowed to fixate the target with the eyes. In that case, the oculomotor signals that have been assumed to guide the hand at the visual goal (e.g., Festinger & Canon, 1965; Gazzaniga, 1969) could not be used. Although their role appears unclear in normal subjects (Prablanc, Echallier, Komilis, & Jeannerod, 1979), oculomotor signals seem critical for controlling hand movements in patients with optic ataxia. Pointing inaccuracy observed in these patients when they are prevented from moving their eyes might be ascribable primarily to the inability to relate hand position, or movement, signals to the visual map. Oculomotor signals due to foveation would compensate for this inability by improving the spatial coding of the target position into a body-centered map of visual space.

In the complementary situation, when the patients are free to direct the gaze at the target but cannot see the hand, a similar impairment in relating signals from hand position or movement to the visual map could be responsible for the inaccuracy of pointing. In this situation the only signals coming from the hand are proprioceptive and do not allow a precise encoding of movement amplitude with respect to target location. As suggested by the improved accuracy shown in normal subjects in the static closed-loop condition, proprioceptive signals need to be calibrated by means of visual cues (Klein & Posner, 1974; Prablanc, Echallier, Jeannerod, & Komilis, 1979). In patients with optic ataxia a mechanism that compares proprioceptive signals from the hand with visual signals from the target's position in space is inefficient for movement control.

According to the results reported here, optic ataxia patients seem to be impaired in those goal-directed hand movements requiring a high level of intermodal integration. Compensation for the deficits created by the parietal lesion can occur through better mutual mapping of somatic and visual representations on egocentric coordinates. However, we cannot state at which stage of motor control between the initial central program and the terminal sensory feedback, intermodal integration is lacking. We have only preliminary indications that in the peripheral vision condition visual feedback mechanisms are the most severely impaired. Indeed, no change in pointing accuracy is observed when the view of the hand is prevented during its displacement. In contrast, pointing accuracy degrades when the view of the hand is prevented before it moves.

Lesions of the posterior parietal lobe do not result in a total disorganization of visually elicited hand movements. The deficit that occurs can be attenuated by compensatory strategies and, even in the critical peripheral vision or open-loop conditions, there remains some capacity for approximating visual targets by hand. Similar findings in the monkey after lesions of areas 5 and 7 (Lamotte & Acuna, 1978) suggest that other structures (such as the cerebellum or the basal ganglia) may also contribute to eye–hand coordination.

References

Baleydier, C., & Mauguiere, F. Pulvinar-latero posterior afferents to cortical area 7 in monkeys demonstrated by horseradish peroxidase tracing technique. *Experimental Brain Research*, 1977, *27*, 501-507.

Balint, R. Seelenlähmung des "Schauens," optische Ataxie, räumliche Störung der Aufmersamkeit. *Monatsschrift für Psychiatrie und Neurologie*, 1909, *25*, 57–81.

Benevento, L. A., & Fallon, J. H. The ascending projections of the superior colliculus in the rhesus monkey (Macaca Mulatta). *Journal of Comparative Neurology*, 1975, *160*, 339-362.

Brain, W. R. Visual disorientation with special reference to lesions of the right cerebral hemisphere. *Brain*, 1941, *64*, 244-272.

Brinkman, J., & Kuypers, H. G. Cerebral control of contralateral and ipsilateral arm, hand and finger movements in the split-brain rhesus monkey. *Brain*, 1973, *96*, 653-674.

Cole, M., Schutta, H. S., & Warrington, E. K. Visual disorientation in homonymous half-fields. *Neurology*, 1962, *12*, 257-263.

Damasio, A. R., & Benton, A. L. Impairment of hand movements under visual guidance. *Neurology*, 1979, *29*, 170-174.

Ettlinger, G., & Kalsbeck, J. E. Changes in tactile discrimination and visual reaching after successive and simultaneous bilateral posterior parietal ablations in the monkey. *Journal of Neurology, Neurosurgery and Psychiatry*, 1962, *25*, 256-268.

Faugier-Grimaud, S., Frenois, C., & Stein, D. Effects of posterior parietal lesions on visually guided behavior in monkeys. *Neuropsychologia*, 1978, *16*, 151-168.

Festinger, L., & Canon, L. K. Information about spatial location based on knowledge about efference. *Psychological Review*, 1965, *72*, 373-384.

Garcin, R., Rondot, P., & de Recondo, J. Ataxie optique localisée aux deux hémichamps homonymes gauches (étude clinique avec présentation d'un film). *Revue Neurologique* (Paris), 1967, *116*, 707-714.

Gazzaniga, M. S. Eye position and visuo-motor coordination. *Neuropsychologia*, 1969, *1*, 209-215.

Godwin-Austen, R. B. A case of visual disorientation. *Journal of Neurology, Neurosurgery and Psychiatry*, 1965, *28*, 453-458.

Haaxma, R., & Kuypers, H. Role of occipito-frontal cortico-cortical connections in visual guidance of relatively independent hand and finger movements in rhesus monkeys. *Brain Research*, 1974, *71*, 361-366.

Hartje, W., & Ettlinger, G. Reaching in light and dark after unilateral posterior parietal ablations in the monkey. *Cortex*, 1973, *4*, 346-354.

Hécaen, H., & de Ajuriaguerra, J. Balint's syndrome (psychic paralysis of visual fixation) and its minor forms. *Brain*, 1954, *77*, 373–400.

Heilman, K. M., The neuropsychological basis of skilled movement in man. In M. S. Gazzaniga (Ed.), *Handbook of behavioral neurobiology, II Neuropsychology*. New York: Plenum Press, 1979, pp. 447–460.

Holmes, G. Disturbance of visual orientation. *British Journal of Ophthalmology*, 1918, *2*, 449–468, 506–516.

Jeannerod, M., & Biguer, B. Visuomotor mechanisms in reaching within extrapersonal space. In D. Ingle, M. Goodale, & R. Mansfield (Eds.), *Advances in the analysis of visual behavior*. Boston: MIT Press, 1982, pp. 387–409.

Jones, E. G., & Powell, T. P. S. An anatomical study of converging sensory pathways within the cerebral cortex of the monkey. *Brain*, 1970, *93*, 793–820.

Kase, C. S., Troncoso, J. F., Court, J. E., Tapia, J. F., & Mohr, J. P. Global spatial disorientation. *Journal of Neurological Sciences*, 1977, *34*, 267–278.

Kimura, D., & Archibald, Y. Motor functions of the left hemisphere. *Brain*, 1974, *97*, 337–350.

Klein, R., & Posner, M. I. Attention to visual and kinesthetic components of skills. *Brain Research*, 1974, *71*, 401–411.

Lamotte, R. H., & Acuna, C. Defects in accuracy of reaching after removal of posterior parietal cortex in monkeys. *Brain Research*, 1978, *139*, 309–326.

Levine, D. N., Kaufman, K. J., & Mohr, J. P. Inaccurate reaching associated with a superior parietal lobe tumor. *Neurology*, 1978, *28*, 609–612.

Liepmann, H. *Drei Aufsätze aus dem Apraxiegebiet*. Berlin: Karger, 1908.

Luria, A. R. Disorders of "simultaneous perception" in a case of bilateral occipito-parietal brain injury. *Brain*, 1959, *82*, 437–449.

Lynch, J. C. The functional organization of posterior parietal association cortex. *Behavior and Brain Sciences*, 1980, *3*, 485–498.

Mesulam, M. M., Van Hoesen, G. W., Pandya, D. N., & Geschwind. N. Limbic and sensory connections in the inferior parietal lobule (PG) in the rhesus monkey: A study with a new method for horseradish peroxidase histochemistry. *Brain Research*, 1977, *136*, 393–414.

Michel, F., Jeannerod, M., & Devic, M. Troubles de l'orientation visuelle dans les trois dimensions de l'espace (à propos d'un cas anatomique). *Cortex*, 1965, *1*, 441–466.

Mountcastle, V. B. Brain mechanisms for directed attention. *Journal of the Royal Society of Medicine*, 1978, *71*, 14–28.

Perenin, M. T., Vighetto, A., Mauguiere, F., & Fischer, C. L'ataxie optique et son intérêt dans l'étude de la coordination oeil-main. *Lyon Médical*, 1979, *242*, 349–358.

Prablanc, C., Echallier, J. F., Jeannerod, M., & Komilis, E. Optimal response of eye and hand motor systems in pointing at a visual target. II. Static and dynamic visual cues in the control of hand movement. *Biological Cybernetics*, 1979, *35*, 183–187.

Prablanc, C., Echallier, J. F., Komilis, E., & Jeannerod, M. Optimal response of eye and hand motor systems in pointing at a visual target. I. Spatio-temporal characteristics of eye and hand movements and their relationships when varying the amount of visual information. *Biological Cybernetics*, 1979, *35*, 113–124.

Riddoch, G. Visual disorientation in homonymous half-fields. *Brain*, 1935, *58*, 376–382.

Rondot, P., & de Recondo, J. Ataxie optique: Trouble de la coordination visuo-motrice. *Brain Research*, 1974, *71*, 367-375.

Rondot, P., de Recondo, J., & Ribadeau Dumas, J. C. Visuomotor ataxia. *Brain*, 1977, *100*, 355-376.

Schneider, G. E. Two visual systems. *Science*, 1969, *163*, 895-902.

Stein, J. Long loop motor control in monkeys. The effects of transient cooling of parietal cortex and of cerebellar nuclei during tracking tasks. In J. E. Desmedt (Ed.), *Cerebral motor control in man: Long loop mechanisms*. Basel: Karger, 1978, pp. 107-122.

Trevarthen, C. B. Two mechanisms of vision in primates. *Psychologische Forschung*, 1968, *31*, 299-337.

Tzavaras, A., & Masure, M. C. Aspects differents de l'ataxie optique selon la latérali-sation hémisphérique de la lésion. *Lyon Médical*, 1976, *236*, 673-683.

Tzavaras, A., Ozonas, G., & Chodkiewicz, J. P. Apraxie ideomotrice à prédominance unilatérale gauche avec ataxie optique lors d'une lésion pariétale gauche. In F. Michel & B. Schott (Eds.), *Les syndromes de disconnexion calleuse chez l'homme*. Lyon: Hopital Neurologique, 1974, pp. 265-285.

Vighetto, A. *Etude neuropsychologique et psychophysique de l'ataxie optique*. Un-published medical thesis, Universite Claude Bernard, Lyon, 1980.

Wyke, M. The effects of brain lesions on the performance of bilateral arm move-ments. *Neuropsychologia*, 1971, *9*, 33-42.

Chapter 18
Multimodal Structure
of the Extrapersonal Space

Otto-Joachim Grüsser

Directly perceived space can be divided into two major parts, *personal space* and *extrapersonal space*. Personal space contains the space of the self (ego space), which is experienced by the "inner senses" within the limits of body space. Ego space, however, is not identical with body space. One can perceive parts of one's own body as objects of the extrapersonal space, as one may analyze the shape and structure of the left index finger in a manner similar to any other object. The subjective localization of the ego remains within the limits of body space in an awake and attentive subject, but it is, within these limits, vague and ill-defined (Kant, 1796).

Compartments of Extrapersonal Space

Extrapersonal space, perceived by the "outer senses," may be subdivided phenomenologically into four different compartments (Figure 18-1; Grüsser, 1978): grasping space, near-distant action space, far-distant action space, and visual background.

Grasping Space

We call the immediate surround of our body "grasping space." Objects in grasping space have multimodal properties. They remain largely invariant when the intensity or the modality of signals is changed. The properties of the perceived objects are derived directly from different sensory signals which are, however, not perceived phenomenologically as sensory stimuli per se. A monomodal sensory stimulus without an object is, as a rule, a laboratory artifact. A cup of tea on my desk is not only invariant with respect to the perceptual modalities and their intensities, but is also experienced as the same object and is known to be in

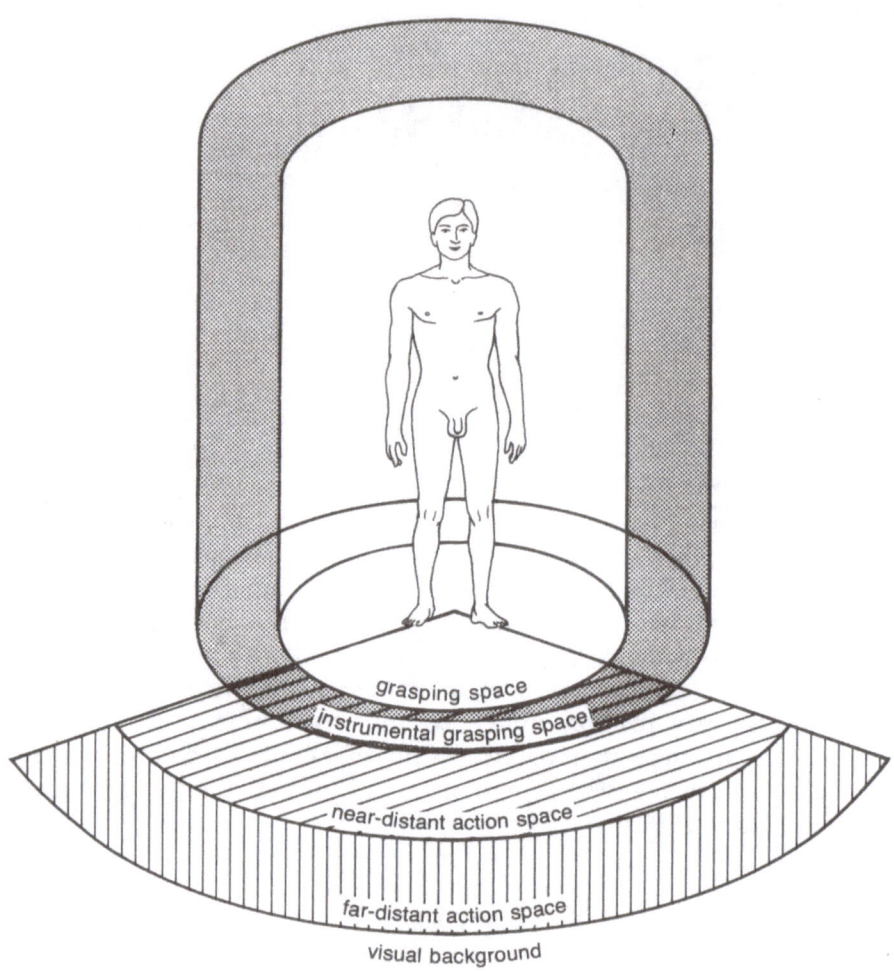

Fig. 18-1. Different compartments of the extrapersonal space. (From Grüsser, O. -J. Grundlagen der neuronalen Informationsverarbeitung in den Sinnesorganen und im Gehirn. In S. Schindler & W. K. Giloi (Eds.), *Informatik-Fachberichte* (Vol. 16). Berlin: Springer, 1978, pp. 234–273.)

its place even when it is totally outside of my sensory field. Thus short-term memory plays a constitutional role in object generation in extra-personal space.

Observations on the ontogenetic development of perception in grasping space lead one to a fourfold subdivision of this space compartment: the intraoral, perioral, manual, and general grasping space. The last includes objects contacted at any point on the body surface. During the first year of life oral grasping is dominant for object recognition, while after the first year hand and fingers become specialized for object recognition (Spitz, 1965). Coordinated oral and manual grasping in eating is present in all primates. It is therefore not surprising that neurons have been found in primate prefrontal cortical areas that have combined tactile and visual receptive fields in the hand, finger, and mouth region (Rizzolatti, Scandolara, Matelli, & Gentilucci, 1981a, 1981b).

The manual grasping range can be extended by instruments. It is important to note that, with skilled instrumental grasping, objects are perceived in their correct position in the extrapersonal space and not at the points where the instruments transmit the signals to the mechanoreceptors of the skin. In the course of the neuronal operations leading to the location of objects in the extrapersonal space, the sensory signals are interpreted as if the instruments were part of the body.

An object that is viewed within grasping space produces a certain expectation of its tactile structure. Obversely, during manipulation of an *invisible object* the tactile signals in combination with active hand movements lead to object recognition and simultaneously to its visualization. In a demonstration of this process, one of the members of our graphic department was blindfolded and asked to palpate bimanually an object unknown to him for a period of 20 seconds. The object was then removed and he was given 20 seconds to draw it under visual control (Figure 18-2a). Next he was permitted to see the object for 20 seconds while he made a second drawing of it (Figure 18-2b). Finally, he was blindfolded again and repeated the drawing (Figure 18-2c). It is evident that an artist is able to draw a realistic picture of an object that he has explored with his fingers, and even when blindfolded he can draw a previously seen object. It is important to note, as Gibson (1950) and Sechenov (1878) did, that active touch is necessary for object recognition. When an object is passively pressed into the hand of a blindfolded observer, he is not able to perceive its shape. Object recognition is also impaired when the object is grasped bimanually, but with crossed hands.

(a)   (b)   (c)

Fig. 18-2. Drawings of an object (a copy of an Etruscan vase) made by artist under different tactile and visual conditions. (a) Drawing made after blindfolded inspection by touching the vase for 20 seconds. (b) Drawing made while seeing the object for 20 seconds. (c) Drawing made while blindfolded for 20 seconds after having touched and seen the object. Note that the shape of the blindfolded drawing c resembles a more than b. In addition, drawing a is more similar to the object than drawing b.

Even for recognizing the texture of a surface, active movements are better than passive. An informal study conducted in our laboratory confirmed this. Blindfolded subjects were asked to judge the texture of different surfaces, and the speed at which the finger tips moved across the surface was measured. It turned out that the speed of movement across fine-grained surfaces was within the range at which the frequency of skin vibrations elicited was optimal with respect to the frequency response maximum of the phasic mechanoreceptors of Meissner corpuscles located in the skin below the rims of the finger tips (more than 60–80 Hz; Darian-Smith, Davidson, & Oke, 1980; Darian-Smith & Oke, 1980). Observations also indicated that active finger movement across the surface permits better texture discrimination than passive movement of the surface at a constant speed across the stationary fingers.

Near-Distant Action Space

Next to the grasping space is the near-distant action space, which may be perceived by visual, auditory, kinesthetic, vestibular, and sometimes olfactory signals. The extent of this space depends on body size and walking speed. Its radius can be estimated by a simple experiment: the subject is instructed to inspect carefully his immediate surroundings. Thereafter he is blindfolded and asked to walk in a certain direction. After 6–15 steps he suddenly becomes unsure about his position in space. Any sensory stimulus within the grasping range, such as a piece of furniture that is touched, immediately improves the spatial orientation and the subject can take some further steps with confidence. As anyone who walks while blindfolded across a large empty area can attest, the uncertainty signaling the limits of the near-distant action space is not dependent on the expectation of bumping into obstacles. Even when evidence is present that no objects are in the way, the uncertainty arises after a few steps forward. The distance at which the uncertainty appears when walking or running blindfolded can be extended to 20–25 m by training 10–15 minutes daily for 8 days. When we move ourselves forward, our nonvisual sense organs and motor memory give us rather poor information about space. For example, when blindfolded and turned either actively or passively, a subject after 5–10 steps will mislocate the direction of a previously viewed object.

Far-Distant Action Space

Beyond a distance of 6–8 m, the near-distant action space is gradually transformed into the far-distant action space. The latter is essentially a visually perceived space, the vertical direction of which is also determined by vestibular signals. Some auditory cues, usually with rather imprecise spatial location signs, contribute to the structure of this space.

Visual Background

The far-distant action space is limited at its outer border by the visual background. As one can see in the sky on a clear night, we do not perceive three-dimensional properties in this outermost limitation of the extrapersonal space. The borders between the far-distant space and visual background depend on the conditions of general visibility and on the speed of the observer.

It is evident that in the perception of objects in the different compartments of the extrapersonal space, the number of different modalities and types of receptors involved decreases from the grasping space to the visual background.

Coordinates of Extrapersonal Space

Subjective Vertical and Prism Adaptation

The vertical axis of extrapersonal space is determined mainly by visual and vestibular signals and also by joint and muscle receptors and tonic mechanoreceptors from the areas of the skin touching the ground. The signals from these receptors contribute to the perceived relationship between the personal and extrapersonal space, which changes from standing to sitting and from sitting to lying. The coordinate directions "forward–backward" and "left–right" depend normally on the egocentric relationship between personal and extrapersonal space. During locomotion, however, these coordinate directions might be related, at least in part, to exocentric landmarks and the sky. Beyond a distance of 6 m or so, binocular vision contributes little to depth perception; depth in the extrapersonal space is mainly perceived by means of the parallactic position changes of the objects relative to the moving observer.

When visual and vestibular signals are contradictory with respect to up and down, as is the case when the subject wears inverting prisms, a slow adaptation occurs. After several hours or days, objects in the extrapersonal space are again perceived in their correct position (Kohler, 1951). Active movements are required for adaptation (Held & Bossom, 1961; Held & Hein, 1958; for review see Gyr, Willey, & Henry, 1979). During the exposure period apparent motion of the extrapersonal space is associated with head and body movements. When left–right reversing prisms are worn, the subjects also experience apparent motion of the extrapersonal space and, initially, disturbances in left–right orientation in the extrapersonal space and of the body are reported (Stratton, 1897; O. Bock, personal communication, 1980). When subjects try to grasp an object appearing in the grasping space, they report feeling the left arm being connected with the right shoulder and the right arm originating from the left shoulder. This indicates that the perceived structures of the personal and the extrapersonal space are not independent of each other and that an interruption in the unity of the perception of body space and grasping space can disturb both percepts.

In the prism experiments one also observes disturbances in hand-mouth coordination. The same is true with an acute paresis of the external eye muscles. Recently I was a subject in experiments in which one eye was immobilized by means of a local anesthetic (5 cm^3 2% mepivacaine-hydrochloride, 0.1% *p*-hydroxybenzoic acid; 0.16 mg nor-epinephrine, 150 IU hyaluronidase) injected into the left orbit (Grüsser, Kulikowski, Pause, & Wollensak, 1981). During the recovery period after the complete immobilization of the left eye, I experienced tilting and shifting of the extrapersonal space for about 2 hours, due to paresis of the medial rectus and inferior oblique muscles of the left eye (the right eye was covered). This apparent tilting of the extrapersonal space was accompanied by disturbances in walking which were greatly reduced when I allowed myself to touch intermittently the wall I was walking along. The coordination between personal and extrapersonal grasping space was also affected: while eating with a fork, I quickly learned to find the plate but then missed my mouth, with the fork deviating toward the left side. These coordination errors disappeared, of course, whenever my eyes were closed. The hand–mouth coordination error decreased when I intentionally compensated for the erroneous hand–mouth coordination, but disappeared only after eight trials.

Nonlinearities in Space Perception

The geometrical structure of the grasping and near-distant action space is perceived as being approximately Euclidean, while the geometrical structure of the far-distant action space is essentially non-Euclidean. An object seen at a distance of more than 20–30 m in the vertical plane appears much smaller than the same object seen in the horizontal plane. This non-Euclidean deformation of space perception probably contributes to the well-known "moon illusion." Only part of this illusion can be attributed to the fact that the observer can see other objects between the horizon and himself in the horizontal plane, while in the vertical plane usually no objects are seen (Kaufman, 1974). The moon illusion disappears when one looks at the moon on the horizon upside down.

Circular Vection and Apparent Movement in Extrapersonal Space

Under normal circumstances the extrapersonal space is perceived as stationary, even when its image on the retinas is moved by *active* eye movements or by active or passive head and body movement. The extrapersonal space seems to move as soon as the vestibular excitation, especially that of the semicircular canal receptors, exceeds the threshold for vestibular nystagmus. The world is seen as rotating in the direction of the quick phases.

The perceived stability of extrapersonal space coordinates does not always include all compartments. When one walks through a forest on a clear night and looks up, the moon and the stars seem to move, while the trees appear stationary, but simultaneously in parallactic movement with respect to each other. The fact that all sensory stimuli that are perceived as having a constant spatial relation to the body are seen as moving as soon as we experience self-movement, however induced, also holds for other illusions such as linear or circular vection (Dichgans & Brandt, 1978; Mach, 1875). To produce the illusion a subject may sit with head stationary inside a striped vertical cylinder that is rotating at an angular speed V_s. If the subject fixates a stationary target he will, after a few seconds, experience *circular vection*, an illusion of rotation in the direction opposite to that of the cylinder (Figure 18-3a). The apparent speed, V_{cv}, of the illusion increases with time and eventually reaches V_s. At that moment the pattern rotating with the cylinder wall is seen as stationary. All visible stationary objects placed between the subject and the cylinder are perceived as rotating with the subject, and the same is true for stationary sources of olfactory, tactile, and auditory stimuli. The stationary visual background behind the rotating cylinder is also seen as rotating with the subject.

When the subject changes his gaze from one stationary object to another, the sequence of saccadic eye movements and fixation periods does not suppress the apparent circular vection. Smooth gaze or pursuit eye movements, however, interfere with circular vection. When the subject carefully fixates one of the moving stripes, thus producing optokinetic nystagmus (*Schaunystagmus*; Ter Braak, 1936), the circular vection and the apparent movement of stationary objects in extrapersonal space is either reduced or eliminated and the pattern rotating with the cylinder wall is again perceived to be moving. This suppression of circular vection is due in part to a reduction in the retinal stimulus angular velocity V_r (the difference between V_s and V_e, the velocity of the slow phase of optokinetic nystagmus). Circular vection is also suppressed when pursuit eye movements are elicited by other targets moving either at a constant speed or sinusoidally in front of the subject. Suppression is independent of target movement direction and speed relative to cylinder movement. Pursuit eye movements performed in the direction opposite or perpendicular to cylinder movement also suppress or reduce circular vection. These results indicate that active gaze commands control visual–vestibular interaction and may suppress circular vection.

Not only the semicircular canals but also the otoliths play an essen-

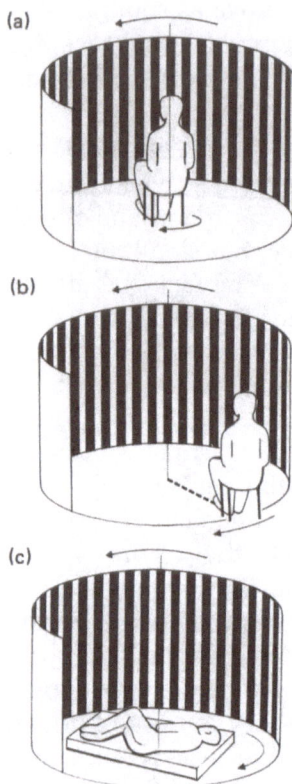

Fig. 18-3(a). The subject is sitting within a vertical cylinder. A stimulus pattern (periodic vertical black and white stripes or random dots) rotates around the subject along the cylinder wall at constant speed to the right. When the subject fixates a small target (e.g., a red light-emitting diode) at the cylinder wall, he feels after a few seconds to be rotated in the direction opposite to the visual pattern movement (circular vection). When full circular vection is reached, the moving visual pattern is seen as stationary. As soon as the subject pursues the moving pattern, the circular vection disappears or is greatly reduced. (b) The subject is sitting eccentrically with respect to the cylinder. The rotation axis of the visual stimulus and of the cylinder are identical. When the subject again fixates a small light-emitting diode in the cylinder wall after a few seconds he has the impression of being rotated around an axis, which corresponds to the rotation axis of the visual stimulus. (c) The same is true when the subject is horizontal or in any other position.

tial role in the perceived direction of extrapersonal space. Tilting the subject to the left, to the right, forward, or backward produces a slight shift in the perceived vertical (Bischof, 1974; Delage & Aubert, 1888). When a subject is placed eccentrically in the cylinder or in a tilting position with respect to its rotation axis, he perceives *pseudocentrifugal* sensations superimposed upon circular vection. The former sensations are normally aroused during real rotation by otolith receptor signals. The subject placed eccentrically in the cylinder experiences a rotation of his whole body in the direction opposite to that of the visual field, and around an axis that is approximately coincident with the axis of the visual stimulus rotation (Figure 18-3b and c).

Apparent Motion in Extrapersonal Space
Elicited by Gaze Pursuit Movements

Whenever the gaze axis is moved in a smooth pursuit fashion by eye, head, or body movements, visual stimuli that are stationary on the retina are perceived as moving in the direction of the eye, head, or body movements. The best example is the Aristoteles-Helmholtz motion illusion seen in a retinal afterimage during active gaze movements. Another is

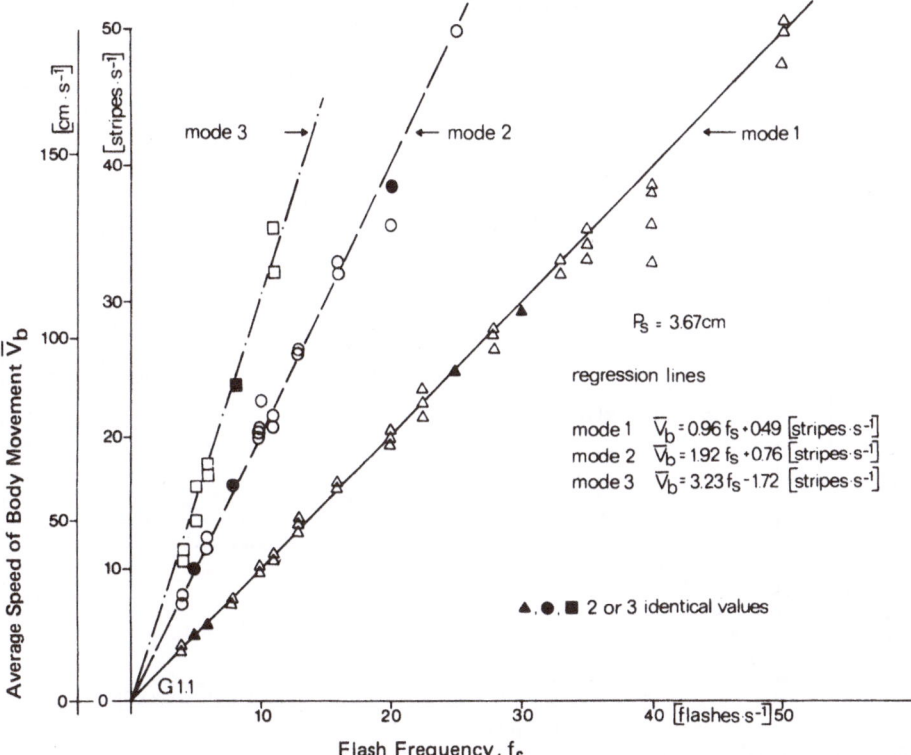

Fig. 18-4. Average speed of body movement during perception of sigma-movement. The subject walks along the inner wall of a cylinder 280 cm diameter and 220 cm high that is covered by vertical black and white stripes of equal widths and a period P_s of 3.67 cm. The stripe pattern is illuminated stroboscopically. If the center of gaze is moved along the wall at a speed corresponding to $(P_s \cdot f_s)$, $(2P_s \cdot f_s)$, or $(3P_s \cdot f_s)$ (called mode 1, 2, or 3), the cylinder is seen in rotation *in the direction of the gaze movement*. The symbols indicate the measurement of the subject's speed, V_b, walking or running along the wall to maintain this sigma-movement illusion. To obtain the single values the time T_{100} needed for the center of gaze to move along 100 periods was measured. The ordinate values are, $V_b = 100\,P_s/T_{100}$.

the apparent movement of a stroboscopically illuminated *spatially peri-odic* visual pattern across which pursuit eye movements are performed. The motion illusion appears when the gaze moves exactly from one stripe to the next during each interflash interval. Thus the gaze axis moves at an angular velocity,

$$V_g = P_s \cdot f_s \quad [\text{deg/sec}] \tag{18-1}$$

where P_s is the period of the pattern and f_s the flash frequency. A smooth, constant movement of the stationary periodic pattern in the direction of the pursuit eye movements is perceived under these conditions (*sigma-movement*; Adler, Collewijn, Curio, Grüsser, Pause, Schreiter, & Weiss, 1981; Adler & Grüsser, 1979; Behrens & Grüsser, 1978, 1979).

Sigma-movement is also present when the wall of a large cylinder

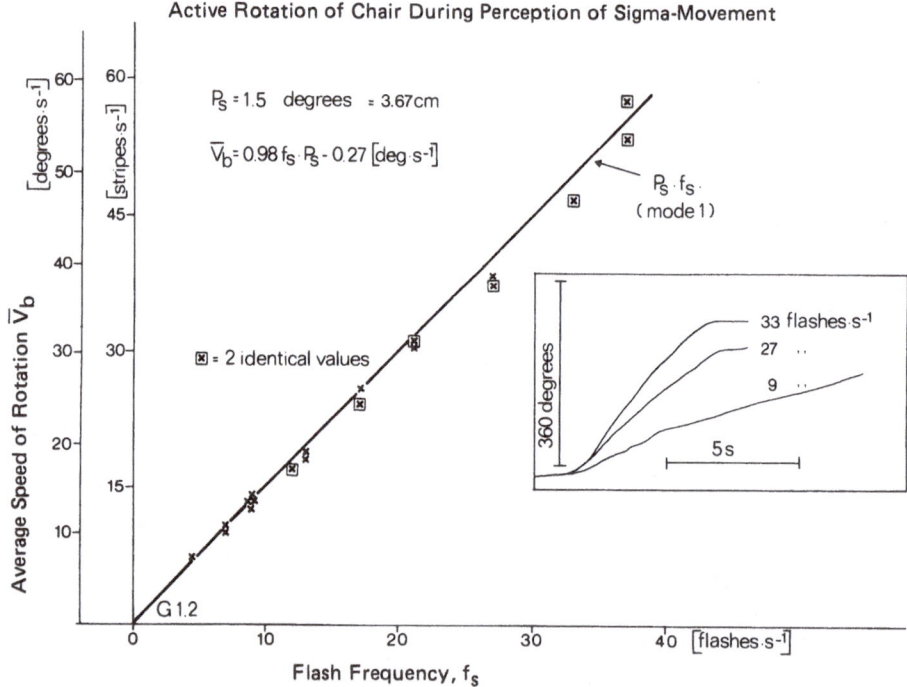

Fig. 18-5. Active rotation of chair during perception of sigma-movement. The subject is sitting in a rotatable chair in the center of the cylinder covered by the same vertical stripes as in Figure 18-4. $P_s = 1.5° = 3.67$ cm. The head of the subject is fixed with respect to the chair axis, and the subject, after initiating sigma-movement, is asked to rotate the chair by his feet to maintain the sigma-movement illusion. The data points represent the average speed of the chair rotation across a defined sector of 150°. The chair position was read out by an angiometer connected to the chair axis (*inset*). *Solid line*, Equation (18-1) in mode 1. The regression line did not deviate significantly from Equation (18-1).

covered by vertical periodic stripes is illuminated stroboscopically and the subject walks along the wall at an appropriate speed. His center of gaze must move during one interflash interval from one stripe to the next or the next but one, the next but two, etc. Experimental data on this sigma-movement perception elicited during active body movement are shown in Figure 18-4.

Apparent movement of a stroboscopically illuminated stripe pattern is also seen when the subject is sitting on a rotating chair inside a striped cylinder and moves the chair actively at an average angular speed corresponding to Equation (18-1). As long as he perceives himself rotating, he also perceives the stroboscopically lit stationary cylinder wall as moving in the same direction as his body movement. Data from such an experiment are shown in Figure 18-5.

Disturbances in the Perception of Objects and Space Following Parietal and Occipital Cortex Lesions

Lesions of the cerebral cortex in the region of the parietal lobe and in the transitional areas between parietal and occipital cortex may induce disturbances in object perception, space perception, and spatial orientation. Similar symptoms appear when the connections between the thalamus, in particular the pulvinar, and the parietal lobe are interrupted. Neurologists have described different disturbances in object and space perception related to the location of the cerebral lesion. Vision that dominates the perception of extrapersonal space has been the principal modality investigated (Hécaen & Albert, 1978), and the following disturbances have been studied: visual object agnosia; visual movement agnosia; visual or general hemineglect (hemi-inattention); disturbances in spatial and depth perception, including stereovision in the near-distant and/or far-distant action space; disturbances in the spatial structure of the grasping space and corresponding errors in visually guided hand movements; simultaneous agnosia; metamorphopsia of objects in the extrapersonal space; and gaze disturbances. The subdivision of extrapersonal space described earlier is useful for the understanding of these disturbances.

Careful clinical examinations of patients with parietal lobe lesions indicate that disturbances in the perception of the grasping and the near-distant action space are frequently associated with disturbances in the perception of body space. To illustrate this I first describe some symptoms in a patient whom I have seen repeatedly over 2 years and who suffered an acute cerebral stroke without losing consciousness. This acute condition was preceded by repeated periods of dizziness, gait disturbances, and spatial disorientation.

Patient A

I first saw this patient 10 days after the stroke (ischemia of cerebral re-
tions supplied by the right medial cerebral artery). This left-handed, highly
educated man suffered from a left-sided hemiplegia, hemihypesthesia, a
right Horner syndrome, and a left hemianopsia with a rather large macular
sparing. I could not detect any signs of aphasia. During the first 4 weeks
the patient admitted that he could not walk, but was otherwise anosog-
nostic for his left hemiplegia. He noted as somewhat strange that his left
hand always seemed to move when he moved his right. In addition, he felt
that his left side was "far away from the self," while the right side was "in-
tensively near to the self." A slight tonic head and eye deviation toward
the right side was also present. He could not move his eyes actively to the
left by means of saccades beyond the midline. Pursuit eye movements to
the left were also impaired beyond the midline position, but eye movements during passive head movements were normal. The patient also suffered a fairly strong left-sided visual and auditory spatial neglect. When I
stood to the left side of his bed, he did not answer my questions. As I
moved to the right side, communication with the patient became normal.
When reading aloud from a newspaper, he disregarded the left half of each
column and added intelligent confabulations to the correct text, thus producing meaningful sentences. Excerpts from short autobiographical notes
that I asked him to write are presented in Figure 18-6. His writing was always confined to the right side of the sheet of paper even when I moved
his right hand to the left (Figure 18-6, *asterisk*). It was not possible to
direct his attention to the left side of the page, even when the whole page
was placed on the right side in front of him.

Naming and recognition of objects in all compartments of the right
extrapersonal space was correct, but perception of the position of objects
relative to each other was impaired. He had difficulty pointing with the
right hand to an object in the grasping space. In addition, he suffered from
severe constructive apraxia. Drawings of a tree and a face attempted by the
patient 10 days after the stroke are shown in Figure 18-7a. Copying of a
simple schematic face was also inadequate. One week later some reduction
in the constructive apraxia was evident (Figure 18-7b). The constructive
apraxia was less severe 26 days after the stroke, but left-sided visual neglect was still present (Figure 18-7c). A further improvement 1 month
later is shown in Figure 18-7d, but signs of constructive apraxia and partial
left-sided neglect are still visible. It is evident from Figure 18-7 that the
left-sided neglect is not only perceptual but also *conceptual* and involves
the constructive concept.

Objects placed in the grasping space have not only a spatial relation
to each other, but also a functional relationship requiring that they be
used in a particular motor sequence. When the mapping between spatial and functional relations of objects is impaired, a new disturbance
emerges called *ideational apraxia*, or, perhaps better, *apraxia of sequential, goal-directed actions*. I will illustrate this by the following case
description.

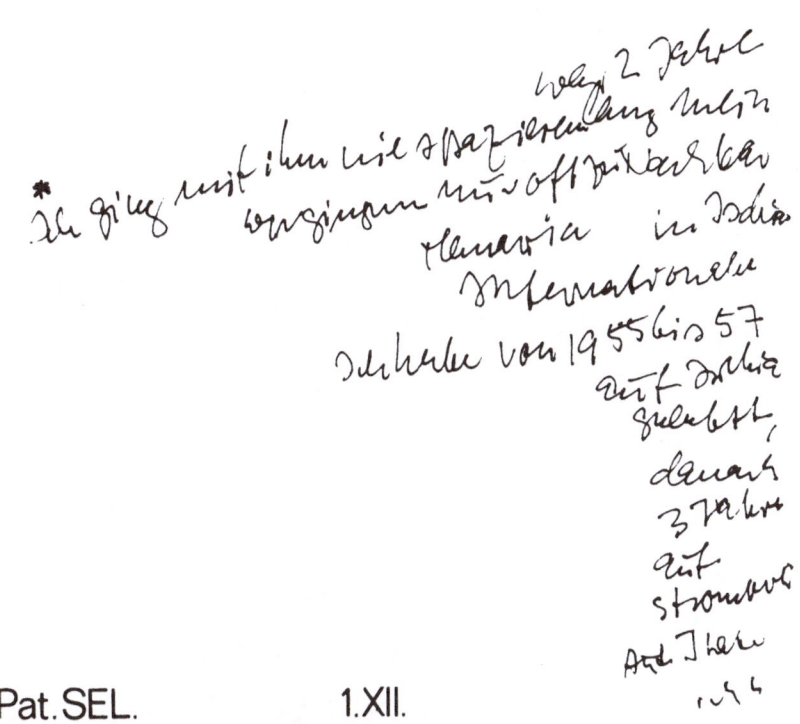

Pat. SEL.   1. XII.

Fig. 18-6. Part of a short autobiography of patient A. He wrote only on the right side of the page (borders of the page marked by arrows). Before the patient wrote the line marked with the asterisk, I moved his hand towards the left border of the page and told him to write across the whole page. From the next line the patient continued to write only on the right side of the page. This neglect of the left half of the page persisted when the whole page was moved towards the right side of the body.

Patient B

A 53-year-old, right-handed woman suffered an acute cerebral stroke that resulted in an extended left parieto-temporal lesion. While at the hospital she showed transient spatial and temporal disorientation, fluent sensory aphasia, and a mild paresis of the right arm and hand. After leaving the hospital she was unable to use the appliances in her kitchen and could not find her way from her home to different stores. She had difficulty getting dressed (dressing apraxia). During the first months of recovery she became aware that she could no longer use numbers correctly, confused left and right, and could not tell time. Difficulties in using simple tools such as a hammer or pliers became apparent; however, eating with knife, fork, and spoon was not impaired. Nearly 10 months after the stroke she reported that she still had great difficulty in preparing a meal because she mixed up the sequence of actions for preparing food.

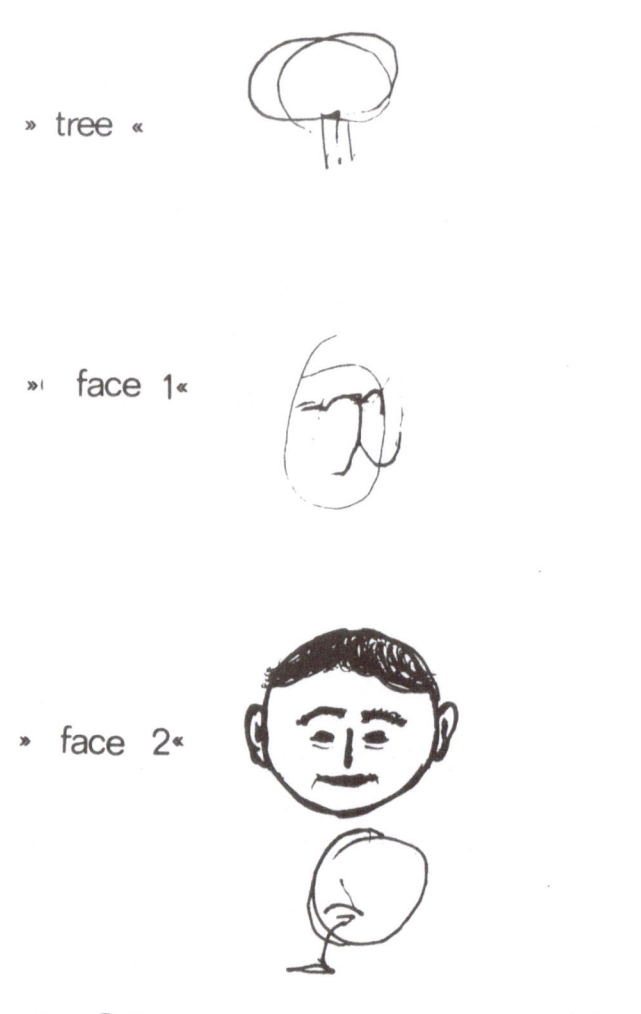

» tree «

»ı face 1«

» face 2«

(a)   Pat. SEL.   14. XI.

Fig. 18-7. Recovery of patient A from severe constructive apraxia. Drawings made 10 days (a), 17 days (b), 26 days (c), and 57 days (d) after the stroke. Except for "face 2" in (a), where the patient was asked to copy the round face, the patient was asked to draw a tree, face, automobile, or bicycle from memory. Note the successive improvement in the drawings. In (b) the automobile is drawn upside down, but the patient did not note this. The left-sided neglect is also present in drawings (c) and (d). The patient was technically highly trained and could describe verbally all technical details of an automobile.

Evidently, nonautomated handling of objects and instruments in the grasping space was severely disturbed, as was the correlation between spatial order of objects visible in the grasping space and the motor acts. She had, however, no trouble with the visuomotor coordination necessary for

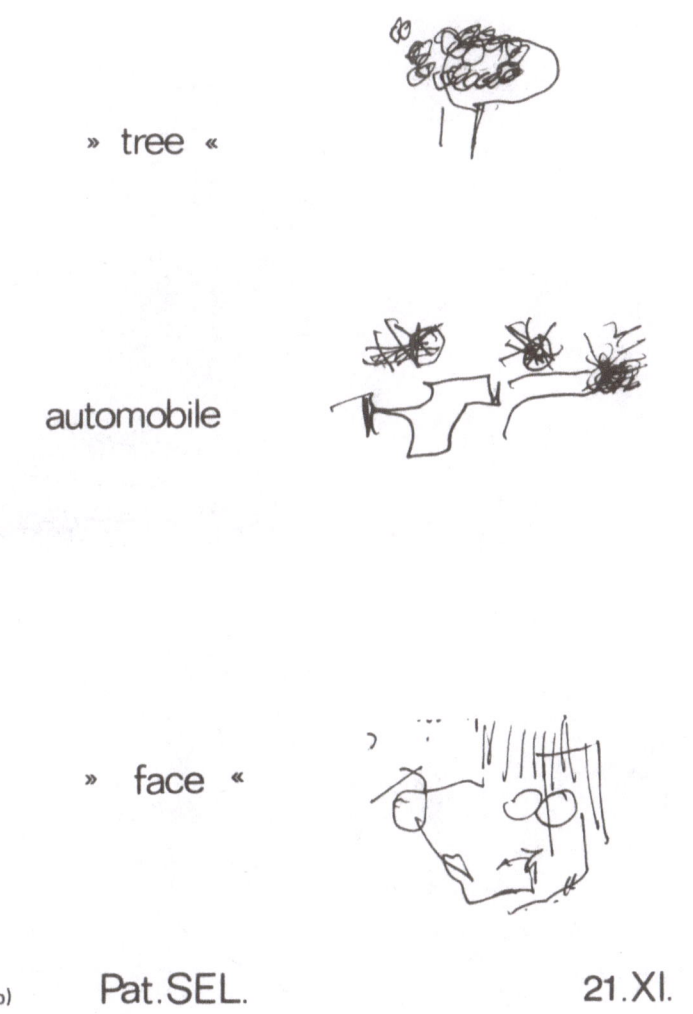

» tree «

automobile

» face «

(b)   Pat.SEL.   21.XI.

Fig. 18-7 (continued)

pointing. In this patient, object recognition and perception of spatial order in the grasping space was not affected, while the mapping of the perceived spatial structure in the near-distant and far-distant action space onto a spatial memory was considerably impaired. While these space perceptual and conceptual disorders were still present 10 months after the stroke, she had recovered considerably from her sensory aphasia.

Simultaneous agnosia and visual movement agnosia are other symptoms indicating impaired perception of the spatial relationship between objects in extrapersonal space, as demonstrated in the following case.

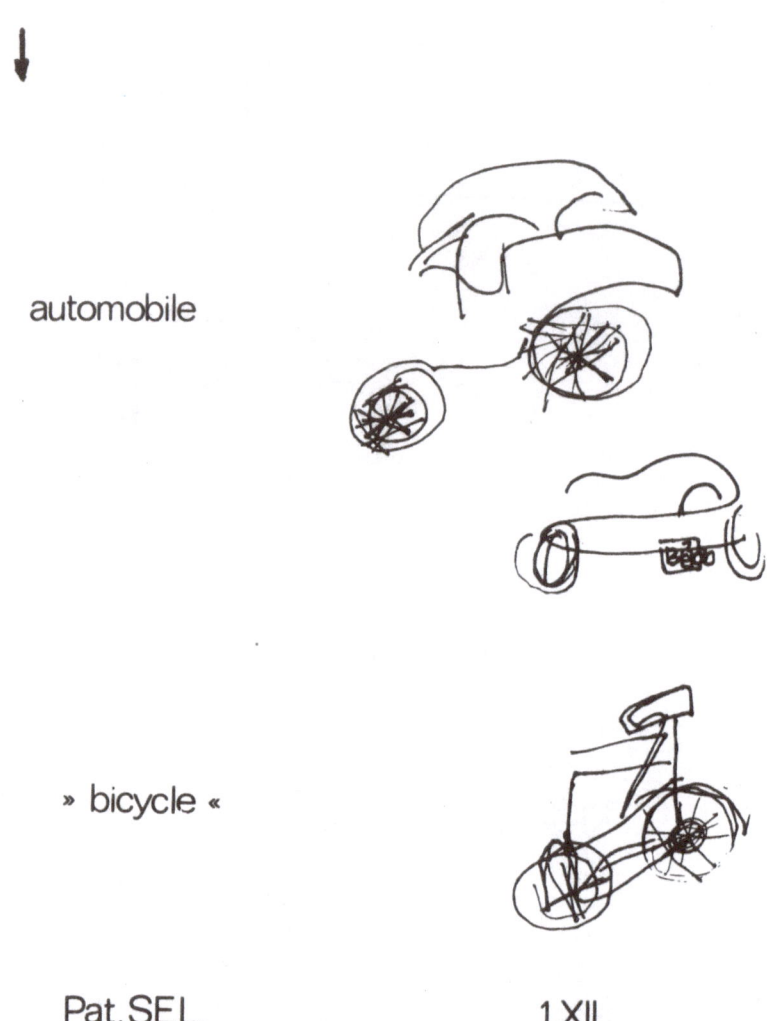

automobile

» bicycle «

(c)   Pat. SEL.   1. XII.

Fig. 18-7 (continued)

Patient C

I examined a patient[1] who had suffered a bilateral lesion in the anterior part of the occipital lobes and the parieto-occipital junction. This right-handed, 42-year-old woman had suffered a bilateral cerebral hemorrhage about 1 year before the examination. She had severe disturbances in the perception of the near-distant and far-distant action space leading to a gross underestimation of object distances and depth perception. After 1

---

[1] Courtesy of Dr. D. von Cramon, Max-Planck-Institut für Psychiatrie, Munich, Germany.

» face «

» bicycle «

automobile

Fig. 18-7 (continued)

year she still had great difficulty in going up or down stairs. In addition, she was not able to pay attention to more than one object at a time in her extrapersonal space. Pointing to targets placed in her grasping range and object recognition (visual or manual) was normal.

When walking on a horizontal plane she experienced apparent motion of stationary objects in the near-distant action space. In addition, she did not perceive movement of objects relative to a complex visual background or to other moving objects. This led to serious problems in finding her way through city traffic. She reported that a car would appear in one part of her visual field, then suddenly in another without the perception of movement or movement direction. However, she could experience the movement of a single dot of light in the grasping or near-distant space, and the same was true for perception of tactile movements across her body surface. When she maintained contact with the investigator's moving hand with her own she was able to perceive its movement and could then easily visually recognize its movement, speed, and direction. When several objects were moved together in her extrapersonal space she reported being "disturbed by the many objects," and the same was true when she looked at a moving two-dimensional random dot pattern. These observations support the idea that the operations necessary for the perception of the far-distant and near-distant action space can be impaired by extrastriate occipital, or parieto-occipital lesions. Impairment of perception in the grasping compartment of the extrapersonal space, however, appears when the central parietal structures are damaged.

Multimodal Control of Eye Movements

From the observations mentioned in the preceding sections it becomes evident that the perception of objects in, and the structure of, extrapersonal space is a task for which multimodal neuronal integration is important, but the visual modality dominates the function in the space beyond the grasping range. The multimodal integration necessary for normal spatial orientation in the grasping range probably develops during a sensitive period in early life. Hyvärinen, Hyvärinen, Farkkila, Carlson, Leinonen (1978) demonstrated that binocular lid closure during the first 6 months in the monkey does not seriously impair the function of the visual cortex or afferent visual pathways, but prevents the visual information from integration with other sensory modalities in area 7. After the eyes are reopened, the monkeys continue to behave as if functionally blind and to rely predominantly on tactile cues for spatial orientation and object recognition.

Evidence from neurophysiological studies in the monkey (see review by Lynch, 1980) indicates that area 7 is a rather complex multimodal region in which large neuronal networks integrate sensory signals for the recognition of objects and for the control of object-directed movement of the eye, head, and hand. In Figure 18-8 the responses of one type of area 7 neuron are shown in which multimodal (visual and tactile) sensory input and goal-directed eye and hand movements modify neuronal activity. A considerable number of the neurons found in this cortical region are neither sensory nor motor but seem to be essentially related to attention mechanisms. Under normal conditions, in directing attention toward a certain object in the extrapersonal space, one gazes toward it. When the object moves, pursuit gaze movements follow it. It is possible, however, for an observer to separate gaze movement and attention. A monkey can also be trained to stabilize his gaze while shift-

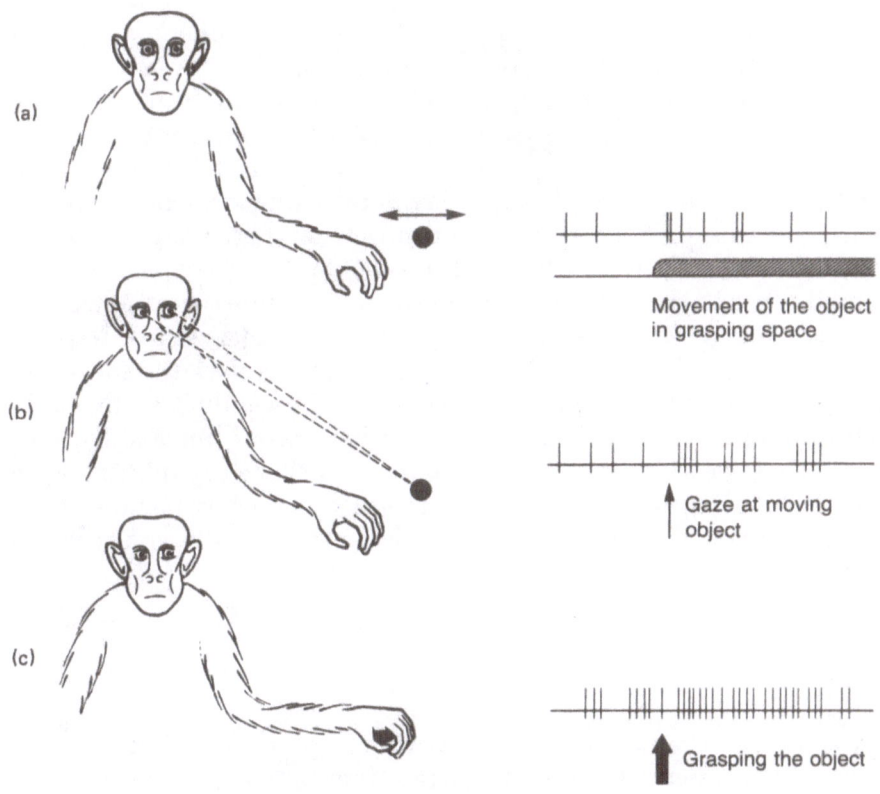

Movement of the object
in grasping space

Gaze at moving
object

Grasping the object

Fig. 18-8. Responses of a neuron in area 7 of the awake rhesus monkey. (a) An object is moved in the grasping space near the hand. (b) Back and forth movement of the object is continued, but now the monkey gazes at the moving object (*arrow*). This leads to an increase in the neuronal activity. (c) The monkey grasps the object. During grasping the neuronal activity increases and is maximal at the moment the monkey touches the object with the hand (*arrow*). (After unpublished observations of Büttner, Grüsser, & Henn, 1975; from Grüsser, O. -J. Grundlagen der neuronalen Informationsverarbeitung in den Sinnesorganen und im Gehirn. In S. Schinder & W. K. Giloi (Eds.), *Informatik-Fachberichte* (Vol. 16). Berlin: Springer, 1978, pp. 234–273.)

ing his attention. In recording from area 7 neurons in such monkeys, Goldberg and Robinson (1977, 1980) demonstrated that a considerable number of the neurons are involved in the shift of attention from one part of the extrapersonal space to another (see also Lynch, 1980). Normally, however, eye movements are a rather sensitive indicator of which part of the extrapersonal space our attention is directed toward.[2] Even

---

[2] Results of systematic experiments on "Spatially selective visual attention and generation of eye pursuit movements" are published by Collewijn, H., Curio, O., and Grüsser, O. -J., *Human Neurobiology*, 1982, *1*, 129–139.

congenitally blind babies direct their eyes toward an auditory stimulus when, for example, they are responding to the mother's voice (Eibl-Eibesfeldt, 1967; see also Chapter 12, this volume).

Eye and gaze movements can be elicited by a variety of stimuli. Because they represent, in an attentive subject, a measurable response toward an object in extrapersonal space, we became interested in the question of how the three large classes of oculomotor behavior (fixations, saccades, and smooth pursuit movements) are triggered by nonvisual stimuli. When the head of the subject is fixed, all nonvisual stimuli activate gaze control mechanisms under "open-loop" conditions, that is, the eye movements performed do not change the respective sensory input. This is not the case for visually guided eye movements because the retinal position of the stimulus changes during a saccade. A "closed-loop" sensory signal is also available from pursuit eye movements. This provides information on the retinal angular velocity V_r of a stimulus moving in the extrapersonal space, which is pursued by the center of gaze. V_r is the difference between the stimulus and gaze angular velocity.

Fixation

An auditory stimulus in the extrapersonal space, an object grasped in darkness, or a part of the body surface that is touched, locally heated, cooled, or otherwise stimulated can be fixated by the eyes in the absence of visual input. During the first 300 msec gaze is as steady as when fixating a small visual stimulus. Afterward, fixation stability declines.

Saccadic Eye Movements

Eye saccades to an auditory stimulus can be performed with a latency only 20–30 msec longer than that of saccades to a visual target. We found that, up to a stimulus distance of $60°$, the auditory saccades are always larger than the angular distance of the auditory signals, but the direction errors are small. The amplitude error is larger when the two sound sources are placed in a vertical, as compared to a horizontal, orientation (Figure 18-9). When the subject (head fixed) is asked to point in darkness toward the sound, he usually points halfway between its viridical locus and the point in extrapersonal space where he is looking. When the subject is asked to move his head toward the sound source, the center of gaze is directed accurately to the target. Under those conditions, however, the subject is acting in a closed-loop condition, using binaural information for adjusting his head.

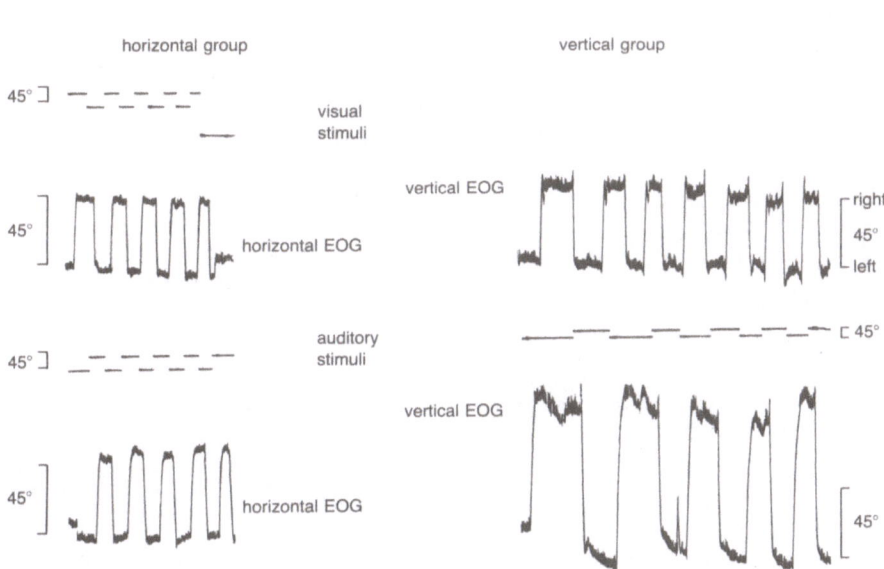

Fig. 18-9. Saccadic eye movements and fixation periods elicited by auditory stimuli from two small loudspeakers located at a distance of 140 cm from the subject, or by successively lit small light spots placed below the loudspeakers. The loudspeakers were separated by 45°. The amplitude of the "auditory" saccades was always larger than the loudspeaker distance. *EOG*: Electrooculogram.

Saccades aroused by two "tactile" stimuli (weak electrical pulses applied to the two index fingers) are directly related to, although larger than, the angular distance and position of the two index fingers with respect to the center of gaze (Figure 18-10). When the finger position is changed either by moving the hand sideways, or bringing the hands to a different distance from the face, the size of the saccades changes. When the hands are crossed, this information is correctly transferred to the gaze control system, but saccade latency is increased. When the sequence of the electrical stimuli applied to the index fingers is higher than .8/sec, the subject trying to fixate the stimulated finger tip as fast as possible produces direction errors. When the sequence is above 1.5/sec, the subject looks predominantly to the left side when the left index finger is stimulated, and to the right side when the right index finger is stimulated, despite the fact that the hands are crossed in front of the body. This observation indicates that the tactile control of eye movements needs at least 100–200 msec additional processing time when the position of the hand, arm, or finger is changed to the contralateral side of the grasping space (Figure 18-10).

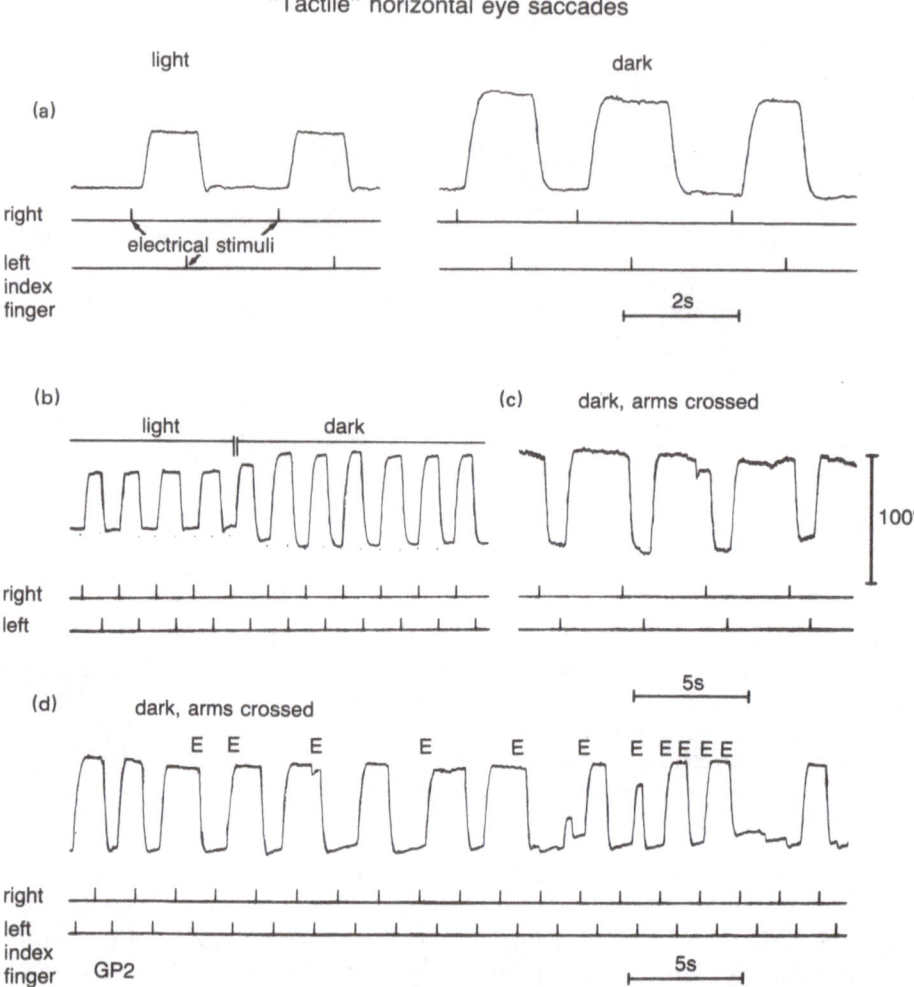

"Tactile" horizontal eye saccades

Fig. 18-10. Eye saccades elicited in a subject asked to gaze in darkness at his left or right index finger tip as fast as possible after he perceived a slight touch at the finger tip (aroused by near threshold, short electrical pulses applied to the index finger tip); head unmoved. (a) The fingers were placed 40 cm apart at eye level, 50 cm from the head. Visually controlled horizontal saccades elicited by tactile stimulation during light and during darkness were recorded. (b) The fingers were brought in darkness to 50 cm from the head (horizontal separation of the index fingers 40 cm). Tactile saccades during light and darkness were recorded. (c) Continuation of recording as in (b), hands now held crossed in darkness; distance between the index fingers 34 cm; distance from the head 37 cm. Tactile saccades were recorded in response to alternating electrical stimulation of the index finger tips at about .6/sec in darkness, variable delay. (d) Increase of alternation frequency to about 1.1/sec elicited many "wrong" responses (E).

Pursuit Eye Movements

Up to a stimulus velocity of 40°–60°/sec the eyes follow a small visual target moving at constant speed in extrapersonal space with an angular velocity corresponding approximately to that of the target. Small correction saccades (less than 1°) interrupt the smooth pursuit movements at intervals of 0.3 to about 3 seconds, depending on the training of the subject. Larger tracking saccades are added to smooth pursuit when angular speed is increased. At an angular speed above 80°–120°/sec tracking of the target is dominated by a sequence of saccades. Nonvisual moving stimuli also arouse tracking eye movements. When an auditory target moves at angular velocities below 80°/sec within the near-distant space, a mixture of smooth pursuit and saccades can be observed. Saccadic tracking increases, while smooth pursuit decreases with the speed of the auditory stimulus (Figure 18-11; Gottschalk, Grüsser, & Lindau, 1978). While the average angular velocity of the eyes (combined smooth pursuit and saccadic components) matches the angular speed of the auditory target, gaze position errors are observed. These increase with the speed of the target.

Tactile stimulation in the dark also permits continuous tracking of a moving target with the eyes when either a stimulus moves along the skin or a moving object is followed with the hand and arm. When a stimulus moves across the fingers or the hands, pursuit eye movements depend on stimulus movement in the grasping space and not on the position of the hands (crossed or uncrossed) relative to the subject's body. Somewhat at variance with the "arthrokinetic nystagmus" of

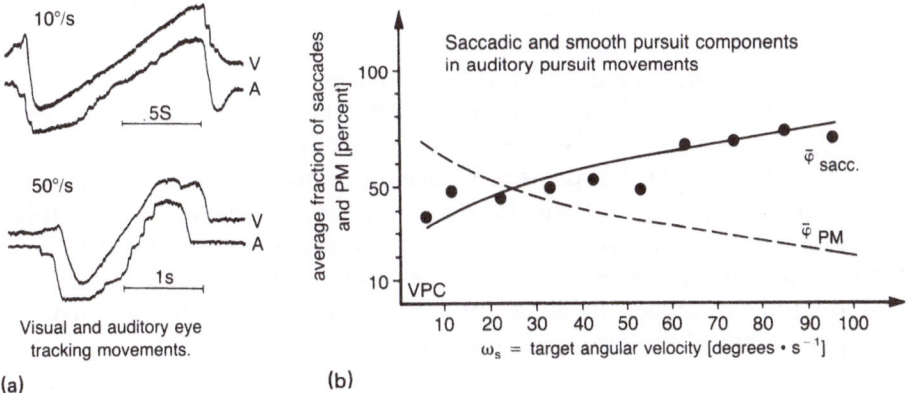

(a) Visual and auditory eye tracking movements.

(b)

Fig. 18-11(a). Recording examples of auditory tracking (A) in darkness and visual tracking (V) of a stimulus moving at constant speed around the subject; two different angular velocities (10°/sec and 50°/sec). (b) The percentage of saccadic tracking increases that of smooth pursuit (SP) tracking decreases when the speed of the auditory target (ω_s) is increased (From Gottschalk, Ch., Grüsser, O. -J., & Lindau, M. Tracking movement of the eyes elicited by auditory stimuli at a constant angular velocity. *Pflügers Archiv*, 1978, *377*, 46.)

Brandt, Büchele, and Arnold (1978), we found in our subjects that eye
tracking aroused by tactile or arthrokinetic stimulation is dominated by
saccadic eye movements with only rare smooth pursuit phases. Tracking
by a sequence of saccades is also possible when thermoreceptors of the
skin are stimulated by an infrared source moving above the arm. The
subject then perceives a small, moving, warm–cold contrast field and
can track this stimulus moving across his skin with the eyes in total
darkness.

In summary, one can demonstrate that eye movements aroused in an
attentive subject by nonvisual sensory stimuli depend on the stimulus
parameters, especially the perceived speed of the stimulus in the extrapersonal space or along the skin surface. In my opinion, measurement
of such nonvisually elicited eye movements could be a useful tool for
understanding the deficiencies of patients suffering from parietal lobe
lesions.

Conclusions

A pragmatic division of extrapersonal space into different compartments and subcompartments has been proposed. These divisions depend
on the contribution of the different sensory and motor modalities to
object recognition and space perception. The neuronal operations that
lead to the recognition, identification, and localization of objects in the
different compartments of extrapersonal space are multimodal. One can
presume from clinical and experimental data that neuronal mechanisms
responsible for the perceptual and motor operations related to the different compartments are dominated by different brain structures: the
oral grasping space by prefrontal neuronal networks; the manual grasping space by "central" parietal mechanisms that also form a link for the
perception between body space and extrapersonal space; the near-
distant action space by posterior parietal regions dominated by visual
and auditory input; and the far-distant action space by the occipital
(peristriate) cortex. Recording of eye movements aroused by visual and
nonvisual stimuli is a useful tool for examining cortical functions that
are related to object localization and identification in the different
compartments of the extrapersonal space.

Acknowledgments

This work was supported by Grants Gr 161 from the Deutsche Forschungsgemeinschaft. I am grateful to Dr. M. Pause for his help in part of the eye movement recordings, Mrs. J. Dames for technical assistance and help in the English translation,
Mrs. U. Saykam for typing the manuscript, Mr. M. Winzer for the drawings, and
Mr. P. Holzner for the photographs. I am also grateful to Professor M. Jeannerod
and Professor A. Hein for their many helpful suggestions regarding the manuscript.

References

Adler, B., Collewijn, H., Curio, G., Grüsser, O. -J., Pause, M., Schrieter, U., & Weiss, L. Sigma-movement and sigma-nystagmus: A new tool to investigate the gaze pursuit system and visual movement perception in man and monkey. *Annals of the New York Academy of Sciences*, 1981, *374*, 284–302.

Adler, B., & Grüsser, O. -J. Apparent movement and appearance of periodic stripes during eye movements across a stroboscopically illuminated random dot pattern. *Experimental Brain Research*, 1979, *37*, 537–550.

Behrens, F., & Grüsser, O. -J. Bewegungswahrnehmung und Augenbewegungen bei Flickerbelichtung unbewegter visueller Muster. In G. Kommerall (Ed.), *Augenbewegungsstörungen, Neurophysiologie und Klinik*. Munich: Bergmann, 1978, pp. 273–283.

Behrens, F., & Grüsser, O. -J. Smooth pursuit eye movements and optokinetic nystagmus elicited by intermittently illuminated stationary patterns. *Experimental Brain Research*, 1979, *37*, 317–336.

Bischof, N. Optic-vestibular orientation to the vertical. In H. H. Kornhuber (Ed.), *Handbook of sensory physiology*, Part 2: *Vestibular system*. (Vol. VI/2). Berlin: Springer, 1974, pp. 155–190.

Brandt, T., Büchele, W., & Arnold, F. Arthrokinetic nystagmus and ego-motion sensation. *Experimental Brain Research*, 1977, *30*, 331–338.

Darian-Smith, I., Davidson, I., & Johnson, K. O. Peripheral neural representation of spatial dimension of a textured surface moving across the monkey's finger pad. *Journal of Physiology* (London), 1980, *309*, 135–146.

Darian-Smith, I., & Oke, L. E. Peripheral neural representation of the spatial frequency of a grating moving across the monkey's finger pad. *Journal of Physiology* (London), 1980, *309*, 117–133.

Delage, Y., & Aubert, H. *Physiologische Studien über die Orientierung*. Tübingen: Laupp, 1888.

Dichgans, J., & Brandt, T. Visual-vestibular interaction: Effects on self-motion perception and postural control. In R. Held, H. W. Leibowitz, & H. -L. Teuber (Eds.), *Handbook of sensory physiology* Part 0: *Perception* (Vol. VIII). Berlin: Springer, 1978, pp. 755–804.

Eibl-Eibesfeldt, I. *Grundriss der vergleichenden Verhaltensforschung. Ethologie*. Munich: Piper, 1967.

Gibson, J. J. *The perception of the visual world*. Boston: Houghton Mifflin, 1950.

Goldberg, M. E., & Robinson, D. L. Visual responses of neurons in inferior parietal lobule: The physiological substrate of attention and neglect. *Neurology*, 1977, *27*, 350–362.

Goldberg, M. E., & Robinson, D. L. The significance of enhanced visual responses in posterior parietal cortex. *Behavior and Brain Science*, 1980, *3*, 503–505.

Gottschalk, Ch., Grüsser, O. -J., & Lindau, M. Tracking movement of the eyes elicited by auditory stimuli at a constant angular velocity. *Pflügers Archiv*, 1978, *377*, 46.

Grüsser, O. -J. Grundlagen der neuronalen Informationsverarbeitung in den Sinnesorganen und im Gehirn. In S. Schindler & W. K. Giloi (Eds.), *Informatik-Fachberichte* (Vol. 16). Berlin: Springer, 1978, pp. 234–273.

Grüsser, O. -J., Kulikowski, J., Pause, M., & Wollensak, J. Optokinetic nystagmus, sigma-optokinetic nystagmus and eye pursuit movements elicited by stimulation of an immobilized eye. *Journal of Physiology* (London), 1981, *320*, 21–22.

Gyr, J., Willey, R., & Henry, A. Motor-sensory feedback and geometry of visual space: An attempted replication. *Behavior and Brain Science*, 1979, *2*, 59–94.

Hécaen, H., & Albert, M. L. *Human neuropsychology*. New York: Wiley, 1978.

Held, R., & Bossom, J. Neonatal deprivation and adult rearrangement: Complementary techniques for analyzing plastic sensory-motor coordinations. *Journal of Comparative Physiological Psychology*, 1961, *54*, 33–37.

Held, R., & Hein, A. Adaptation of disarranged hand–eye coordination contingent upon reafferent stimulation. *Perceptual and Motor Skills*, 1958, *8*, 87–90.

Hyvärinen, J., Hyvärinen, I., Farkkila, M., Carlson, S., & Leinonen, L. Modification of visual functions of the parietal lobe at early age in the monkey. *Medical Biology*, 1978, *56*, 103–109.

Kant, I. Postscript. In S. Th. Soemmering, *Über das Organ der Seele*. Königsberg, 1796, pp. 81–86.

Kaufman, L. *Sight and mind. An introduction to visual perception*. New York: Oxford University Press, 1974.

Kohler, I. Über Aufbau und Wandlung der Wahrnehmungswelt, insbesondere über bedingte Empfindungen. Österreichische/Akademie der Wissenchaften, Philosophische-Historische Klasse, Band 227, 1951, pp. 1–118.

Lynch, J. C. The functional organization of posterior parietal association cortex. *Behavior and Brain Science*, 1980, *3*, 485–534.

Mach, E. *Grundlinien der Lehre von den Bewegungsempfindungen*. Leipzig: Engelmann, 1875.

Rizzolatti, G., Scandolara, C., Matelli, M., & Gentilucci, M. Afferent properties of periarcuate neurons in macaque monkeys. I. Somatosensory responses. *Behavior and Brain Research*, 1981, *2*, 125–146. (a)

Rizzolatti, G., Scandolara, C., Matelli, M., & Gentilucci, M. Afferent properties of periarcuate neurons in macaque monkeys. II. Visual responses. *Behavior and Brain Research*, 1981, *2*, 147–163. (b)

Sechenow, I. [The elements of thought.] Translated in *The lectured works*. Amsterdam: Bonset, 1968 pp. 403–489. (Originally published 1879.)

Spitz, R. *The first year of life. A psychoanalytic study of normal and deviant development of object relations*. New York International University Press, 1965.

Ter Braak, J. W. G. Untersuchungen über optokinetischen Nystagmus. *Arch. Néerl. Physiol.*, 1936, *21*, 309–376.

Author Index

Abel, L. A.   69, 85
Abraham, W.   205, 212
Abramov, I.   169, 172
Abzug, C.   31, 105, 118
Acuna, C.   315, 325
Adler, B.   336, 351
Adler, D.   289, 300
Adler, F. M.   6, 11
Akert, K.   64, 83
Albano, J. E.   64–66, 85
Albert, M. L.   337, 352
Alexeiev, M. A.   17, 31
Allen, G. I.   27, 31
Allman, J. M.   286, 301
Anderson, R. A.   80, 81
Andreyev, A. E.   19, 32
Apter, J. T.   38, 58
Arango, V.   49, 60
Archibald, Y.   316, 325
Arnold, A.   31, 33
Arnold, F.   351
Aronson, E.   175, 195
Asanuma, H.   31, 33
Aslin, R. N.   158, 171, 172, 189, 194
Atkin, A.   155, 173
Atkinson, J.   155, 157, 162, 172
Attneave, F.   231, 241
Aubert, H.   334, 351
Augerinos, G.   52, 60

Babinski, J.   16–18, 31
Bahill, A. T.   267, 277, 289

Baker, K. E.   229, 241
Baker, R. R.   90, 101
Baleydier, C.   315, 324
Balint, R.   324
Ball, W.   198, 211
Baloh, R. W.   169, 172
Banks, M. S.   171, 172
Barlow, H. B.   223, 241
Barnes, W. T.   64, 81
Bartlett, N. R.   290, 302
Batini, C.   128, 132
Battaglini, P. P.   100, 101
Bartz, A. E.   290, 300
Bauer, J. H.   43, 58
Becker, W.   284, 290, 292, 298, 299,
   300
Beeler, G.   265, 277
Behrend, K.   135, 152, 156, 173
Behrens, F.   336, 351
Belenkii, V. E.   17, 18, 32
Bender, M. B.   66, 83
Bender, M. R.   105, 117
Benevento, L. A.   100, 101, 315, 324
Benton, A. L.   306, 315, 324
Bergman, T.   183, 194
Berlucchi, G.   121, 132
Berlyne, D. E.   197, 211
Berman, N.   46, 50, 58, 64, 81, 84, 135,
   136, 152
Berson, D. M.   50
Bertenthal, B. I.   176–180, 193, 194
Bertrand, C.   64, 83
Biguer, B.   1, 11, 315, 325

Birch, E. E.   171, 172
Bischof, N.   334, 351
Bizzi, E.   15, 33, 99, 101, 105, 106, 115, 117, 216, 222
Black, A. H.   52, 60
Blair, S. M.   105, 117
Blakemore, C.   170, 171, 173, 198, 211, 223, 241
Blanc-Garin, J.   299, 301
Bossom, J.   331, 352
Bouisset, S.   17, 32
Bouman, M. A.   285, 301
Bower, T. G. R.   199, 211, 218, 222
Boynton, R, M.   291, 302
Brain, W. R.   306, 324
Brandt, T.   333, 350, 351
Braun, J. J.   156, 174
Brazelton, T. B.   210, 211
Breitmeyer, B. G.   297, 300
Bridgeman, B.   263, 265–268, 271, 272, 275, 277
Brill, S.   164, 172
Brindley, G. S.   6, 10, 11, 261, 262
Brinkman, J.   315, 324
Brodal, A.   105, 117, 118
Bronson, G.   210, 211
Broughton, J. M.   218, 222
Brown, J. F.   233, 241
Brune, F.   265, 269, 270, 278
Büchele, W.   351
Bucher, V.   39, 59
Bueno de Mesquita, A. E.   285, 301
Buisseret, P.   129, 133
Bullinger, A.   175, 193, 194, 216, 218, 222
Burgi, S.   39, 59
Butterworth, G.   175, 194
Buttner, U.   65, 81, 116, 117
Buttner-Ennever, J. A.   65, 81, 105, 115, 117

Cairns, S. J.   39, 60
Callens, M.   265, 278
Campbell, F. W.   5, 11
Campos, J.   175, 178, 193, 194
Campos, R. G.   193, 194
Canon, L. K.   323, 324
Carlson, S.   344, 351

Castle, P.   199, 213
Caviness, V. S., Jr.   46, 58
Chalupa, L. M.   46, 60, 210, 212
Cheal, M.   36, 59
Chodkiewicz, J. P.   316, 326
Churcher, J.   276, 278, 289, 290, 291, 301
Clark, M. E.   267, 277
Cogan, D. G.   155, 172
Cohen, B.   65, 82, 105, 115, 117, 169, 173
Cohen, G. H.   291, 302
Cohen, M. E.   289–291, 300
Cole, M.   306, 324
Collewijn, H.   135, 169, 173, 336, 345, 351
Collins, W. E.   157, 172
Collison, C.   51, 60
Conway, J. L.   66, 84
Cook, J. D.   63, 85
Cooper, R. M.   43, 58
Cordo, P. J.   17, 18, 32
Coren, S.   282, 296, 300
Coulmance, M.   23, 32
Court, J. E.   305, 325
Cowey, A.   285, 288, 289, 300
Craske, B.   8, 11
Crone, R. A.   156, 172
Curio, G.   336, 351
Cynader, M.   46, 58, 64, 81, 84, 135, 152, 156, 168, 172

Dale, R. H. I.   51–55, 58
Damasio, A. R.   306, 315, 324
Darian-Smith, I.   330, 351
Davidson, I.   330, 351
de Ajuriaguerra, J.   305, 325
Dean, P.   43, 44, 54, 58
Delage, Y.   334, 351
Delgado, D.   272, 278
Dell'Osso, L. F.   69, 85
de Rocondo, J.   306, 324
Descartes, R.   264, 278
Devic, M.   305, 325
Dews, P.   121, 133
Diamond, R.   9, 12, 119, 120, 133
Dichgans, J.   105, 115, 117, 333, 351
Di Franco, D.   200, 202, 203, 206, 212

Ditchburn, R.   264, 278
Dixon, J. P.   93, 103
Dizio, P.   36, 59
Dobson, V.   164, 169, 172
Dodwell, P. C.   188, 197, 198, 200, 202,
   203, 206, 205, 212
Drager, U. C.   65, 81
Dufossé, M.   22, 32
Dumas, R.   306
Duysens, J.   265, 278

Eason, R. G.   39, 59, 290, 302
Ebbesson, S.   36, 49, 58
Echallier, J. F.   2, 10, 12, 317, 322, 325
Eckmiller, R.   105, 117
Ectors, L.   89, 101
Edwards, M. W.   243, 262
Edwards, S. B.   64, 65, 81, 88, 101
Eibl-Eibesfeldt, I.   198, 212, 346, 351
Elam, G. W.   157, 172
Eliasson, S. G.   39, 59
Elner, A. M.   17, 18, 32
Emerson, R. C.   6, 13, 261, 262
Emerson, V. F.   43, 58
Engle, W. K.   63, 85
Erickson, R. P.   293, 300
Ettlinger, G.   316, 324
Evarts, E. V.   1, 11

Fallon, J. H.   315, 324
Fantz, R. L.   155, 172, 197, 212
Farkkila, M.   344, 351
Faugier-Grimaud, S.   315, 316, 324
Feldon, P.   285, 300
Feldon, S.   285, 300
Felpel, L. P.   105, 118
Fender, D. H.   300, 301
Fendrich, R.   265, 278
Ferrier, D.   88, 101
Festinger, L.   323, 324
Field, J.   200, 203–205, 212
Filion, M.   105, 118
Findlay, J. M.   282, 286, 287, 290, 292,
   296, 298, 301
Finlay, B. L.   39, 60
Fiorentini, A.   132, 133
Fischer, B.   223, 242

Fischer, C.   306, 325
Fisher, D. F.   294, 296
Foorman, N. P.   36, 59
Forbes, B.   205, 212
Foreman, N. P.   52–54, 58
Frankfurter, A. J.   65, 81, 82, 100, 101
Freedman, S. J.   8, 11
French, J.   162, 172
Frenois, C.   315, 324
Frey, R.   261, 262
Frost, B. J.   227, 241
Frost, D.   3, 12, 290–292, 301
Fuchs, A. F.   65, 69, 83, 89, 102, 105,
   118
Fukuda, V. T.   55, 172
Fukushima, K.   105, 118

Gage, F. H.   54, 60
Gahéry, Y.   20, 22, 24, 27–29, 32
Galletti, C.   100, 101
Garcin, R.   306, 324
Garey, L. J.   170, 173
Gellis, S. S.   155, 172
Gazzaniga, M. S.   123, 124
Gentilucci, M.   329, 352
Gerin, P.   10, 11
Gerstein, G. L.   6, 13
Gerstein, R. L.   261, 262
Geschwind, N.   316, 325
Ghez, C.   31, 33
Gibson, J. J.   198, 199, 212, 224, 227,
   241, 272–274, 278, 329, 351
Gilman, A.   244, 262
Ginsburgh, C. L.   64, 81
Giolli, R. A.   50, 58
Godwin-Austen, R. B.   305, 324
Goldberg, M. E.   64, 65, 82, 85, 297,
   300, 345, 351
Goldman, P. S.   64, 82
Goldstein, H.   224, 241
Goodale, M. A.   36, 38, 39, 42, 43, 53,
   55, 58
Goodman, L. S.   244, 262
Goodwin, G. M.   6, 11, 261, 262
Gordon, B.   65, 82
Gordon, J.   169, 172
Gorman, J. J.   155, 172
Gottschalk, Ch.   349, 351

Gould, J. D. 292, 300
Gower, E. C. 119, 125, 133
Graefe, A., von 6, 11
Graham, C. H. 229, 241
Graham, J. 65, 82
Graham, M. 223, 242
Gray, J. 18, 19, 32
Graybiel, A. M. 50, 58, 64, 65, 82, 105, 118, 115
Gregory, R. L. 3, 11
Gresty, M. 90, 101
Grillner, S. 29, 32
Grüsser, O. J. 275, 327, 328, 332, 336, 345, 349, 351
Guillemot, J. P. 156, 172
Guillery, R. W. 170, 172
Guthrie, B. L. 65, 84
Guthrie, M. D. 50, 58
Gurfinkel, V. S. 17, 17, 32
Gwiazda, J. 164, 171, 172
Gyr, J. 331, 352

Haaxma, R. 312, 315, 324
Haddad, G. 7, 12
Hainline, L. 169, 172
Haith, M. M. 175, 178, 183, 188–191, 194, 195, 197, 220, 222
Hallett, P. E. 2, 11, 63, 69, 83
Halton, A. 200, 213
Halverson, H. M. 200, 212
Hanker, J. S. 136, 152
Hansen, R. M. 4, 12
Harris, K. S. 7, 11
Harris, L. R. 156, 168, 172
Harris, L. S. 205, 212
Harris, P. 189, 194
Harting, J. K. 50, 59, 65, 81, 82, 85, 100, 101
Hartje, W. 316, 324
Hassler, R. 16, 32, 89, 101
Hebb, D. O. 193, 194, 198, 212
Hécaen, H. 305, 325, 337, 352
Hecht, M. 229, 241
Heilman, K. M. 89, 103, 316, 325
Hein, A. 9, 12, 119–121, 125, 127, 133, 175, 195, 331, 352
Heit, G. 263, 277
Held, R. 3, 12, 119, 125, 133, 155, 164,

171–173, 175, 195, 199, 213, 331, 352
Helmholtz, H. von 8, 11, 63, 82, 128, 133, 243, 261, 262, 264, 275, 278
Hendrickson, A. 64, 82, 169, 172
Hendry, D. 265, 278
Henkel, C. K. 64, 65, 81
Hepp, K. 116, 117
Henn, V. 65, 81, 82, 105, 115–117
Henry, A. 331, 352
Hering, E. 275, 278
Hess, S. 39, 59
Hess, W. R. 16, 32
Heywood, S. 276–278, 289–291, 301
Hickey, T. L. 64, 83, 170, 173
Hicks, L. 175, 194
Hoenig, P. 282, 296, 300
Hoffman, K. P. 135, 136, 138, 147, 152, 156, 169, 173
Holland, R. 65, 84
Hollander, H. 135, 153
Hollenberg, M. J. 169, 173
Holmes, G. 18, 32, 305, 325
Holst, E. von 4, 11, 63, 82, 264, 278
Honrubia, V. 169, 172
Horel, J. A. 43, 59
Hou, R. L. 300, 301
Hubel, D. H. 64, 65, 81, 82, 135, 152, 156, 173, 197, 212, 226, 241
Huerta, M. F. 65, 82, 100, 101
Hughes, A. 285, 301
Huizinga, E. 155, 173
Hull, C. L. 198, 212
Hyde, J. E. 39, 59
Hyvärinen, I. 344, 352
Hyvärinen, J. 80, 82, 344, 352

Innis, N. 51, 52, 55, 58
Ioffé, M. E. 19, 29, 31, 32
Iwahori, N. 135, 152
Iwamoto, T. 88, 102

Jackson, J. H. 6, 11
Jacobson, S. 100, 101, 171, 173
Jakway, J. S. 36, 49, 60
Jeannerod, M. 1, 3–5, 10–12, 98, 103, 292, 302, 305, 315, 317, 322, 325

Johnson, C. A.   274, 278
Johnson, K. O.   351
Johnstone, J. R.   5, 11
Jones, E. G.   316, 325
Joseph, J. P.   80, 84, 88, 94, 102, 103
Jung, R.   16, 32
Jürgens, R.   284, 290, 291, 300

Kaas, J. H.   286, 301
Kalil, R. E.   8, 11, 105, 117, 216, 222
Kallos, T.   6, 13, 261, 262
Kalsbeck, J. E.   316, 324
Kant, I.   327, 352
Kasdom, D. L.   100, 101
Kase, C. S.   305, 325
Kase, M.   65, 80, 82
Kaufman, K. J.   306, 325, 332, 352
Kawano, K.   80, 84, 95, 103
Keller, E. L.   65, 83, 105, 107, 114,
    115, 118, 285, 301
Kelso, J. A. S.   7, 11
Kennard, M. A.   89, 101
Kennedy, H.   4, 11, 46, 61, 100, 102
Kessen, W.   188–190, 195
Khambatta, H. J.   256, 262
Kibler, G.   246, 262
Kievit, J.   100, 101
Kimura, D.   316, 325
King, W. M.   65, 83
Klein, R.   323, 325
Koenderink, J. J.   226, 241, 285, 301
Koerner, F. H.   10, 11, 65, 84, 87, 102
Kohler, I.   276, 278, 331, 352
Komilis, E.   3, 10, 12, 17, 322, 325
Kong, R.   265, 278
Kornmuller, A. E.   6, 11, 261, 262
Kotchabhakdi, N.   100, 101
Kowler, E.   237, 242
Kruger, L.   65, 85, 210, 213, 285, 300
Kubota, K.   88, 90, 102
Kulikowski, J. J.   6, 11, 261, 262, 332,
    351
Kunzle, H.   64, 83
Kuypers, H. G. J. M.   64, 83, 100, 101,
    312, 315, 324

La Boissiere, E.   169, 172
Lackner, J. R.   8, 9, 11

Ladpli, R.   105, 118
Lamotte, R. H.   315, 316, 324, 325
Lane, R. H.   286, 301
Lashley, K. S.   38, 43, 52, 57, 59
Lawrence, D. G.   64, 83
Lee, D. N.   175, 195
Leibowitz, H.   229, 241
Leichnetz, G. R.   100, 102
Lehtinen, I.   89, 102.
Leighton, D.   6, 11, 261, 262
Leinonen, L.   344, 352
Lepore, F.   156, 172
Letson, R. D.   171, 172
Leushina, L. I.   292, 301
Le Vay, S.   64, 82
Levi, D. M.   156, 173
Levine, D. N.   306, 316, 325
Lévy-Schoen, A.   282, 286, 291, 292,
    299, 301
Lewis, S.   263, 277
Lewis, T. L.   210, 212
Liepman, H.   316, 325
Lightstone, A. D.   2, 11, 63, 83
Lindau, M.   349, 351
Lloyd, V. V.   229, 241
Lo, A. W.   290, 291, 302
Lobos, E.   210, 213
Longuet-Higgins, H. C.   226, 241
Loomis, J.   223, 242
Lücking, C. H.   265, 278
Lund, R. D.   38, 46, 59, 64, 83
Luria, A. R.   305, 325
Luschei, E. S.   89, 102, 105, 107, 118
Lynch, J. C.   80, 83, 88, 102, 317, 325,
    344, 345, 352

Mac Arthur, R.   243, 262
Mac Farlane, A.   189, 194
Mach, E.   333, 352
Mack, A.   265, 270, 274, 275, 278
Mac Kay, D. M.   238, 241, 266, 275,
    278
Mackel, R. G.   31, 33, 105, 118
Macpherson, J. M.   22, 32
Maeda, M.   31
Maekawa, K.   135, 152, 153
Maes, H.   46, 61
Maffei, L.   132, 133

Magalhaas-Castro, B.   65, 85
Magnin, M.   4, 11, 100, 102
Magnus, R.   15, 32
Magoun, H. W.   64, 81
Maldonado, H.   89, 102
Mann, V. A.   9, 10, 12
Mark, R. F.   5, 11
Marsden, C. D.   17, 18, 32
Marshall, J.   3, 13
Martin, J. P.   16, 18, 32
Massé, D.   2, 12
Massion, J.   22, 29, 32, 33
Masure, M. C.   306, 326
Matelli, M.   329, 352
Matin, E.   246, 262
Matin, L.   3–5, 12, 63, 83, 243, 245,
    246, 261, 262, 275
Matteo, R. S.   256, 262
Mauguiere, F.   306, 315, 324, 325
Maunz, R. A.   31, 33, 105, 118
Maurer, D.   210, 212
Mays, L. E.   2, 12, 63, 64, 66, 69, 75,
    77, 79, 83, 88, 102, 103
Mc Ilwain, J. T.   64, 83, 286, 297, 301
Mc Kee, S. P.   235, 242
Mc Laughlin, S. C.   9, 12
Megaw, E. D.   302
Meltzoff, A. N.   199, 212
Mendelson, M. J.   188, 195
Merton, P. A.   10, 11, 17, 32
Mesulam, M. M.   136, 152, 316, 325
Metz, C. B.   136, 152
Meulen, P.   155, 173
Michard, A.   291, 292, 302
Michel, F.   305, 325
Miezin, F. M.   286, 301
Miles, F. A.   1, 12
Miller, B. D.   89, 103
Miller, D. C.   65, 82
Miller, L. K.   290, 302
Milner, A. D.   36, 38, 39, 42, 59
Mittelstaedt, H.   4, 11, 63, 82, 264, 278
Mizuno, N.   135, 152
Mlinar, E.   39, 42, 60
Mohler, C. W.   66, 67, 83, 90, 102, 297,
    298, 302
Mohindra, I.   164, 172
Mohr, J. P.   305, 306, 325
Montarolo, P. G.   156, 173

Monty, R. A.
Moore, M. J.   183, 194
Moore, M. K.   199, 212, 218, 222
Morasso, P.   99, 101, 105, 115, 117
Mort, E.   39, 43, 60
Morton, H. B.   17, 32
Mountcastle, V. B.   80, 81, 83, 88, 102,
    317, 325
Mouret, J.   10, 11
Movshon, J. A.   171, 173
Muir, D. W.   200, 202, 203, 205, 206,
    212
Munson, J. B.   121, 132
Murison, R. C. C.   36, 39, 42–44, 59

Nadel, L.   50, 54, 60
Naegele, J. R.   155, 158, 173
Nagle, M.   263, 277
Nakamura, Y.   135, 152
Nakayama, K.   223, 226–232, 236–240,
    241, 242
Nashner, L. M.   17, 18, 32
Nauta, W. J. H.   38, 60, 64, 82, 85
Naydel, A. V.   17, 31
Neufeld, G. R.   6, 13, 261, 262
Nichols, C. W.   6, 13, 261, 262
Nieoullon, A.   20, 28, 32
Nissen, M. J.   291, 302
Noback, C. R.   50, 59
Noda, H.   65, 82
Nordby, T.   135, 152

Ogden, W. C.   291, 302
Oke, L. E.   330, 351
O'Keefe, J.   50, 54, 60
Olson, C. R.   9, 12
Olton, D. S.   50, 51, 54, 60
Optican, L. M.   63, 65, 83, 85
Orban, G. A.   46, 61, 265
Ordy, J. M.   155, 172
Orem, J.   89, 102
Oyster, C. W.   169, 173
Ozonas, G.   316, 326

Padel, Y.   29, 32
Paillard, J.   221, 222

Palka, J. 4, 12
Palmer, L. A. 46, 60, 147, 152
Pandya, D. N. 316, 325
Parker, D. M. 297, 302
Partlow, G. D. 65, 85
Pause, M. 332, 336, 351
Pasik, P. 66, 83
Pasik, T. 66, 83
Patton, L. 6, 11
Payne, B. R. 50, 58
Pearce, D. G. 3, 12, 246, 262
Peck, C. K. 80, 84, 94, 103
Perenin, M. T. 3, 6, 7, 12, 306, 325
Peterson, B. W. 31, 33, 105, 118
Petras, J. M. 64, 83
Pettigrew, J. D. 223, 241
Piaget, J. 199, 203, 210, 212
Picoult, E. 243, 246, 262
Pilon, R. 215, 212
Pipp, S. 184, 192, 195
Pitts, N. G. 31, 33, 105, 118
Pleune, J. 265, 278
Poggio, G. F. 223, 240, 242
Poirier, L. J. 64, 83
Pola, J. 246, 262
Polit, A. 15, 33, 99, 103
Pollack, J. G. 64, 83, 85
Pompeiano, O. 105, 118
Pöppel, E. 3, 12, 290–292, 301
Poranen, A. 80, 82
Posner, M. I. 291, 302, 323, 325
Powell, T. P. S. 316, 325
Prablanc, C. 2, 10, 12, 290, 292, 302, 317, 322, 323, 325
Prazdny, K. 226, 241
Precht, W. 135, 152, 156, 173
Pua, E. K. 256, 262

Rabinovitch, H. E. 290, 291, 302
Rademaker, G. G. J. 156, 173
Ranson, S. W. 64, 81
Raphan, T. 169, 173
Regis, H. 29, 33
Reichardt, W. 240, 242
Rezak, M. 100, 101
Rhoades, R. W. 46, 60, 210, 212
Rich, I. 43, 61
Riddoch, G. 306, 325

Riesen, A. H. 121, 133
Rifkin, K. I. 9, 12
Rinrik, E. 100, 101
Riss, W. 36, 49, 60
Ritchie, L. 65, 83
Rizzolatti, G. 121, 132, 329, 352
Roberts, T. D. M. 17, 33
Roberts, W. A. 51, 61
Robinson, D. A. 2, 12, 63, 65–67, 83–85, 99, 102, 107, 109, 116, 118, 141, 153, 281, 284, 285, 302
Robinson, D. L. 4, 5, 12, 64, 82, 345, 351
Rogers, B. J. 223, 238, 242
Rolls, E. T. 285, 288, 289, 300
Ron, S. 65, 84
Rondot, P. 306, 324, 326
Rose, J. 39, 59
Rosenblatt, F. 198, 213
Rosenquist, A. C. 6, 13, 46, 60, 147, 152, 261, 262
Ross, L. E. 289–291, 300
Rothwell, J. C. 18, 33
Roumieu, J. 29, 33
Rovamo, J. 285, 302
Royce, G. J. 65, 82, 100, 101
Ruff, H. 200, 213
Rustioni, A. A. 136, 152

Sakata, H. 80, 84, 95, 102
Salapatek, P. 188, 189, 194, 195
Salinger, W. 120, 133
Salzen, E. A. 297, 302
Samuelson, R. J. 50, 51, 54, 60
Sanders, M. D. 3, 13
Santos-Anderson, R. 100, 101
Saslow, M. G. 290, 302
Scalia, F. 49, 60
Scandolara, C. 329, 352
Scheibel, A. B. 105, 118
Scheibel, M. E. 105, 118
Schiller, P. H. 64–68, 84, 87, 90, 100, 102, 116, 118
Schlag, J. 80, 84, 88, 89, 91, 94, 98, 102, 103
Schlag-Rey, M. 80, 84, 88, 89, 91, 94, 102, 103

Schneider, G. E.   35, 39, 43, 44, 60, 210, 213, 315, 326
Schroeder, D. J.   157, 172
Schoppmann, A. A.   135, 147, 152, 153, 156, 169, 173
Schor, C. M.   156, 167, 173
Schott, B.   98, 103
Schrieter, U.   336, 351
Schutta, H. S.   306, 324
Schwartz, S.   265, 278
Scobey, R. P.   274, 278
Sechenov, I.   329, 352
Seibeck, R.   261, 262
Shanzer, S.   105, 117
Sharpe, J. A.   290, 291, 302
Shebilske, W. L.   63, 84
Sherman, H. B.   46, 58
Sherman, S. M.   136, 152
Sherrington, C. S.   28, 33, 128, 133
Shibutani, H.   80, 84, 95, 102
Shinoda, Y.   31, 33
Siebeck, R.   6, 12
Simpson, J. I.   135, 152
Sinclair, M.   205, 212
Skavenski, A. A.   4, 7, 10, 12, 63, 84, 282, 302
Slappendel, S.   285, 301
Smith, A. M.   29, 33
Sparks, D. L.   2, 12, 63–66, 68, 83, 88, 102, 103, 115, 118
Spear, P. D.   156, 174
Spector, S.   256, 262
Sperry, R. W.   5, 13, 264, 278
Spira, A. W.   169, 173
Spitz, R.   328, 352
Sprague, J. M.   64, 85
Squatrito, S.   100, 101
Stark, L. A.   7, 13, 63, 85, 265–267, 275, 278, 284, 289, 303
Stein, B. E.   64, 65, 81, 85, 93, 103, 210, 212
Stein, D.   315, 324
Stein, J.   316, 326
Steinberg, R.   29, 32
Steinman, R. M.   4, 7, 13, 237, 241, 282, 289, 292, 302
Sterling, P.   46, 61
Sternberg, S.   292, 302
Stevens, J. K.   6, 13, 243, 261, 262

Stevens, R. G.   52–54, 58
Straaten, J. J.   38, 60
Strata, P.   135, 152, 156, 173
Strominger, N. L.   65, 82, 100, 101
Stryker, M.   64, 66, 84, 116, 118
Suzuki, H.   88, 102
Suzuki, S.   52, 60
Svejda, M. J.   193, 194
Swett, J. E.   24, 32, 105, 118

Tagliasco, V.   105, 117, 216, 222
Takanashi, E.   169, 173
Takeda, T.   135, 152
Talbot, W. H.   80, 83, 88, 102
Tapia, J. F.   305, 325
Tauber, E.   155, 169, 173
Teller, D. Y.   164, 172
Ter Braak, J. W. G.   156, 173, 333, 352
Tétard, C.   291, 302
Teuber, H. L.   5, 13
Thomas, A.   18, 33
Thomas, R. K.   54, 61
Thompson, R.   43, 54, 61
Timberlake, G. T.   282, 302
Todd, J. T.   289, 302
Tokita, T.   155, 172
Toyne, M. J.   64, 82
Traub, M. M.   18, 33
Travis, R. P.   115, 118
Trevarthen, C.   263, 279, 315, 326
Troncoso, J. F.   305, 325
Tronick, E.   198, 211
Trouche, E.   29, 33
True, S. D.   66, 84
Tsang, Y. C.   52, 61
Tsukahara, N.   27, 31
Tucker, P.   178, 194
Tuller, B.   7, 11
Tyler, C. W.   229, 233, 238, 242
Tzavaras, A.   306, 315, 316, 326

Udelf, M. S.   155, 172
Updyke, B. V.   135, 152

Van Doorn, A. J.   226, 241
Van Gelder, P.   289, 302

Van Gisburgen, J. A. M.   285, 302
Van Griffen, K.   180–183, 193, 195
Van Hoesen, G. W.   316, 325
Van Hofsten, C.   200, 213
Van Hof-Van Duin, J.   136, 153, 155, 156, 173
Vighetto, A.   306, 320, 325
Virsu, V.   285, 302
Vital-Durand, F.   120, 133, 170, 173
Von Noorden, G. K.   156, 170, 173

Walberg, F.   100, 101, 135, 153
Walk, R. D.   240, 242
Walker, J. A.   54, 60
Wallach, H.   265, 269, 276
Walls, G.   240, 242
Warrington, E. R.   3, 13, 306, 324
Watson, R. T.   89, 103
Weber, J. T.   65, 81, 85
Weir, V. K.   54, 61
Weiskrantz, L.   3, 13
Weiss, L.   336, 351
Werz, M. A.   51, 60
Westheimer, G.   105, 117, 235, 237, 240, 242
Wheeless, L. L.   291, 302
White, B. L.   199, 203, 213
White, C. T.   290, 302
Whitlock, D. G.   64, 85
Wickelgren, B.   46, 61
Wiermsa, C. A. G.   4, 13, 97, 103

Wiesel, T. N.   64, 82, 121, 133, 135, 147, 153, 156, 173, 197, 212, 226, 241
Willey, R.   331, 352
Wilson, M. E.   64, 82
Wilson, V. J.   31
Winston, P. H.   198, 213
Wollensak, J.   332, 351
Wood, C. C.   156, 174
Wurtz, R. H.   4, 5, 12, 64–66, 81, 83, 85, 90, 102, 297, 302
Wyke, M.   316, 326
Wyman, D.   282, 289, 290, 292, 302

Yamaguchi, T.   4, 13
Yates, P. E.   136, 152
Yee, R. D.   169, 172
Yin, T. C. T.   80, 83, 88, 102
Yingchareon, K.   100, 101
Young, D.   243, 262
Young, L. R.   7, 13, 63, 67, 85, 284, 289, 303

Zahn, J. R.   69, 85
Zarzecki, P.   31, 33
Zattara, M.   17, 32
Zee, D. S.   63, 68, 85
Zihl, J.   291, 303
Zoladek, L.   51, 61

Subject Index

Adequate response   91–98
Asynergia   17, 18

Cataract   164, 167, 168
Conditioning of movement patterns   24–27
Corollary discharge   5, 128, 264–266, 275, 276
Constancy, space   255–256, 274–277

Deafferentation of extraocular muscles   128–132
Differential looking   197
Directionality, law of   90

Efference copy   4, 5
Effort of will   243, 264
Extraocular proprioception   128–132
Extrapersonal space   327–337
  far distant action   330–331
  grasping   327–329
  instrumental grasping   329–330
  near distant action   330
  rearrangement   331, 332
  visual background   331
Extraretinal eye position information   243–244, 250, 255, 256, 258–261
Extraretinal signal   4, 5, 7, 10, 121–132
Eye-head coordination   105–117, 120, 223, 224

Eye movement   3–7, 10, 120, 155–171, 175–194, 206–211, 223, 344–350
  saccades   2, 3, 5, 10, 63–81, 132, 221, 265–270, 276, 277, 281–300, 308, 346–348
    amplitude   292–298
    direction   286–289
    latency   289–292, 308
    retinocentric model   67, 68
    spatial model   68–72
  saccadic generation   284, 285
  voluntary   3, 281, 308, 332, 335–337
Eye paralysis   6, 7, 121–128, 241–261
  optic ataxia   305–324
Eye position   1, 6, 10, 247–255

Figure-ground   270, 274, 277
Foveation hypothesis   67, 68
Frontal eye field ablation   100

Gaze direction   7–10, 220, 243–261, 305, 344–346
Guided locomotion   1, 36–50, 119–132, 175
Guided reaching   120, 122–132, 199–201, 310–324

Habituation   176–180
Hand orientation   312–324

Head movement　3, 37, 206–209, 223
Head position　1, 9, 109, 246–250

Image position　1
Internal medullary lamina　88–90, 92, 95–97, 99
　lesions of　89
　recording　89
　　eye position units　89
　　presaccadic units　89
　　visual units　90
　stimulation　89
Inverted vision　5

Lateral geniculate nucleus　135

Modality interaction　35, 57
Motion parallax　223–241
Motor-visual feedback　119
Movement
　anticipation　17, 18
　body　3
　predictive　37
　torsion　22
Multiple visual systems　36

Nativism　197–199, 209–211
Nystagmus　121–132, 333

Ocular paralytic illusion　245–247
Optic
　tectum　49, 227
　tract
　　nucleus of　135–151
　　anatomy　136–138
Optical velocity　223–226
Optokinetic nystagmus　135–151, 155–171
　asymmetry　136, 142–144, 151, 155–171
　cortical lesions　144–147
　development　156–163
　symmetry　156, 162
　visual deprivation　147–151
Orientation behavior　36

Parietal cortex lesions　306–308, 337–344
Past pointing　6, 7
Personal space　327
Place localization　50–57
Pontine reticular formation　105–117
　burst cells　107, 112–114
　burst-tonic cells　106, 107
　omnipauser cells　112, 114, 115
　pause cells　107, 109–112
　　head movement　109
　tonic cells　107, 108
Posture　15–31, 36, 217–221
　central organization　27–30
　diagonal pattern　19–24
　nondiagonal pattern　22–24
Pretectum　48, 49, 156
Prismatic displacement　8, 9
Pulvinar　337

Scotoma　2
Somesthesia　308, 309
Stereognosis　308
Stereopsis　156, 170, 171, 223
Stroboscopic illumination　3, 119, 337
Strabismus　9, 164–168
Superior colliculus
　lesions of　66, 67, 100
　localization mediation　38–43, 210
　projections from　65
　projections to　64
　saccades　64–67, 100, 285
　stimulation　66
　unit recording　65, 66, 75–79

Tonic neck reflex　217, 218
Torsion movements　22
Two visual systems　35, 263, 264

Vestibulo-ocular reflex　1, 123, 132, 333
Vection, linear and circular　333–335
Visual
　anticipation　184–188
　cortex　135
　　acuity　43, 307, 308
　　identification　263

localization   45–48, 263
   pattern discrimination   43–48
fixation   1–3, 184–188, 266–270, 346
localization auditory stimuli   188, 189,

203–206, 256–261
recognition   35
scanning   180–184, 188–194, 201–203
tracking   215–221, 344–346, 349, 350